OT 17:
Operator Theory: Advances and Applications
Vol. 17

Birkhäuser Verlag
Basel · Boston · Stuttgart

Advances in Invariant Subspaces and Other Results of Operator Theory

9th International Conference on Operator Theory,
Timişoara and Herculane (Romania),
June 4–14, 1984

Volume Editors

R. G. Douglas
C. M. Pearcy
B. Sz.-Nagy
F.-H. Vasilescu
D. Voiculescu

Managing Editor

Gr. Arsene

Springer Basel AG

Volume Editorial Office

Department of Mathematics
INCREST
Bd. Păcii 220
79622 Bucharest (Romania)

CIP-Kurztitelaufnahme der Deutschen Bibliothek

Advances in invariant subspaces and other results of operator theory / 9. Internat. Conference on Operator Theory, Timişoara and Herculane (Romania), June 4–14, 1984. Vol. ed. R. G. Douglas . . . Managing ed. Gr. Arsene. – Basel ; Boston ; Stuttgart : Birkhäuser, 1986.
(Operator theory ; 17)
ISBN 978-3-0348-7700-8 ISBN 978-3-0348-7698-8 (eBook)
DOI 10.1007/978-3-0348-7698-8
NE: Douglas, Ronald G. [Hrsg.]; International Conference on Operator Theory <09, 1984, Timişoara; Herculane>; GT

Originally published by Birkhäuser Verlag Basel in 1986
Softcover reprint of the hardcover 1st edition 1986

CONTENTS

FOREWORD

The annual Operator Theory conferences, organized by the Department of Mathematics of INCREST and the University of Timişoara, are intended to promote cooperation and exchange of information between specialists in all areas of operator theory. This volume consists of papers contributed by the participants of the 1984 Conference. They reflect a great variety of topics, dealt with by the modern operator theory, including very recent advances in the invariant subspace problem, subalgebras of operator algebras, hyponormal, Hankel and other special classes of operators, spectral decompositions, aspects of dilation theory and so on.

The research contracts of the Department of Mathematics of INCREST with the National Council for Science and Technology of Romania provided the means for developing the research activity in mathematics; they represent the generous framework of these meetings, too.

It is our pleasure to acknowledge the financial support of UNESCO which also contibuted to the success of this meeting.

We are indebted to Professor Israel Gohberg for including these Proceedings in the OT Series and for valuable advice in the editing process. Birkhäuser Verlag was very cooperative in publishing this volume.

Mariana Bota, Camelia Minculescu and Rodica Stoenescu dealt with the difficult task of typing the whole manuscript using a Rank Xerox 860 word processor; we thank them for the excellent job they did.

Organizing Committee,

Head of Mathematics Department of INCREST, Zoia Ceauşescu

Organizers, F.-H. Vasilescu, Dan Voiculescu

LIST OF PARTICIPANTS*)

ALBRECHT, Ernst	University of Saarland, West Germany
ALTOMARE, Francesco E.	University of Bari, Italy
ARSENE, Grigore	INCREST, Bucharest
ARVERSON, William B.	University of California, Berkeley, USA
BACALU, Ion	Politechnical Institute, Bucharest
BÁKONY, Mihai	University of Timişoara
BALINT, Agneta	University of Timişoara
BALINT, Ştefan	University of Timişoara
BARRÍA, José	Institute of Investig.Sci., Caracas, Venezuela
BEZNEA, Lucian	INCREST, Bucharest
BOBOC, Nicolae	University of Bucharest
BÓGNAR, János	Mathematics Institute, Budapest, Hungary
BRIGOLA, R.	University of Regensburg, West Germany
BRUCKSTEIN, Alfred	Technion Institute of Technology, Israel
BUCUR, Gheorghe	INCREST, Bucharest
BUZULOIU, Ion	IPGGB, Bucharest
CAIN, Brian E.	Iowa State University, Ames, USA
CARTIANU, Dan	ICECHIM, Bucharest
CĂINICEANU, George	ICEFIZ, Bucharest
CEAUŞESCU, Zoia	INCREST, Bucharest
CEAUŞU, Traian	Politechnical Institute, Timişoara
CHEVREAU, Bernard	University of Bordeaux, France
COLOJOARĂ, Ion	University of Bucharest
COLOJOARĂ, Sanda	University of Bucharest
CONSTANTINESCU, Tiberiu	INCREST, Bucharest
CORNEA, Emil	INCREST, Bucharest
COSTINESCU, Roxana	Bucharest
CURTO, Raul E.	University of Iowa, USA
DAZORD, Jean	University of Lyon, France
DĂDĂRLAT, Marius	INCREST, Bucharest
DEACONU, Valentin	INCREST, Bucharest
DOMAR, Yngve	University of Uppsala, Sweden
DOUGLAS, Ronald G.	University of New York, Stony Brook, USA
DRAGOMIR, Achim	University of Timişoara
DURSZT, Endre	University of Szeged, Hungary
ECKSTEIN, Gheorghe	University of Timişoara
ESCHMEIER, Jorg	University of Munster, West Germany
FRUNZĂ, Ştefan	University of Iaşi
GABOR, Mihai	University of Timişoara
GADIDOV, Radu	INCREST, Bucharest
GAŞPAR, Dumitru	University of Timişoara
GĂVRUŢĂ, Paşc	Politechnical Institute, Timişoara

*) Romanian participants are listed only with the name of their institution and the city.

GHEONDEA, Aurelian	INCREST, Bucharest
GODINI, Gliceria	INCREST, Bucharest
GOLOGAN, Radu	INCREST, Bucharest
GRAMSCH, Berhard	Johannes Gutenberg University, West Germany
GRECEA, Valentin	INCREST, Bucharest
GRZĄSŁEWICZ, Richard	Mathematics Institute, Wroclaw, Poland
HĂRĂGUŞ, Dumitru	University of Timişoara
HELSON, Henry	University of California, Berkeley, USA
HESS, Heinz-Ulrich	University of Regensburg, West Germany
HIRIŞ, Viorel	University of Timişoara
HRUŠČEV, Serghei V.	Steklov Institute, Leningrad, USSR
JAFARIAN, A.A.	Sharif University of Technology, Teheran, Iran
JOHN, Kamil	Mathematics Institute, Prague, Czechoslovakia
JONAS, Peter	Mathematics Institute, Berlin, GDR
KAASHOEK, Marinus A.	Free University, Amsterdam, Netherlands
KAILATH, Thomas	Stanford University, USA
KÉRCHY, László	University of Szeged, Hungary
KYMALA, Earl E.	University of California, Sacramento, USA
LANGER, Heinz	University of Dresden, GDR
LEGIŠA, P.	University of Ljubljana, Yugoslavia
LESNJAK, G.	University of Ljubljana, Yugoslavia
LEVY, Roni N.	Mathematics Institute, Sofia, Bulgaria
MAGAJNA, Bojan	University of Ljubljana, Yugoslavia
MEGAN, Mihail	University of Timişoara
MENNIKEN, Reinhard	University of Regensburg, West Germany
MIHALACHE, Georgeta	INCREST, Bucharest
NAGY, Gabriel	University of Bucharest
NEIDHARDT, Hagen	Mathematics Institute, Berlin, GDR
NEUMANN, Michael	University of Essen, West Germany
NICULESCU, Constantin	University of Craiova
OMLADIČ, Matlaž	University of Ljubljana, Yugoslavia
PASNICU, Cornel	INCREST, Bucharest
PEARCY, Carl	University of Michigan, Ann Arbor, USA
PETRESCU, Steliana	INCREST, Bucharest
PIMSNER, Mihai	INCREST, Bucharest
PINCUS, Joel	University of New York, Stony Brook, USA
POPA, Sorin	INCREST, Bucharest
POPESCU, Gelu	INCREST, Bucharest
PREDA, Petre	University of Timişoara
PTAK, Vlastimil	Mathematics Institute, Prague, Czechoslovakia
PUTINAR, Mihai	INCREST, Bucharest
RĂDULESCU, Florin	INCREST, Bucharest
READ, Charles	University of Cambridge, England
REGHIŞ, Mircea	University of Timişoara
ROSENBERG, Johnatan	University of Maryland, College Park, USA
ROŞU, Radu	University of Bucharest
RUDOL,Krzysztof	Mathematics Institute, Kraków, Poland
SINGER, Ivan	INCREST, Bucharest
STACHO, László	University of Szeged, Hungary
STOCHEL, Jan	Mathematics Institute, Kraków, Poland
STRĂTILĂ, Şerban	INCREST, Bucharest
SUCIU, Ion	INCREST, Bucharest
SUCIU, Nicolae	University of Timişoara
SZAFRANIEC, Franciszek H.	University of Kraków, Poland

SZÖKEFALVI-NAGY, Béla	University of Szeged, Hungary
ŞABAC, Mihai	University of Bucharest
TAYLOR, Joseph L.	University of Utah, Salt Lake City, USA
TERESCENCO, Alexandru	Computer Centre, Timişoara
TIMOTIN, Dan	INCREST, Bucharest
TOPUZU, Paul	University of Timişoara
TOROK, A.	University of Timişoara
VALUŞESCU, Ilie	INCREST, Bucharest
VASILESCU, Florian-Horia	INCREST, Bucharest
VOICULESCU, Dan	INCREST, Bucharest
VRBOVA, Pabla	Mathematics Institute, Prague, Czechoslovakia
VUKMAN, J.	University of Maribor, Yugoslavia
VUZA, Dan	INCREST, Bucharest
WANG, Shengwang	University of Nanjing, China
WIDOM, Harold	University of California, Santa Cruz, USA
ZEMÁNEK, Jaroslav	Mathematics Institute, Warsaw, Poland

PROGRAMME OF THE CONFERENCE

Tuesday, June 5

Chairman: B.Szökefalvi-Nagy

9:30-10:10 **C.Pearcy:** Invariant subspaces, dilation theory and the structure of the predual of a dual operator algebra.

10:20-11:00 **H.Langer:** Classes of analytic functions related to operators in π_κ-spaces.

11:10-11:50 **J.D.Pincus:** A local index theory for certain Banach algebras.

Chairman: E.Albrecht

16:00-16:20 **M.Putinar:** Hyponormal operators are subscalar.

16:30-17:00 **R.E.Curto:** Uniform algebras, Hankel operators and invariant subspaces.

Section A

Chairman: P.Jonas

17:20-17:40 **R.Mennicken:** A generalization of a theorem of Keldysh.

17:45-18:05 **J.Barria:** Unicellular operators.

18:10-18:30 **J.Vukman:** A characterization of Hilbert spaces in terms of involution on operator algebra.

18:35-18:55 **A.Terescenco:** On the spectra of some operators acting in interpolation spaces.

Section B

Chairman: M.Reghis

17:20-17:40 **P.Legiša:** Hermitian operators on real Banach space with a hyperorthogonal basis.

17:45-18:05 **M.A.Balint; S.Balint:** On the Born-Oppenheimer approximation.

18:10-18:30 **H.U.Hess:** L^1-preduals and stochastic processes.

Wednesday, June 6

Chairman: C.Pearcy

9:30-10:10 **R.G.Douglas:** Hardy submodules for the polydisk algebra.

10:20-11:00 **M.A.Kaashoek:** Spectral analysis of systems of Wiener-Hopf integral equtions with symbols that are analytic in a strip.

11:10-11:50 **T.Kailath:** Cholesky factorization, generalized Schur algorithms and inverse scattering.

Chairman: H.Langer

16:00-16:30 **Gr.Arsene; T.Constantinescu:** Structure of the Naimark dilation and Gaussian stationary processes.

16:40-17:10 **H.Neidhardt:** The scattering operator of a dissipative scattering theory.

Section A

Chairman: D.Gașpar

17:30-17:50 **B.Magajna:** On the operator equation $ax + (ax)^* - \lambda x = b$.

17:55-18:15 **F.Rădulescu:** Generalized multipliers.

18:20-18:40 **S.Wang:** Duality theorems for closed operators with the SDP.

18:45-19:05 **T.Ceausu:** A reiteration theorem for an abstract interpolation method.

Section B

Chairman: V.Ptak

17:30-17:50 **M.Reghiș; N.Popescu; P.Topuzu:** A Lumer-Phillips type theorem and quasi-dissipativity.

17:55-18:15 **B.Cain:** Inertia theorems for Hilbert space operators.

18:20-18:40 **M.Omladic:** Is this a new spectralness?

18:45-19:05 **P.Găvruță:** On some integral operators.

Thursday, June 7

Chairman: J.D.Pincus

9:30-10:10 **P.Jonas; H.Langer:** A model for π-selfadjoint operators in π_1-spaces and a special linear pencil.

10:20-10:50 **A.A.Jafarian:** Linear maps preserving commutativity.

18:00-19:30 **S.V.Hruščev:** Seminar on De Branges' proof of Bieberbach conjecture.

Friday, June 8

Chairman: R.G.Douglas

9:30-10:10 **E.Albrecht:** Spectral theory on quotient spaces.

10:20-11:00 **F.-H.Vasilescu:** Analytic operators and spectral decomposition.

11:10-11:50 **M.Putinar:** Base change and the Fredholm index.

Chairman: F.-H. Vasilescu

16:00-16:30 **J.Eschmeier:** Local spectral theory and duality.

16:40-17:10 **M.Neumann:** Decomposable operators, divisible subspaces and problems of automatic continuity.

17:20-17:50 **R.N.Levy:** Coherent sheaves, connected with n-tuples of operators.

18:00-18:20 **L.Kérchy:** Approximation and quasisimilarity.

18:30-18:50 **J.Zemanek:** The stability radius of a semi-Fredholm operator.

Saturday, June 9

Chairman: J.L.Taylor

9:30-10:10 **H.Helson:** Differential equation of an inner function.

10:20-11:00 **B.Szökefalvi-Nagy:** On the canonical model of contractions.

11:10-11:40 **N.Boboc; Gh.Bucur:** Operators on Dirichlet spaces.

Monday, June 11

Chairman: H.Helson

9:30-10:10 **S.V.Hruščev:** Unconditional bases of exponentials in $L^2(-T,T)$ and resonance vibration of strings.

10:20-11:00 **B.Gramsch:** Perturbation theory in Ψ-algebras.

11:10-11:50 **H.Widom:** More about pseudodifferential operators on bounded domains.

Chairman: H. Widom

16:00-16:20 **D.Timotin:** On certain higher-dimensional generalizations of a result of Peller.

16:30-16:50 **J.Bógnar:** On the spectral function of definitizable operators.

17:00-17:20 **T.Constantinescu:** $n \times n$ positive operator-matrices.

17:30-17:50 **F.Szafraniec:** Some remarks on unbounded subnormal operators.

18:00-19:30 **J.L.Taylor:** Seminar on new developments on group representations.

Tuesday, June 12

Chairman: W.B.Arveson

9:30-10:10 **D.Voiculescu:** Almost-inductive limit automorphisms and embeddings into AF-algebras.

10:20-11:00 **J.Rosenberg:** The structure of some smooth C^*-crosed products.

11:10-11:50 **M.Pimsner:** Ranges of traces on K-groups of crossed products by free groups.

Chairman: S.-V.Strătilă

16:00-16:20 **C.Pasnicu:** On certain inductive limit C^*-algebras.

16:25-16:45 **R.N.Gologan:** Perturbing the ergodic theorem.

16:50-17:10 **L.Stacho:** Elementary operator-theoretical approach to the subgroups of U(n).

17:15-17:35 **R.Brigola:** A characterization of the Schur property by operator ideals.

17:40-18:35 **D.Vuza:** Applications of principal modules to linear operators on Banach lattices.

18:40-19:10 **S.Frunză:** Jordan operators on Hilbert space.

Wednesday, June 13

Chairman: J.Rosenberg

9:30-10:10 **W.B.Arveson:** Markov operators.

10:20-11:00 **S.Popa:** Derivations into the compacts and some properties of II_1 factors.

11:10-11:50 **J.L.Taylor:** Localization in analysis.

Cahirman: I.Suciu

16:00-16:30 **J.Dazord:** Generic contractions.

16:40-17:20 **V.Ptak; M.Neumann:** Uniform boundedness and automatic continuity.

17:25-17:45 **C.Niculescu:** On an Alfsen-Effros type ordering of L(E,E).

17:50-18:10 **F.Altomare:** Korovkin closures in Banach algebras.

18:15-18:35 **D.Gaspar; N.Suciu:** On Wold decompositions of Hilbert space representations of function algebras.

18:40-19:00 **R.Grzaślewicz:** Isometries of classical Banach spaces.

19:05-19:25 **J.Stochel:** An integral representation of covariance kernels on abelian $*$-semigroups without unit.

Thursday, June 14

Chairman: B.Gramsch

9:30-10:10 **Y.Domar:** Examples of invariant subspace lattices.

10:20-11:00 **B.Chevreau:** On reflexivity of operators in certain classes.

11:10-12:40 **Ch.Read:** Operators without invariant subspaces on the Banach space ℓ_1.

Operator Theory:
Advances and Applications, Vol.17

SOME TOPICS IN THE THEORY OF DECOMPOSABLE OPERATORS

Ernst Albrecht, Jörg Eschmeier, and **Michael M. Neumann**

0. INTRODUCTION AND PRELIMINARIES

A bounded linear operator T on a Banach space X is called *decomposable* if for every open covering $\{\Omega_1, \Omega_2\}$ of the complex plane $\mathbb{C}$ there are closed invariant subspaces X_1, X_2 for T such that $sp(T,X_j) \subset \Omega_j$ for $j = 1,2$ and $X = X_1 + X_2$ (where $sp(T,X_j)$ is the spectrum of T on X_j). We refer to the monographs [7, 16] for the theory of decomposable operators. For bounded linear operators T and S on Banach spaces X resp. Y we introduce the operators L(S), R(T), C(S,T) on the Banach space $L(X,Y)$ of all bounded linear operators from X to Y by

$$L(S)A := SA, \quad R(T)A := AT, \quad C(S,T)A := L(S)A - R(T)A = SA - AT$$

for $A \in L(X,Y)$. We shall see in the following section that the decomposability of R(T) resp. L(S) implies that T resp. S is decomposable (if $X \neq \{0\}$, $Y \neq \{0\}$). Conversely, if T and S are decomposable then L(S), R(T) still have some weaker decomposability properties. In the special case that X and Y are Hilbert spaces, we even obtain decomposability for L(S), R(T), and C(S,T). This kind of questions is related to the problem whether tensor products of decomposable operators are again decomposable. In the case of Hilbert spaces, positive answers to this problem are given in the second section. In the last section we introduce the class of "well"-decomposable operators. This class of operators which is strictly larger than the class of super-decomposable operators introduced in [13] contains most of the interesting examples in the theory of decomposable operators. We shall see that for well-decomposable operators without nontrivial divisible subspaces the algebraic characterization of the spectral maximal spaces given in [13], Proposition 1.5, in the case of super-decomposable operators is still valid. As pointed out in [13] this implies continuity properties for intertwining operators.

This joint work has been started during the 9th International Conference on Operator Theory at Timişoara and Herculane. The authors express their thanks to the organizers of this interesting and stimulating conference.

1. DECOMPOSABLE MULTIPLICATION OPERATORS

Let us first recall some notations and definitions. For an open set $\Omega\subset\mathbb{C}$ and a Banach space X, $H(\Omega,X)$ denotes the Fréchet space of all holomorphic X-valued functions on Ω (endowed with the topology of uniform convergence on all compact subsets of Ω). An operator $T\in L(X)$ is said to have the *single valued extension property* (SVEP) resp. *Bishop's property* (β) if for all open $\Omega\subset\mathbb{C}$ the linear map $\alpha_T^\Omega : H(\Omega,X)\to H(\Omega,X)$ given by $(\alpha_T^\Omega f)(z) := (z-T)f(z)$ for $f\in H(\Omega,X)$, $z\in\Omega$, is injective resp. injective with closed range. For $T\in L(X)$ with the SVEP and $x\in X$, the *local spectrum* $\sigma(x;T)$ *of* T *at* x is by definition the closed set of all $z\in\mathbb{C}$ such that the constant function x is not in the range of α_T^Ω for any open neighborhood Ω of z. Notice that property (β) implies that

$$X_T(F) := \{x\in X \mid \sigma(x;T)\subset F\} = X\cap \operatorname{Im}(\alpha_T^{\mathbb{C}\setminus F})$$

is a closed linear subspace of X for every closed $F\subset\mathbb{C}$. Property (β) is usually stated in the following equivalent form: For every open $\Omega\subset\mathbb{C}$ and every sequence $(f_n)_{n=1}^\infty$ in $H(\Omega,X)$ such that $(z-T)f_n(z)\to 0$ uniformly on all compact subsets of Ω, also $f_n\to 0$ uniformly on all compact subsets of Ω.

Using [7], Proposition 1.3.8, it is easy to see that an operator $T\in L(X)$ is decomposable if and only if T has the SVEP, $X_T(F)$ is closed for every closed $F\subset\mathbb{C}$ and T satisfies the following condition.

(δ) For every $x\in X$ and every open covering $\{\Omega_1,\Omega_2\}$ of $\mathbb{C}$ there are $x_1, x_2\in X$ with $x_1+x_2 = x$ and $\sigma(x_j;T)\subset\Omega_j$ for $j = 1, 2$.

It is well known that every decomposable operator satisfies property (β). Hence, $T\in L(X)$ is decomposable if and only if T has the properties (β) and (δ).

1.1. THEOREM. *Let* U, V, W, *be three Banach spaces,* $T\in L(W)$, *and let* E *be a closed linear subspace of* $L(U,W)$ *which is invariant for L(T). Suppose, that* $B: U\times V\to\mathbb{C}$ *is a not everywhere vanishing continuous bilinear form such that for all* $v\in V$, $w\in W$ *the bounded linear operator* $A(v,w)$ *defined by*

$$A(v,w)u := B(u,v)w \quad \textit{for } u\in U$$

is in E. *Then, if* $L(T)|E$ *is decomposable on* E, *also* T *is a decomposable operator.*

PROOF. Let us first show that T has property (β). Hence, fix an arbitrary open set $\Omega\subset\mathbb{C}$ and let $(f_n)_{n=1}^\infty$ be a sequence in $H(\Omega,W)$ such that $\alpha_T^\Omega f_n\to 0$ uniformly on all compact subsets of Ω. As $B\not\equiv 0$, there are $u\in U$ and $v\in V$ such that $B(u,v) = 1$. The functions $F_n:\Omega\to E$, given by $F_n(z) := A(v,f_n(z))$ for $z\in\Omega$, are obviously in $H(\Omega,E)$.

Moreover, for $n \to \infty$,

$$(\alpha^{\Omega}_{L(T)} F_n)(z) = A(v,(z - T)f_n(z)) \to 0$$

uniformly on all compact subsets of Ω. As $L(T)|E$ has property (β), this implies that also $F_n \to 0$ in $H(\Omega,E)$. Hence, also

$$f_n(z) = B(u,v)f_n(z) = A(v,f_n(z))u \to 0 \quad \text{for } n \to \infty$$

uniformly on every compact subset of Ω. Therefore, T has property (β).

In order to prove the decomposability of T, we still have to verify (δ). Hence, fix an arbitrary $w \in W$. If $u \in U$, $v \in V$ are such that $B(u,v) = 1$, then $A(v,w)u = w$. Therefore, if $\{\Omega_1, \Omega_2\}$ is an open covering of $\mathbf{C}$ then, by the decomposability of $L(T)|E$, there are $A_1, A_2 \in E$ such that $A_1 + A_2 = A(v,w)$ and $\sigma(A_j;L(T)|E) \subset \Omega_j$ $(j = 1, 2)$. With $w_j := A_j u$ we have $w_1 = w_2 = w$ and we obtain $\sigma(w_j;T) \subset \Omega_j$ for $j = 1, 2$. Indeed, as $K_j := \sigma(A_j;L(T)) \subset \subset \Omega_j$, there are $G_j \in H(\mathbf{C} \setminus K_j, E)$ satisfying $(z - T)G_j(z) \equiv A_j$ on $\mathbf{C} \setminus K_j$. Therefore, the functions $z \to G_j(z)u$ are in $H(\mathbf{C} \setminus K_j, W)$ and satisfy $(z - T)G_j(z)u \equiv A_j u = w_j$ on $\mathbf{C} \setminus K_j$. Thus, $\sigma(w_j;T) \subset \Omega_j$, for $j = 1, 2$.

1.2. COROLLARY. *Let* X *and* Y *be two non-trivial Banach spaces and* $T \in L(X)$, $S \in L(Y)$.

(a) *If the operator* $L(S)$ *is decomposable on* $L(X,Y)$, *then also* S *is decomposable.*

(b) *If the operator* $R(T)$ *is decomposable on* $L(X,Y)$, *then also* T *is decomposable.*

PROOF. (a) is an immediate consequence of Theorem 1.1, taking $U = X$, $V = X^*$, $W = Y$, $E = L(X,Y)$, S instead of T, and $B(x,x^*) := x^*(x)$ for $x \in U = X$ and $x^* \in V = X^*$.

(b) Let $J : L(X,Y) \to L(Y^*,X^*)$ be the canonical isometric monomorphism given by $J(A) := A^*$ for $A \in L(X,Y)$ and let E be the range of J. Then E is invariant for $L(T^*)$ and $L(T^*)|E = JR(T)J^{-1}$. Suppose now that $R(T)$ is decomposable on $L(X,Y)$. Then also $L(T^*)|E$ must be decomposable. If we now take $U := Y^*$, $V := Y$, $W := X^*$, T^* for T, and $B(y^*,y) := y^*(y)$ for $y^* \in U = Y^*$, $y \in V = Y$, then we see from Theorem 1.1 that T^* is decomposable. By [8], Corollary 3, T is decomposable.

Part (a) of Corollary 1.2 gives a positive answer to a question raised in [13].

1.3. COROLLARY. *Let* X *and* Y *be two non-trivial Banach spaces. Then, for* $T \in L(X)$ *and* $S \in L(Y)$, *the operators* $L(S)$ *and* $R(T)$ *may also be considered as bounded linear operators on the space* $C(X,Y)$ *of all compact linear operators from* X *to* Y.

(a) *If* $L(S)$ *is decomposable on* $C(X,Y)$ *then* S *is decomposable on* Y.

decomposable on Y.

(b) *If* R(T) *is decomposable on* C(X,Y) *then* T *is decomposable on* X.

For the proof we only have to replace $L(X,Y)$ by $C(X,Y)$ and $L(Y^*,X^*)$ by $C(Y^*,X^*)$ in the proof of 1.2.

The converse problem, namely whether the decomposability of S on Y resp. of T on X implies the decomposability of L(S) resp. of R(T) on $L(X,Y)$ or $C(X,Y)$ seems to be more complicated. We are only able to prove a weak decomposability property for L(S) resp.R(T).

1.4. DEFINITION. Let X be a Banach space and let τ be a locally convex topology on X. An operator $T \in L(X)$ will be called τ-*quasi-decomposable* if

(i) T has the SVEP.

(ii) For all closed $F \subset \mathbf{C}$ the linear space $X_T(F)$ is norm closed in X.

(iii) For every finite open covering $\{U_1,\dots,U_n\}$ of $\mathbf{C}$ the τ-closure of $X_T(\bar{U}_1) + \dots + X_T(\bar{U}_n)$ coincides with X.

If τ is the topology induced by the norm of X then we say simply quasi--decomposable instead of τ-quasi-decomposable. It is known [2] that there exist quasi--decomposable operators which are not decomposable.

1.5. THEOREM. *Suppose that* $S \in L(Y)$ *and* $T \in L(X)$ *are decomposable.*

(a) If X *or* Y *has the approximation property then the operators* L(S) *and* R(T) *are* τ_c*-quasi-decomposable on* $L(X,Y)$, *where* τ_c *denotes the topology of uniform convergence on all compact subsets of* X. *For all closed* $F \subset \mathbf{C}$ *we have (with* $L := L(X,Y)$*)*

(1) $\quad L_{L(S)}(F) = \{A \in L \mid A(X) \subset Y_S(F)\}$

(2) $\quad L_{R(T)}(F) = \{A \in L \mid A(X_T(H)) = \{0\}$ for all closed $H \subset \mathbf{C} \setminus F\}$ *and these spaces are closed in* L.

(b) *If* X^* *or* Y *has the approximation property then the operators* L(S) *and* R(T) *are quasi-decomposable on* C(X,Y). *With* $C := C(X,Y)$ *we have for all closed* $F \subset \mathbf{C}$

(3) $\quad C_{L(S)}(F) = \{A \in C \mid A(X) \subset Y_S(F)\}$

(4) $\quad C_{R(T)}(F) = \{A \in C \mid A(X_T(H)) = \{0\}$ for all closed $H \subset \mathbf{C} \setminus F\}$ *and these spaces are closed in* C.

PROOF. By Proposition 2.2 in [11] the operators L(S) and R(T) have the SVEP on L and hence on C (as C is a closed linear subspace of L which is invariant for L(S) and

R(T)). (1) and (2) follow from Theorem 2.5 in [11]. Because of

$$\mathcal{C}_{L(S)}(F)\subset\mathcal{L}_{L(S)}(F)\cap\mathcal{C} \quad\text{and}\quad \mathcal{C}_{R(T)}(F)\subset\mathcal{L}_{R(T)}(F)\cap\mathcal{C}$$

the inclusions "$\subset$" in (3) and (4) are obvious. Conversely, if $A\in\mathcal{C}$ such that $A(X)\ \ Y_S(F)$, then for all $z\in\mathbb{C}\setminus F$ the operator $G(z):=(z-S|Y_S(F))^{-1}A$ exists and the function $G:\mathbb{C}\setminus F\to\mathcal{C}$ is analytic with values in $\mathcal{C}$. Because of $(z-L(S))G(z)=(z-S)\cdot$ $\cdot(z-S|Y_S(F))^{-1}A=A$ for all $z\in\mathbb{C}\setminus F$ we see that $\sigma(A;L(S))\subset F$ i.e. $A\in\mathcal{C}_{L(S)}(F)$. This proves (3). Let now A be an operator in $\mathcal{C}$ such that $A(X_T(H))=\{0\}$ for all closed $H\subset\mathbb{C}\setminus F$ and let w be an arbitrary point in $\mathbb{C}\setminus F$. Fix an open neighborhood U of w with $\bar{U}\cap F=\emptyset$ and two open sets $H, V\subset\mathbb{C}$ such that $\bar{U}\subset H\subset\bar{H}\subset\mathbb{C}\setminus F$ and $\mathbb{C}\setminus H\subset V\subset\bar{V}\subset\mathbb{C}\setminus\bar{U}$. Then $\mathbb{C}=H\cup V$ and thus $X=X_T(\bar{H})+X_T(\bar{V})$ by the decomposability of T. For $z\in U$ and $x\in X$ we now define $B(z)x:=A(z-T|X_T(\bar{V}))^{-1}x_2$ where $x=x_1+x_2$ with $x_1\in X_T(\bar{H})$ and $x_2\in X_T(\bar{V})$. Because of $A(X_T(\bar{H}))=\{0\}$ the mapping B(z) is well-defined. Moreover,

$$((z-R(T))B(z))x=B(z)(z-T)x=A(z-T|X_T(\bar{V}))^{-1}(z-T)x_2=Ax_2=Ax_1+Ax_2=Ax,$$

i.e. $B(z)(z-R(T))=A$ for all $z\in U$. By the open mapping theorem there exists a constant $C>0$ such that for all x in the unit ball of X we have a decomposition $x=x_1+x_2$ with $x_1\in X_T(\bar{H})$, $x_2\in X_T(\bar{V})$ and $\|x_1\|$, $\|x_2\|\leq C$. With this choice and because of the boundedness of the operator $(z-T|X_T(\bar{V}))^{-1}$ we see that the set $M:=$ $:=\{(z-T|X_T(\bar{V}))^{-1}x_2\,|\,x\in X,\ \|x\|\leq 1\}$ is bounded in $X_T(\bar{V})$ and hence in X. As A is a compact operator the set $B(z)(\{x\in X\,|\,\|x\|\leq 1\})\subset A(M)$ must be relatively compact in Y. This shows that $B(z)\in\mathcal{C}$. From the definition of $B:U\to\mathcal{C}$ we see that B is analytic in the strong operator topology. By a standard argument one shows easily that this implies the analyticity of B with respect to the operator norm. Thus, $B:U\to\mathcal{C}$ is in $H(U,\mathcal{C})$ and satisfies $(z-R(T))B(z)=A$ on U. This shows that $w\in\mathbb{C}\setminus\sigma(A;R(T))$. Hence $\sigma(A;R(T))\subset$ $\subset F$ i.e. $A\in\mathcal{C}_{R(T)}(F)$ and the proof of (4) is complete.

We have proved that the operators L(S) and R(T) in (a) and (b) satisfy (i) and (ii) in Definition 1.4. Notice that in (a) the set $X^*\otimes Y$ of all finite rank operators is dense in $\mathcal{L}$ with respect to τ_c (cf. [12], p.232) and that in (b) the set $X^*\otimes Y$ is norm dense in $\mathcal{C}$ (cf. [12], p.235). Thus, our proof will be complete if we can show that for every finite open covering $\{U_1,\ldots,U_n\}$ of $\mathbb{C}$ and every $A\in X^*\otimes Y$ there are $A_1,\ldots,A_n\in X^*\otimes Y$ with $A=A_1+\ldots+A_n$ and $A_j\in\mathcal{C}_{L(S)}(\bar{U}_j)$ for $j=1,\ldots,n$ (resp. $A_j\in\mathcal{C}_{R(T)}(\bar{U}_j)$ for $j=1,\ldots,n$). Obviously it suffices to give the proof for operators A of the form $A=$ $x^*\otimes y$ with $x^*\in X$, $y\in Y$. As S is decomposable, we find $y_j\in Y_S(\bar{U}_j)$ $(j=1,\ldots,n)$ with $y=y_1+\ldots+y_n$. Then we have $A_j:=x^*\otimes y_j\in\mathcal{C}_{L(S)}(\bar{U}_j)$ for $j=1,\ldots,n$ and

$A = A_1 + \dots + A_n$. Also, the decomposability of S implies that S^* is decomposable (cf. [16], Proposition IV.5.6) so that there are $x_j^* \in X^*_{T^*}(\bar{U}_j)$ $(j = 1,\dots,n)$ with $x^* = x_1^* + \dots + x_n^*$. Put $A_j := x_j^* \otimes y$ for $j = 1,\dots,n$. If $H \subset \mathbf{C} \setminus \bar{U}_j$ is closed and $x \in X_T(H) = X_T(H \cap sp(T,X))$ then $A_j x = x_j^*(x) y = 0$ (by [16], Proposition IV.5.6). Hence, $A_j \in C_{R(T)}(\bar{U}_j)$ for $j = 1,\dots,n$ and $A = A_1 + \dots + A_n$ and the proof of the theorem is complete.

In the situation of Hilbert spaces we can prove more:

1.6. THEOREM. *Let* H *and* K *be two non-trivial Hilbert spaces.*

(a) *For* $S \in L(K)$ *are equivalent:*

(i) S *is decomposable;*

(ii) L(S) *is decomposable on* $L(H,K)$;

(iii) L(S) *is decomposable on* $C(H,K)$.

(b) *For* $T \in L(H)$ *are equivalent:*

(i) *T is decomposable;*

(ii) R(T) *is decomposable on* $L(H,K)$;

(iii) R(T) *is decomposable on* $C(H,K)$.

PROOF. Because of the Corollaries 1.2 and 1.3 we have only to show that (i) implies (ii) and (iii) in (a) and (b).

(a) Suppose that $S \in L(K)$ is decomposable. By Theorem 1.5, L(S) has the SVEP and the spaces $L_{L(S)}(F)$ and $C_{L(S)}(F)$ (with $L = L(H,K)$, $C = C(H,K)$) are closed and have the form (1) resp. (3). If U, $V \subset \mathbf{C}$ are open with $U \cup V = \mathbf{C}$ then, by the decomposability of S we have the direct decomposition $K = K_S(\bar{U}) + K_S(\bar{V}) = K_S(\bar{U}) \oplus (K_S(\bar{V}) \ominus K_S(\bar{U} \cap \bar{V}))$. Let $P \in L(K)$ be the corresponding projection onto $K_S(\bar{U})$ with kernel $K_S(\bar{V}) \ominus K_S(\bar{U} \cap \bar{V})$. For all $A \in L$ (resp. for all $A \in C$) we then have $A = PA + (1-P)A$ and $(PA)(H) \subset K_S(\bar{U})$, $((1-P)A)(H) \subset K_S(\bar{V}) \ominus K_S(\bar{U} \cap \bar{V}) \subset K_S(\bar{V})$. Because of (1) (resp.(3)) this implies $PA \in L_{L(S)}(\bar{U})$ and $(1-P)A \in L_{L(S)}(\bar{V})$ (resp. $PA \in C_{L(S)}(\bar{U})$ and $(1-P)A \in C_{L(S)}(\bar{V})$). Hence ($\delta$) is satisfied and L(S) is decomposable on L and C.

(b) For $A \in L(H,K)$ let $J(A) \in L(K^*,H^*)$ be the transposed operator. Then $J : L(H,K) \rightarrow L(K^*,H^*)$ (resp. $J : C(H,K) \rightarrow C(K^*,H^*)$) is an isometric isomorphism and $R(T) = J^{-1} L(J(T)) J$. As the decomposability of T implies that of J(T) (cf. [16], Proposition IV.5.6) the result now follows from (a).

In the following section the statements (i)$\Longrightarrow$(ii) and (i)$\Longrightarrow$(iii) will be improved.

2. THE HILBERT SPACE CASE AND TENSOR PRODUCTS

In this part we shall give two closely related applications of a perturbation result derived for spectral decompositions in [9]. We shall show that for decomposable operators T, $S \in L(H)$ on a complex Hilbert space H all analytic functions in the commuting pair $M = (L(T), R(S))$ are decomposable, in particular this applies to the commutator

$$C(T,S) : L(H) \rightarrow L(H); \qquad A \rightarrow TA - AS$$

and to the multiplier

$$M(T,S) : L(H) \rightarrow L(H); \qquad A \rightarrow TAS.$$

Instead of giving a direct proof we shall first show that operator decomposability is compatible with topological tensor products between Hilbert spaces and then use the duality between $L(H)$ and the completed projective tensor product $H\tilde{\otimes}_{\pi}H'$.

By a topological tensor product between Hilbert spaces we shall mean a rule assigning to each pair of Hilbert spaces H, K a Banach space $H\tilde{\otimes}K$ and a continuous bilinear map $H \times K \twoheadrightarrow H\tilde{\otimes}K$ such that

(1) the linear span of $\{x\otimes y \,;\, x \in H,\, y \in K\}$ is dense in $H\tilde{\otimes}K$;

(2) for each continuous linear map $T : H_1 \rightarrow H_2$ between Hilbert spaces there are continuous linear maps $T\otimes 1 : H_1\tilde{\otimes}K \rightarrow H_2\tilde{\otimes}K$, and $1\otimes T : K\tilde{\otimes}H_1 \rightarrow K\tilde{\otimes}H_2$ with $T\otimes 1(x\otimes y) = Tx\otimes y$ and $1\otimes T(y\otimes x) = y\otimes Tx$ for $x \in H_1$, $y \in K$;

(3) for T as in (2) we have $\|T\otimes 1\| \leq \|T\|$, $\|1\otimes T\| \leq \|T\|$;

(4) the maps $H \rightarrow H\tilde{\otimes}\mathbf{C}$; $x \rightarrow x\otimes 1$, $H \rightarrow \mathbf{C}\tilde{\otimes}H$, $x \rightarrow 1\otimes x$ are topological isomorphisms.

Canonical examples satisfying these conditions are the ε - tensor product $\tilde{\otimes}_{\varepsilon}$, the π-tensor product $\tilde{\otimes}_{\pi}$ and the Hilbert tensor product $\bar{\otimes}$.

We shall need the following elementary and well known fact.

2.1. LEMMA. Let E, F *be topological vector spaces. For a continuous linear map* $T : E \rightarrow F$ *(* $S : F \rightarrow E$ *) there is a continuous linear map* $S : F \rightarrow E$ *(resp.* $T : E \rightarrow F$*) with* $TS = I$ *if and only if* T *(resp.*S*) is a surjective (injective) homomorphism with continuously projected kernel (range).*

In this case $E = \operatorname{Ker} T \oplus \operatorname{Im} S$ *and* ST *is the projection of* E *onto* $\operatorname{Im} S$ *along* $\operatorname{Ker} T$.

Here of course a homomorphism is a continuous linear map, which is open onto its range. Now, consider an arbitrary topological tensor product $\tilde{\otimes}$ between Hilbert spaces H, K.

If $P : H \to H$ is a continuous projection with range M, if $\pi_M : H \to M$ denotes the same map, but with image space M, and if $i_M : M \to H$ denotes the embedding, then

$$M\tilde{\otimes}K \xrightarrow{i_M \otimes 1} H\tilde{\otimes}K \xrightarrow{\pi_M \otimes 1} M\tilde{\otimes}K$$

is the identity map on $M\tilde{\otimes}K$. Therefore

$$M\tilde{\otimes}K \xrightarrow{i_M \otimes 1} (i_M \otimes 1)(M\tilde{\otimes}K) = (P\otimes 1)(H\tilde{\otimes}K)$$

is a topological isomorphism onto a continuously projected subspace of $H\tilde{\otimes}K$. If $T : H_1 \to H_2$ is a surjective bounded operator between Hilbert spaces, then by Lemma 2.1 for each Hilbert space K

$$T\otimes 1 : H_1\tilde{\otimes}K \to H_2\tilde{\otimes}K$$

is a surjective bounded operator with continuously projected kernel

$$\mathrm{Ker}(T\otimes 1) \simeq \mathrm{Ker}(T)\tilde{\otimes}K.$$

For each closed subspace M of H we regard $M\tilde{\otimes}K$ as a closed subspace of $H\tilde{\otimes}K$ in the above sense. If M is invariant for $T \in L(H)$, then for arbitrary $S \in L(K)$ the space $M\tilde{\otimes}K$ is invariant for $T\otimes S$, and $(T|M)\otimes S$ is similar to $T\otimes S|M\tilde{\otimes}K$. If M, N are closed subspaces of H with $H = M + N$, then $H\tilde{\otimes}K = M\tilde{\otimes}K + N\tilde{\otimes}K$. This is due to the fact, that the tensor product of the map $M\oplus N \to H$, $x\oplus y \to x + y$ with the identity operator on K remains surjective.

For each $k \in K$ define $u_k : K \to \mathbf{C}$ by $u_k(y) = \langle y,k\rangle$. The composition f_k of $1\otimes u_k$ and of the topological isomorphism

$$H\tilde{\otimes}\mathbf{C} \to H, \quad x\otimes\alpha \to \alpha x$$

is the unique continuous linear map $H\tilde{\otimes}K \to H$ with

$$f_k(x\otimes y) = \langle y,k\rangle x \quad (x \in H,\ y \in K).$$

Condition (3) can be used to prove:

2.2. LEMMA. *If* $z \in H\tilde{\otimes}K$, *then* $f_k(z) = 0$ *for all* $k \in K$ *implies* $z = 0$.

PROOF. For each subset M of K we denote by P_M the orthogonal projection of K onto the orthogonal complement $M^\perp$ of M in K. For each $k \in K$ it follows that

$$\mathrm{Ker}\, f_k = 1\otimes P_{\{k\}}(H\tilde{\otimes}K).$$

Choose an orthonormal basis B of K and notice that $P_C P_D = P_{C\cup D}$ for arbitrary subsets C, D of B. Denote by J the system of all finite subsets of B ordered by inclusion. Then the net $(P_C)_{C\in J}$ converges strongly to the zero operator. Since $(1\otimes P_C)_{C\in J}$ is norm-bounded by (3) and converges to zero pointwise on a dense subset of $H\tilde{\otimes}K$, it follows that $(1\otimes P_C)_{C\in J}$ converges to zero in the strong operator topology. The observation, that

$$\bigcap \{\mathrm{Ker}\, f_k\, ;\, k\in C\} = (1\otimes P_C)(H\tilde{\otimes}K)$$

holds for each $C\in J$ completes the proof.

Now we have gathered all pieces needed to prove the first of our main results.

2.3. THEOREM. *If* $T\in L(H)$ *is decomposable, then* $T\otimes 1$ *is decomposable. Moreover for each closed set* F *in* $\mathbb{C}$

$$H\tilde{\otimes}K_{T\otimes 1}(F) = H_T(F)\tilde{\otimes}K.$$

PROOF. Let $\mathbb{C} = U_1\cup U_2$ be an open cover. If M_1, M_2 are closed invariant subspaces for T with

$$H = M_1 + M_2;\quad sp(T|M_i)\subset U_i \quad (i = 1, 2),$$

then $H\tilde{\otimes}K = M_1\tilde{\otimes}K + M_2\tilde{\otimes}K$ is a decomposition into a sum of closed subspaces invariant for $T\otimes 1$ with

$$sp(T\otimes 1\,|M_i\tilde{\otimes}K) = sp((T|M_i)\otimes 1)\subset sp(T|M_i)\subset U_i \quad (i = 1, 2).$$

By Corollary 2 of [3] the operator $T\otimes 1$ is decomposable. The inclusion $H_T(F)\tilde{\otimes}K\subset$ $\subset H\tilde{\otimes}K_{T\otimes 1}(F)$ is obvious because of

$$sp(T\otimes 1\,|H_T(F)\tilde{\otimes}K) = sp((T|H_T(F))\otimes 1)\subset sp(T|H_T(F))\subset F.$$

To prove the opposite inclusion let P denote the orthogonal projection of H onto $H_T(F)^{\perp}$ and notice that

$$f_k\circ (S\otimes 1) = S\circ f_k$$

holds for each $k\in K$ and each $S\in L(H)$. For S = T we get

$$f_k(H\tilde{\otimes}K_{T\otimes 1}(F))\subset H_T(F)$$

for all $k\in K$. For S = P we obtain

$$f_k(P\otimes 1(H\tilde{\otimes}K_{T\otimes 1}(F)))\subset PH_T(F) = \{0\}$$

for all $k \in K$. Hence we have $H\tilde{\otimes}K_{T\otimes 1}(F) \subset Ker(P\otimes 1) = H_T(F)\tilde{\otimes}K$.

If $S \in L(K)$ is decomposable, then of course also $1\otimes S \in L(H\tilde{\otimes}K)$ is decomposable with spectral subspaces $H\tilde{\otimes}K_{1\otimes S}(F) = H\tilde{\otimes}K_S(F)$. Using a result of [9] we can prove a much stronger result.

2.4. THEOREM. If $T \in L(H)$, $S \in L(K)$ *are decomposable, then the commuting pair* $(T\otimes 1, 1\otimes S)$ *has the local decomposition property.*

PROOF. For the definition of the local decomposition property and its properties see [10]. By Theorem 1.7 of [9] it suffices to prove that for each closed set F in $\mathbf{C}$ the operators induced by $1\otimes S$ on $H\tilde{\otimes}K_{T\otimes 1}(F)$ and on $(H\tilde{\otimes}K)/(H\tilde{\otimes}K_{T\otimes 1}(F))$ are decomposable. The first is decomposable, because it is similar to $(1|H_T(F))\otimes S$. The second is decomposable, because the exactness of the following short sequence

$$0 \to H_T(F)\tilde{\otimes}K \xrightarrow{i\otimes 1} H\tilde{\otimes}K \xrightarrow{q\otimes 1} (H/H_T(F))\tilde{\otimes}K \to 0,$$

where $i : H_T(F) \to H$ denotes the embedding and $q : H \to H/H_T(F)$ the quotient map, yields a similarity with

$$(H/H_T(F))\tilde{\otimes}K \xrightarrow{1\otimes S} (H/H_T(F))\tilde{\otimes}K.$$

The local decomposition property of $(T\otimes 1, 1\otimes S)$ implies the decomposability of $f(T\otimes 1, 1\otimes S)$ for all $\mathbf{C}$ - valued functions f which are analytic in a neighborhood of $sp(T\otimes 1, 1\otimes S)$. In particular we obtain the following corollaries:

2.5. COROLLARY. *If* $T \in L(H)$, $S \in L(K)$ *are decomposable, then* $T\otimes S \in L(H\tilde{\otimes}K)$ *is decomposable.*

In the following, $C_1(H)$ (resp. $C_2(H)$) denotes the set of all trace-class (resp. Hilbert-Schmidt) operators on H.

2.6. COROLLARY. *Let* $S, T \in L(H)$ *be decomposable operators. Consider* $L(T)$, $R(S)$ *as multiplication operators on the Banach space* $J \in \{C_1(H), C_2(H), C(H), L(H)\}$. *In each case we have*

$$sp(L(T), R(S)) = sp(T)\otimes sp(S).$$

For all f *analytic in a neighborhood of* $sp(T)\otimes sp(S)$ *the operator* $f(L(T),R(S))$ *is decomposable on* J *for each choice of* J.

PROOF. Theorem 5.14 of [14] shows that $L(H)$ is isometrically isomorphic to the norm dual of the completed projective tensor product $H\tilde{\otimes}_\pi H'$. An operator $A \in L(H)$ acts on an elementary tensor (proof of Theorem 3.1 in [14]) as

$$\langle x\otimes u, A\rangle = \langle Ax, u\rangle.$$

For $T, S \in L(H)$ the system $M = (L(T), R(S))$ on $L(H)$ is the adjoint of $(1\otimes T', S\otimes 1)$ acting on $H\tilde{\otimes}_\pi H'$. This follows from

$$\langle TAx, u\rangle = \langle(1\otimes T')(x\otimes u), A\rangle,$$

$$\langle ASx, u\rangle = \langle(S\otimes 1)(x\otimes u), A\rangle.$$

The natural identification between the algebraic tensor product $H\otimes H'$ and the set of continuous finite rank operators on H assigning to the elementary tensor $x\otimes u \in H\otimes H'$ the rank one operator

$$x\otimes u(\xi) = u(\xi)x \qquad (\xi \in H),$$

entends to an isometric isomorphism between $H\tilde{\otimes}_\pi H'$ and $C_1(H)$, $H\bar{\otimes}H'$ and $C_2(H)$, $H\tilde{\otimes}_\varepsilon H'$ and $C(H)$ respectively.

Relative to this identification the system M is similar to $(T\otimes 1, 1\otimes S')$. Since the local decomposition property is compatible with duality (Corollary 1.7 of [10]), it follows that M has the local decomposition property on J for each choice of J, whenever $T, S \in L(H)$ are decomposable.

Due to Theorem 3.1 of [15] we have

$$sp(M, C_2(H)) = sp(T)\otimes sp(S') = sp(T)\otimes sp(S)$$

for arbitrary operators $T, S \in L(H)$. To deduce this equality in the remaining cases one can use that $C_2(H)$ is continuously embedded in J for $J = C(H), L(H)$ and that $M|C_2(H)$ has the local decomposition property. It follows that $C_2(H)_M(F) = \{0\}$ for each closed set F in $\mathbb{C}$ with $F\cap sp(M, J) = \emptyset$, and hence

$$sp(M, C_2(H))\subset sp(M, J) \qquad (J = C(H), L(H)).$$

The opposite inclusion is obvious and the case $J = C_1(H)$ follows by duality.

3. AN ALGEBRAIC CHARACTERIZATION OF SPECTRAL MAXIMAL SPACES

The present section is motivated by some recent results from [13], where an algebraic description of the spectral maximal spaces is obtained for a certain class of decomposable operators. Apart from certain applications to problems of automatic continuity, this characterization seems to be of some intrinsic value. Here, we shall extend some basic results from [13] to a somewhat larger class of decomposable operators.

Let X be a complex Banach space. In [13] an operator $T \in L(X)$ is called *super--decomposable* if for every open covering $\{U,V\}$ of $\mathbf{C}$ there exists an $R \in L(X)$ commuting with T such that $sp(T|\overline{R(X)}) \subset U$ and $sp(T|\overline{(I-R)(X)}) \subset V$. Now, let us define $T \in L(X)$ to be *well-decomposable* if for every open covering $\{U,V\}$ of $\mathbf{C}$ there exist an $R \in L(X)$, an integer $n \in \mathbf{N}$ as well as closed T-invariant linear subspaces Y and Z of X such that $sp(T|Y) \subset U$, $sp(T|Z) \subset V$, $R(X) \subset Y$, $(I-R)(X) \subset Z$, and finally $C(T)^n R = 0$, where $C(T) := L(T) - R(T)$ denotes the commutator of T acting on $L(X)$.

Every super-decomposable operator is certainly well-decomposable. On the other hand, well-decomposable operators are decomposable and in fact strongly decomposable (see [16], V.6.26 for the definition), since the restriction of a well-decomposable operator to an arbitrary spectral maximal space is easily seen to be well-decomposable again; in this connection let us note that the spectral maximal spaces of a decomposable operator $T \in L(X)$ are invariant under every $R \in L(X)$ satisfying $C(T)^n R = 0$ for some $n \in \mathbf{N}$, see Corollary 2.3.4 of [7]. The class of well-decomposable operators is strictly larger than the class of super-decomposable operators:

3.1. PROPOSITION. *Let* $M \in L(X)$ *be super-decomposable and* $N \in L(X)$ *be nilpotent such that* $MN = NM$. *Then* $T := M + N$ *is well-decomposable, but not necessarily super-decomposable.*

PROOF. (1) By Theorem 2.2.1 of [7], T is decomposable and satisfies $X_T(F) = X_M(F)$ for all closed $F \subset \mathbf{C}$. Now, given an arbitrary open covering $\{U,V\}$ of $\mathbf{C}$, we choose an $R \in L(X)$ commuting with M such that $R(X) \subset X_M(\overline{U})$ and $(I-R)(X) \subset X_M(\overline{V})$. Then we have

$$C(T)^{2n}R = C(N)^{2n}R = \sum_{k=0}^{2n} \binom{2n}{k} (-1)^k N^{2n-k} R N^k = 0,$$

where $n \in \mathbf{N}$ is chosen so that $N^n = 0$. It follows that T is well-decomposable.

(2) We shall use the example from [1] to show that such an operator T need not be super-decomposable. Let D be the open unit disc in the complex plane, consider the Banach space X_0 of all continuous complex-valued functions on $\overline{D}$ endowed with the sup-norm $\|\cdot\|_0$, and let X_1 consist of all $g \in X_0$ such that the derivative $\bar{\partial} g$ in the sense of distributions belongs to X_0 as well. X_1 is a Banach space with respect to the norm $\|\cdot\|_1$ given by

$$\|g\|_1 := \|g\|_0 + \|\bar{\partial} g\|_0 \qquad \text{for all } g \in X_1.$$

On the Banach space $X := X_0 \oplus X_1$, we consider the operators M, N $\in L(X)$ given by

$$M(f,g) := (zf\ ,\ zg), \quad N(f,g) := (\bar{\partial} g\ ,\ 0) \quad \text{for all } (f,g) \in X,$$

where z stands for the identity function on $\bar{D}$. By Proposition 2.2 of [1], M is a generalized scalar operator and satisfies

$$X_M(F) = \{(f,g) \in X : \text{supp } f, \text{ supp } g \subset F\} \quad \text{for all closed } F \subset \mathbf{C}.$$

In particular, M is super-decomposable by Proposition 2.2 of [13]. Obviously N is nilpotent and commutes with M. But T := M + N is not super-decomposable. Indeed, given an arbitrary R $\in L(X)$ commuting with T, we obtain from Proposition 2.6 of [1] the representation

$$R(f,g) = (af + b\bar{\partial} g + cg\ ,\ ag) \qquad \text{for all } (f,g) \in X,$$

where a, b, c $\in X_0$ and a is analytic on D. Hence the identity theorem for analytic functions tells us that T cannot be super-decomposable.

Given an operator $T \in L(X)$ with the single-valued extension property and an arbitrary closed subset F of $\mathbf{C}$, let $E_T(F)$ denote the largest algebraic linear subspace Y of X such that $(T - \lambda)Y = Y$ holds for all $\lambda \in \mathbf{C} \setminus F$. Obviously $X_T(F)$ is contained in $E_T(F)$. In order to obtain the algebraic representation $X_T(F) = E_T(F)$ for all closed $F \subset \mathbf{C}$, one certainly has to assume that T has no *divisible* linear subspace different from $\{0\}$, which means that every algebraic linear subspace Y of X with the property $(T - \lambda)Y = Y$ for all $\lambda \in \mathbf{C}$ has to be trivial. In the case of a well-decomposable operator $T \in L(X)$, this necessary condition turns out to be sufficient as well, which generalizes Proposition 1.5 of [13]:

3.2. PROPOSITION. *Assume that* $T \in L(X)$ *is well-decomposable and has no divisible linear subspace different from* $\{0\}$. *Then* $X_T(F) = E_T(F)$ *holds for all closed* $F \subset \mathbf{C}$.

PROOF. Since $X_T(\cdot)$ preserves countable intersections, it suffices to prove that $E_T(F)$ is contained in $X_T(\bar{V})$, where V denotes an arbitrary open neighborhood of a given closed subset F of $\mathbf{C}$. We choose an open subset U of $\mathbf{C}$ such that $F \subset U \subset \bar{U} \subset V$. Since T is well-decomposable, there exist an $R \in L(X)$ and an integer $n \in \mathbf{N}$ such that $R(X) \subset X_T(\mathbf{C} \setminus U)$, $(I - R)(X) \subset X_T(\bar{V})$, and $C(T)^n R = 0$. Obviously we shall obtain the desired inclusion $E_T(F) \subset X_T(\bar{V})$ as soon as R is seen to vanish on $E_T(F)$. Let Z denote the largest algebraic linear subspace of $X_T(\mathbf{C}\setminus U)$ such that $(T-\lambda)Z = Z$ holds for all $\lambda \in \mathbf{C}\setminus F$. Since $sp(T|X_T(\mathbf{C}\setminus U)) \subset \mathbf{C}\setminus U \subset \mathbf{C}\setminus F$, the maximality of the space Z implies that Z is actually divisible for T. Hence $Z = \{0\}$ by our assumption on T. On the other hand, let us introduce the algebraic linear subspace

$$Y := R(E_T(F)) + TR(E_T(F)) + \ldots + T^{n-1}R(E_T(F))$$

of X. Since $R(E_T(F)) \subset X_T(\mathbf{C}\setminus U)$ and since the latter space is invariant under T, we conclude that $Y \subset X_T(\mathbf{C}\setminus U)$. Moreover, from $C(T)^n R = 0$ one easily deduces that Y is invariant under T. In particular, it follows that $(T-\lambda)Y \subset Y$ holds for all $\lambda \in \mathbf{C}\setminus F$. In order to prove the opposite inclusion, let $\lambda \in \mathbf{C}\setminus F$ and $y \in Y$ be arbitrarily given. Then we have

$$y = \sum_{k=0}^{n-1} T^k R a_k \qquad \text{for suitable } a_0, \ldots, a_{n-1} \in E_T(F).$$

Obviously, the system of linear equations

$$\sum_{j=k}^{n-1} \binom{j}{k} (-\lambda)^{j-k} b_j = a_k \qquad \text{for } k = 0, 1, \ldots, n-1$$

has a unique solution $b_0, \ldots, b_{n-1} \in E_T(F)$. For this solution we obtain

$$\sum_{j=0}^{n-1} (T-\lambda)^j R b_j = \sum_{j=0}^{n-1} \sum_{k=0}^{j} \binom{j}{k} (-\lambda)^{j-k} T^k R b_j = \sum_{k=0}^{n-1} T^k R a_k = y.$$

Hence $y = (T-\lambda)u + Rb_0$ for some $u \in Y$. Now, by the defining property of $E_T(F)$, there exists some $c_0 \in E_T(F)$ such that $b_0 = (T-\lambda)^n c_0$. Since

$$0 = C(T)^n R = C(T-\lambda)^n R = \sum_{k=0}^{n} \binom{n}{k} (-1)^k (T-\lambda)^{n-k} R (T-\lambda)^k,$$

we obtain the representation

$$Rb_0 = R(T-\lambda)^n c_0 = \sum_{k=0}^{n-1} \binom{n}{k} (-1)^{n+k+1} (T-\lambda)^{n-k} R (T-\lambda)^k c_0$$

and therefore $Rb_0 = (T-\lambda)v$ for some $v \in Y$. We have shown that $y \in (T-\lambda)Y$. Hence $Y \subset X_T(\mathbf{C}\setminus U)$ satisfies $(T-\lambda)Y = Y$ for all $\lambda \in \mathbf{C}\setminus F$ and consequently $Y \subset Z = \{0\}$. In particular, it follows that $R = 0$ on $E_T(F)$, which completes the proof.

Two operators $S, T \in L(X)$ are said to be *nilpotent equivalent* if there exists some $n \in \mathbf{N}$ such that $C(S,T)^n I = 0 = C(T,S)^n I$. From the computation on p. 11 in [7] it is immediate that this notion actually defines an equivalence relation on $L(X)$, which is certainly stronger than the quasinilpotent equivalence in the sense of Definition 1.2.1 in [7]. It turns out that the algebraic representation of spectral maximal spaces is preserved by nilpotent equivalence. Note that this result ceases to be true for the case of quasinilpotent equivalence, since quasinilpotent operators may well have non-trivial divisible linear subspaces; see for instance Remark 1.6 of [13].

3.3. PROPOSITION. *Let* $S, T \in L(X)$ *be nilpotent equivalent and consider a closed subset* F *of* $\mathbf{C}$. *Then* S *has the* SVEP *(resp. is decomposable) and satisfies* $X_S(F) = E_S(F)$ *if and only if* T *has the* SVEP *(resp. is decomposable) and satisfies* $X_T(F) = E_T(F)$.

PROOF. It is well-known that the SVEP as well as the decomposability are both preserved by quasinilpotent equivalence and hence, in particular, by nilpotent equivalence; see Theorem 1.2.3 and 2.2.1 of [7]. Moreover, if $S, T \in L(X)$ are quasinilpotent equivalent and have the SVEP, then for all closed $F \subset \mathbf{C}$ the identity $X_S(F) = X_T(F)$ holds by Theorem 1.2.4 of [7]. Hence it remains to show that $E_S(F) \subset E_T(F)$ whenever S, $T \in L(X)$ satisfy $C(T,S)^n I = 0$ for some integer $n \in \mathbf{N}$. To this end, let us introduce the linear subspace

$$Y := \{\sum_{k=0}^{m-1} T^k a_k : m \in \mathbf{N} \text{ and } a_o, \ldots, a_{m-1} \in E_S(F)\}$$

of X. This space is certainly invariant under T, which implies in particular that $(T - \lambda)Y \subset Y$ for all $\lambda \in \mathbf{C} \setminus F$. In order to prove the opposite inclusion, let $\lambda \in \mathbf{C} \setminus F$ and $y \in Y$ be arbitrarily given. Thus

$$y = \sum_{k=0}^{m-1} T^k a_k \qquad \text{for suitable } a_o, \ldots, a_{m-1} \in E_S(F).$$

Similar to the proof of 3.2, we obtain the representation

$$y = \sum_{j=0}^{m-1} (T - \lambda)^j b_j = (T - \lambda)u + b_o,$$

where $b_o, \ldots, b_{m-1} \in E_S(F)$ and $u \in Y$. Since $b_o \in E_S(F)$, there exists some $c_o \in E_S(F)$ such that $b_o = (S - \lambda)^n c_o$. Now, from

$$0 = C(T,S)^n I = C(T - \lambda, S - \lambda)^n I = \sum_{k=0}^{n} \binom{n}{k} (-1)^k (T - \lambda)^{n-k} (S - \lambda)^k$$

we conclude that $b_o = (S - \lambda)^n c_o = (T - \lambda)v$ holds for some $v \in Y$. Hence we have the identity $(T - \lambda)Y = Y$ for every $\lambda \in \mathbf{C}\setminus F$, which implies by maximality $E_S(F) \subset Y \subset E_T(F)$. The assertion follows.

Next we show that the algebraic representation of the spectral maximal spaces we have in mind does hold for a somewhat more concrete and fairly large class of decomposable operators. An operator $T \in L(X)$ is said to be *$C^{\infty}(\mathbf{C})$-decomposable of finite type*, if T is nilpotent equivalent to a generalized scalar operator, i.e. to an operator $S \in L(X)$ having a continuous functional calculus on $C^{\infty}(\mathbf{C})$. Again, the example of quasinilpotent operators with non-trivial divisible subspaces reveals that the following result does not carry over to the case of an arbitrary $C^{\infty}(\mathbf{C})$-decomposable operator.

3.4. THEOREM. *If* $T \in L(X)$ *is* $C^{\infty}(\mathbf{C})$*-decomposable of finite type, then* $X_T(F) = E_T(F)$ *holds for all closed* $F \subset \mathbf{C}$.

PROOF. According to Theorem 1.2 of [17], a generalized scalar operator has no divisible linear subspace different from $\{0\}$. Moreover, by Proposition 2.2 of [13], every generalized scalar operator is super-decomposable and therefore well-decomposable. Hence the assertion follows immediately from 3.2 in combination with 3.3.

In the following, the algebraic description of the spectral maximal spaces will be related to the concept of multiplication operators from Section 1.

3.5. LEMMA. *Assume that* $T \in L(X)$ *has the* SVEP *and that the spectral maximal spaces* $X_T(F)$ *are closed for all closed* $F \subset \mathbf{C}$. *Then* $L(T) \in L(L(X))$ *has the* SVEP *as well, and for the spectral maximal spaces of* $L(T)$ *we have*

$$L_{L(T)}(F) = \{S \in L(X) : S(X) \subset X_T(F)\}$$

$$sp(L(T)|L_{L(T)}(F)) \subset F$$

$$X_T(F) = E_T(F) \Leftrightarrow L_{L(T)}(F) = E_{L(T)}(F)$$

for every closed $F \subset \mathbf{C}$. *In particular,* T *has no non-trivial divisible linear subspace if and only if* $L(T)$ *has no non-trivial divisible linear subspace.*

PROOF. The first assertions are proved in Lemma 3.5 of [13]. Now suppose that the equality $X_T(F) = E_T(F)$ holds and consider the algebraic linear span Y of the set $\{Sx : S \in E_{L(T)}(F),\ x \in X\}$ in X. One easily verifies that $(T - \lambda)Y = Y$ holds for all $\lambda \in \mathbf{C}\setminus F$. This implies $Y \subset E_T(F) \subset X_T(F)$ and hence $S(X) \subset X_T(F)$ for all $S \in E_{L(T)}(F)$.

According to the representation of $L_{L(T)}(F)$ stated above, we arrive at $E_{L(T)}(F) \subset \subset L_{L(T)}(F)$, and the opposite inclusion is obvious. Conversely assume that $L_{L(T)}(F) = E_{L(T)}(F)$ is fulfilled and consider the algebraic linear span Z of all operators $S \in L(X)$ of the type $Su = \langle x^*, u\rangle x$ for $x^* \in X^*$ and $x \in E_T(F)$. Again, it is not hard to see that $(L(T) - \lambda)Z = Z$ holds for all $\lambda \in \mathbf{C} \setminus F$, which implies $Z \subset E_{L(T)}(F) \subset L_{L(T)}(F)$. We conclude that $E_T(F) \subset X_T(F)$, which completes the proof.

3.6. PROPOSITION. *If* $T \in L(X)$ *is well-decomposable, then* $L(T) \in L(L(X))$ *is also well-decomposable.*

PROOF. Given an arbitrary open covering $\{U,V\}$ of $\mathbf{C}$, there exist an $R \in L(X)$ and an $n \in \mathbf{N}$ such that $R(X) \subset X_T(\bar{U})$, $(I - R)(X) \subset X_T(\bar{V})$, and $C(T)^n R = 0$. Obviously $\tilde{R} := L(R)$ satisfies $C(L(T))^n \tilde{R} = 0$, and from 3.5 we conclude that $\tilde{R}(L(X)) \subset L_{L(T)}(\bar{U})$ as well as $(I - \tilde{R})(L(X)) \subset L_{L(T)}(\bar{V})$. Moreover, since $sp(L(T)|L_{L(T)}(F)) \subset F$ for all closed $F \subset \mathbf{C}$, it follows that L(T) is well-decomposable.

It is not known to the authors whether the converse of the preceding result is true. From 1.2 we conclude that T is at least decomposable if L(T) is supposed to be well-decomposable, but one might expect that there is a counterexample to show that T need not be well-decomposable again. In this case, the following result would be a strict generalization of 3.2.

3.7. THEOREM. *Let* $T \in L(X)$ *and assume that* L(T) *is well-decomposable and has no divisible linear subspace different from* $\{0\}$. *Then* T *is decomposable and satisfies* $X_T(F) = E_T(F)$ *for all closed* $F \subset \mathbf{C}$.

The proof is an immediate consequence of 1.2, 3.2 and 3.5.

We finally turn to some aspects of automatic continuity, from which our interest in the present investigations actually arose. Given a pair of decomposable operators $T \in L(X)$ and $S \in L(Y)$ on Banach spaces X and Y, one is interested in the automatic continuity of all linear transformations $\theta : X \to Y$ which intertwine the pair (T,S) in the sense that $\theta T = S\theta$. In this situation, it is decisive to know whether $\theta X_T(F) \subset Y_S(F)$ holds for all closed $F \subset \mathbf{C}$. This problem was posed by Jewell in [6] and has recently found a negative answer in [13]. It is still open, however, if the answer is positive as soon as one adds the natural assumption that S has no divisible linear subspace different from $\{0\}$. By the preceding results, there is some strong indication that the answer to this modified version of the Jewell problem should be positive, since the algebraic representation $Y_S(F) = E_S(F)$ certainly implies $\theta X_T(F) \subset Y_S(F)$ for closed $F \subset \mathbf{C}$. Actually the following somewhat stronger result turns out to be true:

3.8. THEOREM. *Let* $T \in L(X)$ *and* $S \in L(Y)$ *be decomposable, assume that the representation* $Y_S(F) = E_S(F)$ *holds for all closed* $F \subset \mathbf{C}$, *and consider a linear transformation* $\theta : X \to Y$ *which almost intertwines the pair* (T,S) *in the sense that*

$$C(S,T)^n\theta = \sum_{k=0}^{n} \binom{n}{k} S^{n-k}\theta(-T)^k = 0$$

for some $n \in \mathbf{N}$. *Then* $\theta X_T(F) \subset Y_S(F)$ *holds for all closed* $F \subset \mathbf{C}$. *Furthermore, there exists a finite subset* Λ *of* $\mathbf{C}$ *such that the separating space*

$$\sigma(\theta) := \{y \in Y : \exists x_n \in X \text{ such that } x_n \to 0, \theta x_n \to y \text{ as } n \to \infty\}$$

is contained in $Y_S(\Lambda)$, *and* $\theta | X_T(F)$ *is continuous for every closed subset* F *of* $\mathbf{C}$ *satisfying* $F \cap \Lambda = \emptyset$.

PROOF. Similar to former considerations in 3.2 and 3.3, let us introduce the linear subspace Z of Y given by

$$Z := \{ \sum_{k=0}^{m-1} S^k\theta a_k : m \in \mathbf{N} \text{ and } a_o, \ldots, a_{m-1} \in E_T(F)\}.$$

Again, it is not hard to see that $(S - \lambda)Z = Z$ holds for all $\lambda \in \mathbf{C} \setminus F$, which implies $\theta X_T(F) \subset \theta E_T(F) \subset Z \subset E_S(F) = Y_S(F)$. Now the remaining assertions follow from standard automatic continuity theory; see Theorem 3.7 of [4] or Theorem 4.3 of [5].

Of course, the most interesting applications of the preceding result arise for $n = 1$ in combination with 3.4 or 3.7. In this connection, let us also refer to the various examples of super-decomposable operators in [13] to which this theorem applies. To give the proper aspect of 3.8, let us note that a *continuous* linear operator $\theta : X \to Y$ satisfies $\theta X_T(F) \subset Y_S(F)$ for all closed $F \subset \mathbf{C}$ if and only if the asymptotic condition $\|C^n(S,T)\theta\|^{1/n} \to 0$ as $n \to \infty$ is fulfilled; see Theorem 2.3.3 of [7]. It will become obvious from the following final result that the singularity set Λ in 3.8 need not be empty at all.

3.9. THEOREM. *Assume that* $T \in L(X)$ *is decomposable and* $S \in L(Y)$ *is well--decomposable. Then the following assertions are equivalent:*

(a) *Every linear transformation* $\theta : X \to Y$ *satisfying* $\theta T = S\theta$ *is necessarily continuous.*

(b) *The pair* (T,S) *has no critical eigenvalue in the sense that* $\text{codim}(T - \lambda)(X) < \infty$

for every eigenvalue λ of S, *and either* T *is algebraic or* S *has no divisible linear subspace different from* {0}.

In Theorem 4.3 of [13], this result was obtained for the special case of a super--decomposable S ∈ L(Y). In view of 3.7 and 3.4, the former proof immediately carries over to the case of an operator S ∈ L(Y) such that L(S) is well-decomposable and to the case of a $C^{\infty}(\mathbb{C})$-decomposable operator S ∈ L(Y) of finite type. We refer to [13] for various applications of automatic continuity results of this type.

REFERENCE

1. **Albrecht, E. :** An example of a $C^{\infty}(\mathbb{C})$-decomposable operator which is not $C^{\infty}(\mathbb{C})$-spectral, *Rev. Roumaine Math. Pures Appl.* **19** (1974), 131-139.
2. **Albrecht, E. :** An example of a weakly decomposable operator which is not decomposable, *Rev. Roumaine Math. Pures Appl.* **20** (1975), 855-861.
3. **Albrecht, E. :** On decomposable operators, *Integral Equations Operator Theory* **2** (1979), 1-10.
4. **Albrecht, E. ; Neumann, M.M. :** Automatic continuity of generalized local linear operators, *Manuscripta Math.* **32** (1980), 263-294.
5. **Albrecht, E. ; Neumann, M.M. :** Automatic continuity for operators of local type, in *Radical Banach Algebras and Automatic Continuity*, Springer Lecture Notes in Math. **975** (1983), 342-355.
6. **Bekken, O.B. ; Øksendal, B.K. ; Stray,A.** (Eds.) : *Spaces of analytic functions*, Springer Lecture Notes in Math. **512** (1976).
7. **Colojoară, I. ; Foiaş, C. :** *Theory of generalized spectral operators*, Gordon and Breach, New York, 1968.
8. **Eschmeier, J. :** Some remarks concerning the duality problem for decomposable systems of commuting operators, in *Spectral Theory of Linear Operators and Related Topics*, Birkhäuser Verlag, Basel, 1984, pp. 115-123.
9. **Eschmeier, J. :** Spectral decompositions and decomposable multipliers, *Preprint*, Universität Münster, 1984.
10. **Eschmeier,J. ; Putinar, M. :** Spectral theory and sheaf theory. III, *INCREST Preprint Series in Math.* **59** (1983).
11. **Foiaş, C. ; Vasilescu, F.-H. :** On the spectral theory of commutators, *J. Math. Analysis Appl.* **31** (1970), 473-486.
12. **Köthe, G. :** *Topological vector spaces.II*, Springer-Verlag, New York--Heidelberg–Berlin, 1979.
13. **Laursen, K.B. ; Neumann, M.M. :** Decomposable operators and automatic continuity, *J. Operator Theory* **15** (1986).
14. **Schatten, R. :** *A theory of cross spaces,*, Princeton University Press, 1950.
15. **Vasilescu, F.-H. :** On pairs of commuting operators, *Studia Math.* **62** (1978), 203-207.

16. **Vasilescu, F.-H. :** *Analytic functional calculus and spectral decompositions*, D.Reidel Publ. Comp., Dordrecht and Editura Academiei, Bucureşti, 1982.

17. **Vrbová, P. :** Structure of maximal spectral spaces of generalized scalar operators, *Czech. Math. J.* **23** (1973), 493-496.

E. Albrecht

Fachbereich Mathematik der
Universität des Saarlandes
D-6600 Saarbrücken
West Germany.

J. Eschmeier

Mathematisches Institut der
Universität Münster
Einsteinstrasse 64
D-4400 Münster
West Germany.

M.M. Neumann

Fachbereich Mathematik der
Universität Essen - Gesamthochschule
D-4300 Essen
West Germany.

Operator Theory:
Advances and Applications, Vol.17

KOROVKIN CLOSURES IN BANACH ALGEBRAS

Francesco Altomare*)

INTRODUCTION

The main problem arising from Korovkin's original theorems can be stated as follows:

Let A be a normed space and L a class of continuous linear operators on A. If S is a subset of A, we call a net $(L_i)^{\leq}_{i\in I}$ of operators in L S - *admissible* if $(L_i)^{\leq}_{i\in I}$ is norm bounded and $\lim_{i\in I \leq} L_i(x) = x$ for all $x \in S$.

We define the L - Korovkin closure of S as

$$\mathrm{Kor}^{L}(S) = \{y \in A \mid \lim_{i\in I \leq} L_i(y) = y \text{ for all S - } \textit{admissible} \text{ nets } (L_i)^{\leq}_{i\in I} \text{ in } L\}.$$

Then the problem is, given a subset S of A and a class L of continuous linear operators on A, to determine the L - Korovkin closure $\mathrm{Kor}^{L}(S)$ and to characterize those sets of A which have the property that $\mathrm{Kor}^{L}(S) = A$.

In this paper we give a short survey and some new results about this problem, and some others related to it, in the context of particular Banach algebras A and for special classes L of operators on A.

More precisely, we consider function algebras on compact Hausdorff spaces and, in this case, L is the class L^1 of all linear contractions on A; moreover we study the case of C^*-algebras and of the classes L^+ and $L^{1,+}$ of all continuous positive operators and of all positive linear contractions on A.

Finally some open questions are discussed.

KOROVKIN CLOSURES

We begin with some preliminary remarks and notations. Let A be a complex commutative Banach algebra. We denote by $\Delta(A)$ the set of all multiplicative linear functionals on A and by $\Delta'(A)$ the set $\Delta(A) \cup \{0\}$. It is well-known that $\Delta'(A)$ is a

---

*) Attendance of the Conference was made possible by support from the Ministero Pubblica Istruzione (M.P.I.).

weak*-compact subset of the dual A' of A and $\Delta(A)$ is weak*-locally compact; if A has an identity, then $\Delta(A)$ is weak*-compact.

For every $x \in A$ let us denote by $\hat{x} : \Delta(A) \to \mathbf{C}$ the Gelfand transform of x defined by putting $\hat{x}(m) = m(x)$ for all $m \in \Delta(A)$. Then $\hat{x}$ belongs to the Banach algebra $C_o(\Delta(A), \mathbf{C})$ of all continuous complex-valued functions on $\Delta(A)$ vanishing at infinity.

For all $x \in A$ we put $\|x\|_\infty = \sup_{m \in \Delta(A)} |m(x)|$; $\|x\|_\infty$ coincides with the spectral radius of x. If $\|x\|_\infty = \|x\|$ for all $x \in A$, the algebra A is also called a function algebra.

We shall denote by F(A) the Silov boundary of $\Delta(A)$. F(A) is a closed subset of $\Delta(A)$ (but not of A', unless A has an identity) and $\|x\|_\infty = \sup_{m \in F(A)} |m(x)|$ for all $x \in A$.

Moreover we put $B_1(A) = \{\mu \in A' \mid \|\mu\| \leq 1\}$; if A has an involution we put $A'_+ = \{\mu \in A' \mid \mu \text{ positive}\}$ and $B_1^+(A) = B_1(A) \cap A'_+$.

If X is a closed subset of $\Delta(A)$, we shall use $\mathcal{R}(X)$ (resp. $\mathcal{R}^1(X)$) to indicate the space of all complex Radon measures on X (resp. the space of all probability measures on X).

If S is a subset of A, we shall denote by $Kor^1(S)$, $Kor^+(S)$ and $Kor^{1,+}(S)$ respectively, the Korovkin closures $Kor^{\mathcal{L}^1}(S)$, $Kor^{\mathcal{L}^+}(S)$ and $Kor^{\mathcal{L}^{1,+}}(S)$ defined as in the introduction (of course, $Kor^+(S)$ and $Kor^{1,+}(S)$ are defined if A has an involution).

The problem in determining these Korovkin closures or in seeing when they coincide with A seems very difficult and, indeed, has, so far, remained unsolved.

Together with the above defined closures it seems interesting to study the so-called universal Korovkin closures defined as follows: *the $\mathcal{L}^1$-universal Korovkin closure* (resp. if A has an involution, *the $\mathcal{L}^+$-universal Korovkin closure, $\mathcal{L}^{1,+}$-universal Korovkin clousre*) of S is, by definition, the subspace $Kor_u^1(S)$ (resp. $Kor_u^+(S)$, $Kor_u^{1,+}(S)$) of all $y \in A$ such that, if B is a function algebra (resp. a commutative C^*-algebra), $T : A \to B$ an algebraic homomorphism (resp. an algebraic $*$-homomorphism) and $(L_i)_{i \in I}^{\leq}$ a net of linear contractions (resp. of continuous positive linear operators, positive linear contractions) from A into B and if $\lim_{i \in I}^{\leq} L_i(x) = T(x)$ for all $x \in S$, then $\lim_{i \in I}^{\leq} L_i(y) = T(y)$.

If A has an identity, the subspaces $Kor_u^1(S)$, $Kor_u^+(S)$ and $Kor_u^{1,+}(S)$ are defined by requiring that B has an identity and T is unital.

Clearly if A is a function algebra, then $Kor_u^1(S) \subset Kor^1(S)$ and hence $Kor^1(S) = A$ if $Kor_u^1(S) = A$.

If A is a commutative C^*-algebra, $Kor_u^+(S) \subset Kor^+(S)$ and $Kor_u^{1,+}(S) \subset Kor^{1,+}(S)$ (in fact are the same, see the next Theorem 1).

While an extensive analysis of these closures and of some others can be found in [3] and [5], here we shall illustrate only the connection among them in some, perhaps important, cases.

It will be convenient for the sequel to use the following notation: If S is a subset of A, and A and B two subsets of A', we put

(1) $U(S, A, B) = \{y \in A \mid \text{If } \delta \in A \text{ and } \mu \in B \text{ and } \delta(x) = \mu(x) \text{ for all } x \in S, \text{ then } \delta(y) = \mu(y)\}$.

We have the following theorem:

THEOREM 1. 1. *If* $A = C(X, \mathbf{C})$, *for* X *compact, then for all subsets* S *of* A *containing the constant function* 1, *we have*

$$\mathrm{Kor}^+(S) = \mathrm{Kor}^1(S) = \mathrm{Kor}^{1,+}(S) = \mathrm{Kor}_u^+(S) = \mathrm{Kor}_u^1(S) = \mathrm{Kor}_u^{1,+}(S) = U(S, \Delta(A), B_1^+(A)) =$$

$$= \{g \in A \mid \text{If } x \in X \text{ and } \mu \in R^1(X) \text{ and if } \textstyle\int f d\mu = f(x) \text{ for all } f \in S, \text{ then } \int g d\mu = g(x)\}.$$

2. *If* $A = C_o(X, \mathbf{C})$, *for* X *locally compact and non compact, then for all subsets* S *of* A *we have*

(i) $$\mathrm{Kor}^{1,+}(S) = \mathrm{Kor}_u^{1,+}(S) = U(S, \Delta'(A), B_1^+(A)) =$$

$$= \{g \in A \mid \text{If } x \in X \text{ and if } \mu \text{ is a positive bounded Radon measure on } X, \text{ with } \|\mu\| \le 1, \text{ such that } \textstyle\int f d\mu = f(x) \text{ for all } f \in S, \text{ then } \int g d\mu = g(x)\}.$$

(ii) $$\mathrm{Kor}^+(S) = \mathrm{Kor}_u^+(S) = U(S, \Delta'(A), A'_+) =$$

$$= \{g \in A \mid \text{If } x \in X \text{ and if } \mu \text{ is a positive Radon measure on } X, \text{ such that } \textstyle\int f d\mu = f(x) \text{ for all } f \in S, \text{ then } \int g d\mu = g(x)\}.$$

(iii) $\mathrm{Kor}^1(S) = \mathrm{Kor}_u^1(S) =$ the closed subspace generated by S.

The statement 1) can be proved combining the results of [1], Theorem 1 and Corollary 2.2, [3], Theorem 2.1 and [5], Theorem 4.1, 3). The proof of the statement 2), i) can be found in [5], Theorem 4.2; the proof of ii) is analogous to one of i) and iii) is trivial, since, by using the same proof of Theorem 4.2 of [5], we have $\mathrm{Kor}^1(S) = \mathrm{Kor}_u^1(S) = U(S, \Delta'(A), B_1(A))$ and this last subspace coincides with the closed subspace generated by S by virtue of the Hahn-Banach theorem.

Theorem 1 can be considered the key result of the Korovkin approximation theory in complex function algebras.

From Theorem 1, part 1, it follows in particular that, for a subset S of $C(X, \mathbf{C})$ (X compact), we have $\mathrm{Kor}^+(S) = C(X, \mathbf{C})$ if an only if the Choquet boundary of the subspace generated by S coincides with X.

This last result allows us to determine the Korovkin subsets of $C(X, \mathbf{C})$ (i.e. those subsets S of $C(X, \mathbf{C})$ for which $Kor^+(S) = C(X, \mathbf{C})$) very easily.

We refer to [1], [2], [7], [9], [16] for some significant examples and applications in Approximation theory and Harmonic Analysis.

We take from [4], Theorem 12, and [5], Corollary 4.3, the following results which show the close connection between the theorems of Korovkin type and the Weierstrass-Stone theorem for function algebras (see also [7], Corollary 3; [11], Theorem 2.8; [12], Example 5, (ii); [14], Corollary p.207).

THEOREM 2. 1. *Let* X *be a compact Hausdorff space and* S *a subset of* $C(X, \mathbf{C})$. *Let denote by* A(S) *the closed* $*$*-subalgebra of* $C(X, \mathbf{C})$ *generated by* S. *Then, if we put* $|S|^2 = \{|f|^2 \mid f \in S\}$,

$$Kor^+(\{1\} \cup S \cup |S|^2) = A(S) = \{f \in C(X, \mathbf{C}) \mid \text{If } x,y \in X \text{ and } h(x) = h(y) \text{ for all } h \in S \text{ then } f(x) = f(y)\}.$$

In particular, if S *separates* X, *then* $Kor^+(\{1\} \cup S \cup |S|^2) = C(X, \mathbf{C})$.

2. *Let* X *be a locally compact Hausdorff space and* $S \subset C_0(X, \mathbf{C})$. *If* S *separates* X *and if for all* $x \in X$ *there exists* $f \in S$ *such that* $f(x) \neq 0$, *then* $Kor^{1,+}(S \cup |S|^2) = C_0(X, \mathbf{C})$.

Among other things, this corollary is useful to find examples of universal Korovkin subsets in other Banach algebras as $\ell_1(\mathbf{C})$, the big disk algebra and the group algebra $L^1(G)$, G being a locally compact abelian group ([3], Example 3.1, 3.2, 3.4; [5], Corollary 3.6, 4.5 and 4.6, Example 4.7).

By returning to the problem to determine the L^1-Korovkin closures in an arbitrary Banach algebra, we point out that not many results are known in this direction; moreover, in general, the L^1-Korovkin closures and the L^1-universal Korovkin closures are different also in unitary function algebras, as the following example shows ([5], Corollary 3.3 and Remark 3.3).

EXAMPLE 3. Let $\mathbf{D} = \{z \in \mathbf{C} \mid |z| \leq 1\}$, $n \in \mathbf{N}$, $n \geq 1$ and let us denote by A the Banach algebra of all functions belonging to $C(\mathbf{D}^n, \mathbf{C})$ that are holomorphic on the interior of $\mathbf{D}^n$, endowed with the sup-norm. If for every $i \in \{1, \ldots, n\}$ we denote by $pr_i : \mathbf{D}^n \to \mathbf{C}$ the i-th projection and if we put $S = \{1, pr_1, \ldots, pr_n\}$ then $Kor^1(S) = A$ whereas $Kor^1_u(S) = U(S, \Delta(A), B_1(A)) \neq A$.

At this point we formulate the following conjecture

CONJECTURE 4. If A is a unitary function algebra and S a subset of A containing the identity, then

(2) $$\mathrm{Kor}^1(S) = U(S, F(A), B_1(A)) =$$

$$= \{y \in A \mid \text{If } m \in F(A) \text{ and } \mu \in R(F(A)),\ \|\mu\| \leq 1, \text{ and if } \int_{F(A)} \hat{x}\,d\mu = m(x) \text{ for every } x \in S, \text{ then } \int_{F(A)} \hat{y}\,d\mu = m(y)\}.$$

Clearly the equality (2) is true, for $A = C(X, \mathbf{C})$, X compact, since in this case $F(A) = \Delta(A) = X$; moreover it holds in the algebra A of the Example 2 ([5], Corollary 3.3). We also note that the second equality of (2) is always true ([5], Remark 3.2); moreover $U(S, F(A), B_1(A)) \subset \mathrm{Kor}^1(A)$ since in [5], Theorem 3.1, we prove that $U(S, F(A), B_1(A))$ coincides with the subspace of all $y \in A$ such that, if B is a function algebra, $T : A \to B$ a linear map such that for all $m \in \overline{F(B)}$, $m \circ T \in \overline{F(A)}$, and $(L_i)_{i \in I}^{\leq}$ a net of linear contractions of A in B, and if $\lim_{i \in I \leq} L_i(x) = T(x)$ for all $x \in S$, then $\lim_{i \in I \leq} L_i(y) = T(y)$.

We are able to show the validity of (2) in another particular case.

THEOREM 5. *Let X be a compact Hausdorff space and A a closed subalgebra of $C(X, \mathbf{C})$ containing the constant functions, separating X and not antisymmetric *). If $S \subset A$ with $1 \in S$ and if Re h and Im h are in A for all $h \in S$, then $\mathrm{Kor}^1(S) = U(S, F(A), B_1(A))$.*

PROOF. We shall make use of the methods of the convexity theory as developed in [6]. Let $\partial_A X$ be the Choquet boundary of A with respect to X. Then F(A) can be identified with $\overline{\partial_A X}$.

Let us consider on $\partial_A X$ the so-called M-topology T_M ([8], Chapter 4, n.8). Because of Theorem 8.2 of [8], Chapter 4, this topology is extendable in the sense of [15], Definition 55, and so there exists an uniformizable topology T^*, which is weaker than T_M; moreover the real-valued functions on $\partial_A X$ continuous for T^* are continuous for T_M and vice versa.

To prove the inclusion $\mathrm{Kor}^1(S) \subset U(S, F(A), B_1(A))$, we fix $f \in \mathrm{Kor}^1(S)$ and consider $x \in \overline{\partial_A X}$ and $\mu \in B_1(A)$ such that $\mu(h) = h(x)$ for all $h \in S$. Let $(y_\gamma)_{\gamma \in \Gamma}^{\leq}$ a net in $\partial_A X$ such that $\lim_{\gamma \in \Gamma \leq} y_\gamma = x$. Moreover let denote by $U(S)$ the set of non-empty finite subsets of S, ordered by inclusion.

For all $h \in S$ Re h and Im h are continuous for T_M ([8], Chapter 4, Corollary 8.4) and, hence, for T^*; so for all $B \in U(S)$ and $\gamma \in \Gamma$ there exists a neighbourhood $U_{B,\gamma}$ of y_γ for T^* such that

*) i.e. the subalgebra $A_{\mathbf{R}} = \{f \in A \mid f \text{ is real-valued}\}$ does not consist only of constants.

$$|h(y) - h(y_\gamma)| \leq (1/\text{card } B) \quad \text{for all } y \in U_{B,\gamma} \text{ and } h \in B .$$

Since T^* is uniformizable there exists a function $f_{B,\gamma} : \partial_A X \to \mathbf{R}$ which is continuous for T^* and hence for T_M, such that

$$0 \leq f_{B,\gamma} \leq 1, \quad f_{B,\gamma}(y_\gamma) = 1, \; f_{B,\gamma} = 0 \text{ on } \partial_A X | U_{B,\gamma}$$

(cf. [10], IX.7, n.5, Theorem 2).

Consequently, by virtue of Corollary 8.4 of [8], Chapter 4, there exists a (unique) real-valued extension $\tilde{f}_{B,\gamma} \in A$ of $f_{B,\gamma}$. By Bauer's maximum principle we have $0 \leq \tilde{f}_{B,\gamma} \leq 1$ too.

Let us consider the linear map $L_{B,\gamma} : A \to A$ defined by putting

$$L_{B,\gamma}(g) = \mu(g)\tilde{f}_{B,\gamma} + (1 - \tilde{f}_{B,\gamma})g \text{ for all } g \in A.$$

$L_{B,\gamma}$ is a contraction; in fact, if $g \in A$ and $y \in \partial_A X$

$$|L_{B,\gamma}(g)(y)| \leq \|\mu\| \, \|g\| \tilde{f}_{B,\gamma}(y) + (1 - \tilde{f}_{B,\gamma}(y))|g(y)| \leq \|g\|$$

since $\|\mu\| \leq 1$. Therefore, $\|L_{B,\gamma}(g)\| \leq \|g\|$.

By endowing the set $U(S) \times \Gamma$ with the componentwise defined order $<$, one can show, as in [6], that $\lim_{<,\,(B,\gamma)\in U(S)\times\Gamma} L_{B,\gamma}(g) = g$ for all $g \in S$. Consequently since $f \in \text{Kor}^1(S)$, $\lim_{<,\,(B,\gamma)\in U(S)\times\Gamma} L_{B,\gamma}(f) = f$; in particular $\lim_{<,\,(B,\gamma)\in U(S)\times\Gamma} L_{B,\gamma}(f)(y_\gamma) - f(y_\gamma) = 0$ since $|L_{B,\gamma}(f)(y_\gamma) - f(y_\gamma)| \leq \|L_{B,\gamma}(f) - f\|$ for all $(B,\gamma) \in U(S) \times \Gamma$. But $L_{B,\gamma}(f)(y_\gamma) = \mu(f)$ for all $(B,\gamma) \in U(S) \times \Gamma$ and so $\mu(f) = \lim_{\gamma\in\Gamma \leq} f(y_\gamma) = f(x)$.

Finally we present a positive result concerning the determination of $\text{Kor}^+(S)$ in the context of C^*-algebras with identity. Also in this case few results are known.

If A is a C^*-algebra with identity, we denote by P(A) the set of all pure states on A. $\overline{P(A)}$ is the closure of P(A) in A' for the weak*-topology.

In this context we suggest the following conjecture.

CONJECTURE 6. If A is a C^*-algebra with identity 1 and S a subset of A with $1 \in S$, then

$$\text{Kor}^+(S) = \text{Kor}^{1,+}(S) = \text{Kor}^1(S) = U(S, \overline{P(A)}, A'_+) = \tag{3}$$

$$= \{y \in A \mid \text{If } \delta \in \overline{P(A)} \text{ and } \mu \in A'_+ \text{ and } \delta(x) = \mu(x) \text{ for all } x \in S, \text{ then } \delta(y) = \mu(y)\} .$$

Only the inclusion $U(S, \overline{P(A)}, A'_+) \subset Kor^+(S)$ is known ([5], Corollary 5.2; see also [17], Theorem 3.4). Analogously one can prove that $U(S, \overline{P(A)}, A'_+) \subset Kor^1(S)$. Obviously $Kor^+(S) \subset Kor^{1,+}(S)$ and $Kor^1(S) \subset Kor^{1,+}(S)$.

In some cases we can prove all the equalities.

THEOREM 7. *Let* A *be a* C^**-algebra with identity and with a non trivial centre. Let* S *be a subset of the algebraic centre of* A *). *Then* $Kor^+(S) = Kor^{1,+}(S) = Kor^1(S) = U(S, \overline{P(A)}, A'_+)$.

The proof of the Theorem 7 is based on the fact that the partially ordered real Banach space A_h of the Hermitian elements of A is isometrically and order isomorphic to the space A(K) of all continuous affine functions on the state space K of A, on the fact that the algebraic centre of A_h coincides with the centre of A(K) and, finally, on a result establihed in [6] in the spaces A(K), for an arbitrary convex compact subset K of a locally convex space. We refer to this article for the details.

Another intersting result, partially analogous to Theorem 2 in the non commutative case, can be formulated by considering the class $L_*^{1,+}$ of all positive linear operators L on A such that $L(1) \leq 1$ and the class S of all Schwarz maps L on A (i.e. $L(x)^* L(x) \leq L(x^* x)$ for all $x \in A$).

We recall that a J^*-subalgebra of A is a norm-closed $*$-subspace of A closed under the Jordan product $\circ$ defined by $x \circ y = \frac{1}{2}(xy + yx)$ for all $x,y \in A$.

We have then

THEOREM 8. *Let* A *be a* C^**-algebra with identity* 1 *and* S *a subset of* A . *Let us denote by* J(S) *and* A(S) *the* J^**-subalgebra and the* C^**-subalgebra generated by* S. *If we put* $S \circ S^* = \{x \circ x^* \mid x \in S\}$, *then*

1) $$J(S) \subset Kor^{L_*^{1,+}}(\{1\} \cup S \cup S \circ S^*) ;$$

2) $$A(S) \subset Kor^{S}(\{1\} \cup S \cup S \circ S^*).$$

This theorem follows directly from Theorem 2 of [12], which improves Theorem 1.3 of [13] and the result of [14].

It would be very interesting to know if the inclusions 1) and 2) are strict or not (as in the commutative case, where $L_*^{1,+} = L^{1,+}$, $S = L^+$ and J(S) = A(S)).

---

*) The algebraic centre of A is the set of those elements of A which commute with all elements of A.

REFERENCES

1. **Altomare, F. ; Boccaccio, C. :** On Korovkin-type theorems in spaces of continuous complex-valued functions, *Bollettino U.M.I.* **6**:1B (1982), 75-86.
2. **Altomare, F. :** Quelques remarques sur les ensemble de Korovkin dans les espaces des functions continues complexes, in *Séminaire Initation à l'Analyse: G.Choquet - M.Rogalski - J.Saint Raymond*, 20e année, 1980/81, Exp. no. 4, 12 pp., Publ. Math. Univ. Pierre et Marie Curie.
3. **Altomare, F. :** On the Korovkin approximation theory in commutative Banach algebras, *Rendiconti di Matematica (4)* **2**(1982), 755-767.
4. **Altomare, F. :** Frontières abstraites et convergence de familles filtrées de formes linéaires sur les algèbres de Banach commutatives, in *Séminaire Initation à l'Analyse: G.Choquet - M.Rogalski - J.Saint Raymond*, 21e année, 1981/82, Publ. Math. Univ. Pierre et Marie Curie.
5. **Altomare, F. :** On the universal convergence sets, *Annali di Matematica (IV)* **138**(1984), 223-243.
6. **Altomare, F. :** Nets of positive operators in spaces of continuous affine functions, Preprint, 1984.
7. **Arveson, W.B. :** An approximation theorem for function algebras, Preprint, 1970.
8. **Asimow, L. ; Ellis, A.J. :** *Convexity theory and its applications in functional analysis*, London Mathematical Society Mon., **16**, Academic Press, 1980.
9. **Bloom, W.R. ; Sussich, J.F. :** Positive linear operators and the approximation of continuous functions on locally compact abelian groups, *J. Austral. Math.* **30** (1980), 180-186.
10. **Bourbaki, N. :** *Topologie générale*, Ch. 5-10, Herman, Paris, 1974.
11. **Limaye, B.V. ; Shirali, S. :** Korovkin's theorem for positive functionals on $*$normed algebras, *J. Indian Math. Soc.* **40**(1976), 163-172.
12. **Limaye, B.V. ; Namboodiri, M.N.N. :** Korovkin-type approximation on C^*-algebras, *J. Approx. Theory* **34**(1982), 237-246.
13. **Priestley, W.M. :** A non commutative Korovkin theorem, *J. Approx. Theory* **16** (1976), 251-260.
14. **Robertson, A.G. :** A Korovkin theorem for Schwarz maps on C^*-algebras, *Math. Z.* **156**(1977), 205-207.
15. **Rogalski, M. :** Topologies faciales dans les convexes compacts; calcul functionnel et décomposition spectrale dans le centre d'un espace A(X), in *Séminaire Choquet, Initation à l'Analyse*, 9e année, 1969/70, Fasc.1, n.3, 56 pp., Publ. Math. Univ. Pierre et Marie Curie.
16. **Sussich, J.F. :** *Korovkin's theorem for locally compact abelian groups*, Ph. D. Dissertation, Murdoch University, October 1982.
17. **Takahasi, S.E. :** Korovkin's theorems for C^*-algebras, *J. Approx. Theory* **27**(1979), 197-202.

Francesco Altomare

Dipartimento di Matematica, Università degli Studi di Bari
Via G.Fortunato, 70125 Bari
Italia.

Operator Theory:
Advances and Applications, Vol.17

ON THE THEORY OF THE CLASS $A_{\aleph_0}$ WITH APPLICATIONS TO INVARIANT SUBSPACES AND THE BERGMAN SHIFT OPERATOR

C. Apostol, H.Bercovici, C.Foiaş and C.Pearcy

1. INTRODUCTION

Let H be a separable, infinite dimensional, complex Hilbert space, let $L(H)$ denote the algebra of all bounded linear operators on H, and let $C_1(H)$ denote the ideal of trace-class operators in $L(H)$ under the trace norm. The spectrum of an operator T in $L(H)$ will be denoted by $\sigma(T)$. Let $\mathbf{D}$ be the open unit disc in $\mathbf{C}$, and let $\mathbf{T} = \partial\mathbf{D}$. The spaces $L^p = L^p(\mathbf{T})$, $1 \leq p \leq \infty$, are the usual Lebesgue spaces. One knows that L^∞ is the dual space of L^1 under the bilinear functional

$$\langle f,g\rangle = (1/2\pi)\int_0^{2\pi} f(e^{it})g(e^{it})dt, \quad f \in L^\infty, g \in L^1,$$

and that $L(H)$ is the dual space of $C_1(H)$ under the bilinear functional

$$\langle T,K\rangle = \mathrm{tr}(TK), \quad T \in L(H), K \in C_1(H).$$

Let $H^p = H^p(\mathbf{T})$, $1 \leq p \leq \infty$, be the subspace of L^p consisting of those functions $f \in L^p$ whose negative Fourier coefficients vanish. One knows that H^∞ is a weak*--closed subspace of L^∞, and thus it follows from general principles that H^∞ can be identified with the dual space of L^1/H^1_0 where H^1_0 is the space consisting of those functions g in H^1 whose analytic extension $\hat{g}$ to $\mathbf{D}$ satisfies $\hat{g}(0) = 0$ (and, of course, H^1_0 is the preannihilator ${}^\perp(H^\infty)$ in L^1 of H^∞).

A *dual algebra* is, by definition, a subalgebra of $L(H)$ that is closed in the weak* topology and contains 1_H. If $T \in L(H)$, the dual algebra A_T *generated by* T is the smallest dual algebra containing T, and consists, of course, of the weak*-closure of the algebra of all polynomials p(T). It follows from the same general principle applied above that if $A \subset L(H)$ is a dual algebra, then A can be identified with the dual space of $C_1(H)/{}^\perp A = Q_A$, where ${}^\perp A$ denotes the preannihilator of A in $C_1(H)$. If x and y are vectors in H, we denote by $x \otimes y$ the rank-one operator defined by $(x\otimes y)(u) = (u,y)x$, $u \in H$. If A is a given dual algebra under consideration, we denote by $[x\otimes y]_A$, or simply $[x\otimes y]$ when no confusion will arise, the image in Q_A of the trace-class operator $x\otimes y$. If one wishes to study the invariant subspace problem for operators T on Hilbert space, no generality is lost by supposing that T satisfies $||T|| = 1$ and that T is an ***absolutely continuous*** contraction, meaning by this that in the canonical decomposition $T = U \oplus T'$, where U is a unitary operator and T' is a completely nonunitary contraction, either U acts on the space (0) or U is an absolutely continuous unitary operator. We denote the class of all absolutely continuous contractions in $L(H)$ by (ACC). One knows that if $T \in$ (ACC), then T has an H^∞-functional calculus Φ_T, which is a norm-decreasing, weak*-continuous, algebra homomorphism of H^∞ into A_T with the properties that $\Phi_T(1) = 1_H$, $\Phi_T(\zeta) = T$, and the range of Φ_T is weak* dense in A_T (cf. [13]). It therefore follows from general principles that Φ_T is the adjoint of a bounded, linear, one-to-one mapping $\phi_T : Q_T \to L^1/H^1_0$. In this situation, one knows from [1] the following theorem of Apostol:

THEOREM 1.1. *If* $T \in$ (ACC), $\sigma(T) \supset \mathbf{T}$, *and* Φ_T *is not an isometry, then* T *has a nontrivial hyperinvariant subspace.*

In view of this theorem we make the following definition.

DEFINITION 1.2. We denote by **A** the class of all T in (ACC) such that Φ_T is an isometry.

If $T \in \mathbf{A}$, then one knows that the range of Φ_T is all of A_T and that Φ_T is a weak*-homeomorphism of H^∞ onto A_T (cf. [13]). Furthermore it follows easily in this case that ϕ_T is an isometry of Q_T onto L^1/H^1_0. Thus if $T \in \mathbf{A}$, the pairs of spaces $\{H^\infty, L^1/H^1_0\}$ and $\{A_T, Q_T\}$ may be identified via the isometries Φ_T and ϕ_T.

DEFINITION 1.3. Let n be a fixed cardinal number satisfying $1 \leq n \leq \aleph_0$. We denote by $\mathbf{A}_n$ the set of all those T in **A** for which every system of simultaneous equations

$$[x_i \otimes y_j] = [L_{ij}], \qquad 0 \leq i, j < n,$$

has a solution $\{x_i\}_{0 \leq i < n}$, $\{y_j\}_{0 \leq j < n}$, consisting of sequences of vectors form H. (Here the array $\{[L_{ij}]\}_{0 \leq i,j < n}$ is an arbitrary $n \times n$ array of elements of Q_T.)

One observes, then, from the definition, that

$$\mathbf{A} \supset \mathbf{A}_1 \supset \ldots \supset \mathbf{A}_n \supset \ldots \supset \mathbf{A}_{\aleph_0} .$$

It is easy to see (cf. [13]) that if $T \in \mathbf{A}_1$, then T has nontrivial invariant subspaces, and in [3] the authors made the following conjecture.

CONJECTURE 1.4. $\mathbf{A} = \mathbf{A}_1$.

This conjecture is interesting because of the following result.

PROPOSITION 1.5. *If Conjecture 1.4 is true, then every contraction* T *in $L(H)$ such that* $\sigma(T) \supset \mathbf{T}$ *has nontrivial invariant subspaces.*

PROOF. If T has a unitary part, the conclusion is obvious, so we may suppose that T is a completely nonunitary contraction. By Theorem 1.1, we may suppose that $T \in \mathbf{A}$, and if the conjecture is true, then $T \in \mathbf{A}_1$. But, as noted above, this implies that T has nontrivial invariant subspaces.

We showed in [7] that if $n \in \mathbf{N}$ and U_n is a unilateral shift operator of multiplicity n, then $U_n \in \mathbf{A}_n$ but $U_n \notin \mathbf{A}_{n+1}$, so the sequence of classes $\{\mathbf{A}_n\}_{n=1}^{\infty}$ is strictly decreasing. In [7] we also developed a dilation theory for the classes $\mathbf{A}_n$, $n = 1,2,\ldots,\aleph_0$, and the dilation theory of the class $\mathbf{A}_{\aleph_0}$ gives considerable information about the invariant subspace lattice Lat(T) of any operator T in $\mathbf{A}_{\aleph_0}$.

The purpose of this expository note is, first, to briefly summarize this dilation theory of the class $\mathbf{A}_{\aleph_0}$ and its consequences for invariant subspace lattices, and, secondly, to set forth some recently obtained sufficient conditions (cf. [3]) for an operator to belong to $\mathbf{A}_{\aleph_0}$. We then conclude by making various applications of these results to obtain new invariant subspace theorems and new results on the structure of the Bergman shift operator (as well as other shift operators).

2. THE DILATION THEOREM AND SOME CONSEQUENCES

The dilation theory for the class $\mathbf{A}_{\aleph_0}$ can be briefly summarized as follows.

THEOREM 2.1. ([7]). *Suppose* $T \in \mathbf{A}_{\aleph_0}$, *and let* A *be any contraction (acting on a separable Hilbert space) that is either*

(a) *a strict contraction,*

(b) *a (finite or countable) direct sum of strict contractions, or*

(c) *a unilateral or bilateral weighted shift operator with the property that the infinite product of its weights is equal to zero.*

Then there exist $M = M_A$ *and* $N = N_A$ *in* Lat(T) *with* $M \supset N$ *such that the compression* $T_{M \ominus N} = P_{M \ominus N} T \mid M \ominus N$ *of* T *to the semi-invariant subspace* $M \ominus N$ *is unitarily equivalent to* A.

On the basis of this theorem, one can deduce (cf. [3], [5]) the following consequences.

THEOREM 2.2. *Suppose* $T \in \mathbf{A}_{\aleph_0}$. *Then* Lat(T) *has the following properties:*

1) Lat(T) *contains a lattice that is isomorphic to* Lat(H), *the lattice of all subspaces of* H,

2) T *is reflexive, and, moreover,* Alg Lat(T) = A_T,

3) *For every cardinal number* n, $1 \leq n \leq \aleph_0$, *there exists a subspace* $L_n \in$ Lat(T) *such that the cyclic multiplicity* $\mu_{T|L_n}$ *of* $T|L_n$ *equals* n,

4) *If* T *has trivial kernel, closed range, and satisfies* $\bigcap_{n=1}^{\infty} T^n H = (0)$, *then there exists a family* $\{M_\lambda\}_{\lambda \in \mathbf{C}}$ *of nonzero subspaces contained in* Lat(T) *such that for* $\lambda_1 \neq \lambda_2$, $M_{\lambda_1} \cap M_{\lambda_2} = (0)$. *Furthermore, there exists another family* $\{N_A\}_{A \in 2^{\mathbf{N}}}$ *of nonzero subspaces contained in* Lat(T) *such that* $A_1 \subset A_2$ *implies* $N_{A_1} \subset N_{A_2}$ *and* $A_1 \cap A_2 = \emptyset$ *implies* $N_{A_1} \cap N_{A_2} = (0)$.

3. SUFFICIENT CONDITIONS FOR MEMBERSHIP IN $\mathbf{A}_{\aleph_0}$

It turns out that the class $\mathbf{A}_{\aleph_0}$ is quite large. The following theorem, which is taken from [3] and [8], sets forth some sufficient conditions that an operator belong to $\mathbf{A}_{\aleph_0}$. Recall that a set $\Lambda \subset \mathbf{D}$ is said to be *dominating* for $\mathbf{T}$ if almost every point of $\mathbf{T}$ is a nontangential limit of a sequence of points from Λ, and a contraction T belongs to C_{00} if both sequences $\{T^n\}_{n=1}^{\infty}$ and $\{T^{*n}\}_{n=1}^{\infty}$ converge to zero in the strong operator topology.

THEOREM 3.1. *If* $T \in$ (ACC) *and any one of the following conditions is satisfied, then* $T \in \mathbf{A}_{\aleph_0}$:

1) *For some* θ, $0 \leq \theta < 1$,

$$\sigma_e(T) \cap \mathbf{D} \cup \{\lambda \in \mathbf{D} \setminus \sigma_e(T) : \theta \| (\pi(T) - \lambda I)^{-1} \| \geq (1 - |\lambda|)^{-1}\}$$

is dominating for $\mathbf{T}$, *where* $\sigma_e(T)$ *is the essential spectrum of* T *and* $\pi(T)$ *is the image of* T *in the Calkin algebra,*

2) $T \in C_{00}$ *and* $(\sigma_e(T) \cup \sigma(T)') \cap \mathbf{D}$ *is dominating for* $\mathbf{T}$, *where* $\sigma(T)'$ *is the derived set of* $\sigma(T)$,

3) $T \in C_{00} \cap \mathbf{A}_1$,

4) $T \in C_{00} \cap \mathbf{A}$ *and the weak* topology and weak operator topology coincide on* A_T,

5) $T \in C_{00} \cap \mathbf{A}$ *and* $\{[x \otimes y] \in Q_T : \|x\|, \|y\| \leq 1\}$ *is* ***sequentially*** *weakly dense in the unit ball of* Q_T,

6) $T \in C_{00} \cap \mathbf{A}$, $\sigma(T) = \mathbf{T}$, *and the closure in the weak operator topology of the algebra of polynomials* $\{p(T) : p(0) = 0\}$ *does not contain* 1_H,

7) $T \in C_{00} \cap \mathbf{A}$ *and* T *is subnormal.*

4. APPLICATIONS

In this section we set forth some applications of the results in Sections 2 and 3. Let $A_2(\mathbf{D})$ denote the Hilbert space consisting of all holomorphic functions on $\mathbf{D}$ that are square integrable with respect to planar Lebesgue measure on $\mathbf{D}$. One knows that the family $\{(1/\sqrt{n+1})\zeta^n\}_{n=0}^{\infty}$ is an orthonormal basis for $A_2(\mathbf{D})$. Furthermore, if $B = M_\zeta$ is the operator of multiplication by the position function on $A_2(\mathbf{D})$, then B is the much studied *Bergman shift* operator. It is, indeed, a unilateral weighted shift relative to the above-mentioned basis, whose weight sequence is $\{\beta_n = \sqrt{(n+1)/(n+2)}\}_{n=0}^{\infty}$. It is well known that $\sigma(B) = \mathbf{D}^-$ and it is an easy consequence of the theory of analytic functions that if $(0) \neq M \in \mathrm{Lat}(B)$, then also $\sigma(B|M) = \mathbf{D}^-$. Since $\prod_{n=0}^{\infty} \beta_n = 0$ and $\bigcap_{n=1}^{\infty} B^n(A_2(\mathbf{D})) = (0)$, we have from 2) or 7) of Theorem 3.1 and Theorem 2.2 the following result.

THEOREM 4.1. *If* $(0) \neq M \in \mathrm{Lat}(B)$, *then* $\bigcap_{n=0}^{\infty} (B|M)^n|M = (0)$ *and* $B|M \in \mathbf{A}_{\aleph_0}$. *Therefore* $B|M$ *is reflexive and* $\mathrm{Lat}(B|M)$ *contains a lattice that is isomorphic to* $\mathrm{Lat}(A_2(\mathbf{D}))$. *Moreover* $\mathrm{Lat}(B|M)$ *contains a family of nonzero subspaces* $\{N_A\}_{A \in 2^{\mathbf{N}}}$ *indexed by the subsets of the positive integers such that if* $A_1 \subset A_2$, *then* $N_{A_1} \subset N_{A_2}$ *and if* $A_1 \cap A_2 = \emptyset$, *then* $N_{A_1} \cap N_{A_2} = (0)$. *Finally, for every* n *satisfying* $1 \leq n \leq \aleph_0$, *there exists a subspace* $L_n \in \mathrm{Lat}(B|M)$ *such that the cyclic multiplicity of* $(B|M)|L_n$ *is* n.

These results generalize considerably previously known results about the structure of Lat(B) (see [17], for example). It is clear that the above theorems will apply to any weighted unilateral shift U_α such that $\|U_\alpha\| = 1 = |\sigma(U_\alpha)|$ and such that the infinite product of the weights in the weight sequence α is equal to zero. With some additional argument one arrives at the following theorem from [3], which improves the current state of knowledge (see [17]).

THEOREM 4.2. *If* W *is a unilateral or bilateral, forward or backward, weighted* *shift operator (with scalar weights) in $L(H)$ and W satisfies $\|W\| = |\sigma(W)|$ (where $|\sigma(W)|$ is the spectral radius of W), then*

1) *A_W is closed in the weak operator topology,*

2) *The weak* and weak operator topologies coincide on A_W,*

3) *W is reflexive.*

Finally, we set forth some consequences of Theorems 1.1 and 3.1 for the invariant subspace problem.

THEOREM 4.3. *If $T \in L(H)$, $\|T\| = 1$, and $\sigma(T) \supset \mathbf{T}$, then any one of the following is a sufficient condition that T have a nontrivial invariant subspace:*

1) *$T \in C_{00}$ and the weak* and weak operator topologies coincide on A_T,*

2) *$T \in C_{00}$, $\sigma(T) = \mathbf{T}$, and the closure in the weak operator topology of the algebra of polynomials $\{p(T) : p(0) = 0\}$ does not contain 1_H,*

3) *$\sigma(T) \cap \mathbf{D}$ is dominating for $\mathbf{T}$.*

As mentioned earlier, the results discussed herein are mostly from [7] and [3]. For a recent exposition of these and other results in this area, see [9].

REFERENCES

1. **Apostol, C.** : Ultraweakly closed operator algebras, *J. Operator Theory* **2**(1979), 49-61.

2. **Apostol, C. ; Chevreau, B.** : On M-spectral sets and rationally invariant subspaces, *J. Operator Theory* **7**(1982), 247-266.

3. **Apostol, C. ; Bercovici, H. ; Foiaş, C. ; Pearcy, C.** : Invariant subspaces, dilation theory, and the structure of the predual of a dual algebra.I., *J. Functional Analysis*, to appear ; II., submitted to *Indiana Univ. Math. J.*

4. **Bercovici, H.** : A reflexivity theorem for weakly closed subspaces of operators, to appear in *Trans. Amer. Math. Soc.*

5. **Bercovici, H. ; Chevreau, B. ; Foiaş, C. ; Pearcy, C.** : Dilation theory and systems of simultaneous equations in the predual of an operator algebra. II, *Math. Z.* **187**(1984), 97-103.

6. **Bercovici, H. ; Foiaş, C. ; Langsam, J. ; Pearcy, C.** : (BCP)-operators are reflexive, *Michigan Math. J.* **29**(1982), 371-379.

7. **Bercovici, H. ; Foiaş, C. ; Pearcy, C.** : Dilation theory and systems of simultaneous equations in the predual of an operator algebra. I, *Michigan Math. J.* **30**(1983), 335-354.

8. **Bercovici, H. ; Foiaş, C. ; Pearcy, C.** : Factoring trace-class operator-valued functions with applications to the class $\mathbf{A}_{\aleph_0}$, submitted to *J. Operator Theory.*

9. **Bercovici, H. ; Foiaş, C. ; Pearcy, C.** : *Dual algebra with applications to invariant subspaces and dilation theory*, CBMS/NSF Lecture Notes No.56, American Math. Society, 1985.

10. **Bercovici, H. ; Foiaş, C. ; Pearcy, C. ; Sz.-Nagy, B. :** Functional models and generalized spectral dominance, *Acta Sci. Math. (Szeged)* **43**(1981), 243-254.

11. **Bercovici, H. ; Foiaş, C. ; Pearcy, C. ; Sz.-Nagy, B. :** Factoring compact operator--valued functions, to appear in *Acta Sci. Math. (Szeged).*

12. **Brown, S. :** Some invariant subspaces for subnormal operators, *Integral Equations and Operator Theory* **1**(1978), 310-333.

13. **Brown, S. ; Chevreau, B. ; Pearcy, C. :** Contractions with rich spectrum have invariant subspaces, *J. Operator Theory* **1**(1979), 123-136.

14. **Chevreau, B. ; Pearcy, C. ; Shields, A. :** Finitely connected domains G, representations of $H^{\infty}(G)$, and invariant subspaces, *J. Operator Theory* **6**(1981), 375-405.

15. **Foiaş, C. ; Pearcy, C. :** (BCP)-operators and enrichment of invariant subspace lattices, *J.Operator Theory* **9**(1983), 187-202.

16. **Olin, R. ; Thomson, J. :** Algebras of subnormal operators, *J. Functional Analysis* **37**(1980), 271-301.

17. **Shields, A. :** Weighted shift operators and analytic function theory, in *Topics in Operator Theory*, pp. 49-128, A.M.S., Providence, 1974.

C.Apostol
Department of Mathematics
Arizona State University
Tempe, Arizona 85287
U.S.A.

H.Bercovici
Department of Mathematics
Indiana University
Bloomington, IN 47405
U.S.A.

C.Foiaş
Department of Mathematics
Indiana University
Bloomington, IN 47405
U.S.A.

C.Pearcy
Department of Mathematics
University of Michigan
347 West Engineering Building
Ann Arbor, Michigan 48109
U.S.A.

Operator Theory:
Advances and Applications, Vol.17

EXAMPLES OF CHAINS OF INVARIANT SUBSPACES

José Barría and **Kenneth R.Davidson**

In [2] the authors have proven that for any countable ordinal (with last element) there exists an operator whose lattice of invariant subspaces is order isomorphic to the given ordinal. In this note an account is given of certain chains related to Volterra operators and weighted shifts.

Given two operators T_1 and T_2, $\mathbf{Lat}\, T_1 + \mathbf{Lat}\, T_2$ will denote the ordinal sum of the lattices of invariant subspaces $\mathbf{Lat}\, T_1$ and $\mathbf{Lat}\, T_2$. The following theorem is analogous to Theorem 2.1 in [2], and its proof can be adapted to this case also.

THEOREM 1. *Let* S *be a cyclic operator. If there exists a unilateral (forward) weighted shift* A *with weights decreasing monotonely to zero and such that*

$$\sum_{n=1}^{\infty} \| S^n \| \,/\, \| A^{n+k} \| < \infty$$

for all $k \geq 0$, *then there exists an operator* C *such that*

$$\mathbf{Lat} \begin{bmatrix} S & C \\ 0 & A \end{bmatrix} \simeq \mathbf{Lat}\, S + \mathbf{Lat}\, A.$$

REMARK. The assumption on A implies that $\mathbf{Lat}\, A \simeq 1 + \omega^*$, where ω is the ordinal of the nonnegative integers. This is a recent result of B.V.Yakubovič.

LEMMA 2. *Let* T_1, T_2 *be operators such that* $\| T_i^n \| = O((n!)^{-p})$ *for* $i = 1,2$ *and* $p > 0$. *If*

$$T = \begin{bmatrix} T_1 & C \\ 0 & T_2 \end{bmatrix}$$

then $\| T^n \| = o((n!)^{-q})$ *for all* $0 < q < p$.

PROOF. From the expression of T^n we have

$$\| T^n \| \leq \max\{ \| T_1^n \| , \| T_2^n \| \} + \| \sum_{j=0}^{n-1} T_1^{n-1-j} C T_2^j \| .$$

Now the conclusion follows easily from the inequalities

$$\|T_1^{n-1-j}\| \, \|T_2^j\| \leq K[(n-1-j)!\, j!]^{-p}$$

and

$$(n-1)!\,[(n-1-j)!\, j!]^{-1} \leq 2^{n-1}.$$

■

Let μ be a measure on $[0,1]$ such that μ is Lebesgue measure on a finite disjoint union of open subintervals, and on the complement of this union μ is purely atomic with a finite number of atoms. Define the operator V_μ on $L^2(\mu)$ by

$$(V_\mu f)(x) = \int_{[o,x)} f(t)d\mu(t).$$

Then the lattice of invariant subspaces of V_μ is a chain [1]. If V denotes the usual Volterra operator on $L^2[0,1]$, then the powers of V_μ have the expression

$$V_\mu^k = \sum_{i=0}^{r} \lambda_i M_i V^{k-r} A_i \, ,$$

where λ_i (scalar), M_i, A_i (operators) are independent of k, and r is the number of atoms of μ [1]. From this it follows that

$$\|V_\mu^k\| = O(\|V^{k-r}\|)\, ,$$

and therefore

$$\|V_\mu^k\| = O((k!)^{-q}) \quad \text{for all } 0 < q < 1.$$

THEOREM 2. *Given* $0 < p < 1$, *and* m,n *nonnegative integers, there exists an operator* T *such that*

$$\textbf{Lat}\, T \simeq \omega m + \textbf{Lat}\, V_\mu + \omega^* n\, ,$$

and

$$\|T^k\| = O((k!)^{-p})\, .$$

PROOF. Let $p < p_m < p_{m-1} < \ldots < p_1 < p_0 = 1$. Let A_j be the weighted shift with sequence of weights $(k+1)^{-p_j}$ $(k \geq 0)$. Then $\|A_j^k\| = (k!)^{-p_j}$. Let $T_0 = V_\mu$. For $j = 1, 2, \ldots, m$ let T_j be an operator of the form

$$T_j = \begin{bmatrix} V_\mu^* & & & * \\ & A_1 & & \\ & & \ddots & \\ 0 & & & A_j \end{bmatrix}$$

From Lemma 2 it follows that $\|T_j^k\| = O((k!)^{-q})$ for any q such that $0 < q < p_j$. Therefore

$$\sum_{k=1}^{\infty} \|T_j^k\| \,/\, \|A_{j+1}^{k+\ell}\| < \infty$$

for all $\ell \geq 0$, and $j = 0, 1, \ldots, m-1$. Note that all this is true for any particular choice of the (nonzero) off diagonal elements in T_j. From Theorem 1, there are operators C_1, $C_2, \ldots, C_m$ such that

$$T_1 = \begin{bmatrix} V_\mu^* & C_1 \\ 0 & A_1 \end{bmatrix}, \quad T_2 = \begin{bmatrix} T_1 & C_2 \\ 0 & A_2 \end{bmatrix}, \ldots, T_m = \begin{bmatrix} T_{m-1} & C_m \\ 0 & A_m \end{bmatrix}$$

are all unicellular operators. Therefore

$$\textbf{Lat}\, T_m^* \simeq \omega m + \textbf{Lat}\, V_\mu .$$

Since $\|T_m^{*k}\| = \|T_m^k\| = O((k!)^{-q})$ for any q such that $0 < q < p_m$, we can proceed as in the first part of the proof in order to get a unicellular operator T of the form

$$\begin{bmatrix} T_m^* & & & * \\ & B_1 & & \\ & & \ddots & \\ 0 & & & B_n \end{bmatrix}$$

where B_j is a weighted shift with sequence of weights $(k+1)^{-q_j}$ $(k \geq 0)$ $(j = 1, 2, \ldots, n)$, and $p < q_n < q_{n-1} < \ldots < q_1 < p_m$. From Lemma 2, $\|T^k\| = O((k!)^{-p})$. Clearly we have

$$\textbf{Lat}\, T \simeq \textbf{Lat}\, T_m^* + \omega^* n \simeq \omega m + \textbf{Lat}\, V_\mu + \omega^* n .$$ ■

REFERENCES

1. **Barría, J.** : The invariant subspaces of a Volterra operator, *J. Operator Theory*, **6** (1981), 341-349.

2. **Barría, J. ; Davidson, K.R.** : Unicellular operators, *Trans. Amer. Math. Soc.* **284**(1984), 229-246.

J.Barria
Instituto Venezolano
de Investigationes Cientificas
Matematicas
1010 A, Caracas
Venezuela

K.R.Davidson
Department of Mathematics
University of Waterloo
Waterloo, N2L 3G1 Ontario
Canada

Operator Theory:
Advances and Applications, Vol.17

ISOMETRIC DILATIONS OF COMMUTING CONTRACTIONS. IV

Zoia Ceauşescu and I. Suciu

In this paper we continue the study developed in [4] of the set of all Ando dilations of a given pair of commuting contractions.

We show that any Ando dilation can be produced in a canonical way by an appropriate adequate isometry. Considering the labelling by choice sequences of the set of all adequate isometries we establish the obstructions in choice, in order that the corresponding adequate isometry produces an Ando dilation (Section 2).For a special class of Ando dilations we exhibit a system of free parameters which produces it (Section 3). In Section 1 we recall some known facts on adequate isometries (cf.[2], [3]). Considering only a particular case, of interest for our purpose, we give to the known general results a form which will be convenient in what follows.

1. ADEQUATE ISOMETRIES

For j = 1, 2 let T_j be a contraction on the Hilbert space H_j and let A be a contraction from H_1 to H_2 such that $AT_1 = T_2A$. In [3] the concept of A-*adequate isometry* was introduced and used in the labelling of the set of all contractive intertwining dilations of A.

In this section we shall recall some results about A-adequate isometries in the particular case when $H_1 = H_2 = K_0$, $T_1 = T_2 = 0_{K_0}$ – the null operator on K_0 and $A = V_0$ – a contraction on K_0.

In this case a V_0 - *adequate isometry* is an isometry V on a Hilbert space K containing K_0 as a (closed) subspace such that

(1.1) $$K = \bigvee_{n \geq 0} V^n K_0,$$

(1.2) $$V_0 = P^K_{K_0} V | K_0.$$

Hence, a V_0 - adequate isometry is in fact a minimal isometric dilation (not necessarly *power dilation*) of V_0. Since the term "minimal isometric dilation" is generally used for the minimal power dilation, we adopt the term "V_0 - adequate isometry" in order to make this distinction. It is in concordance with the term introduced in [3] in the general context.

We say that two V_o - adequate isometries (K,V) and (K',V') coincide if there exists a unitary operator X from K on K' such that $X|K_o = I_{K_o}$ – the identity operator on K_o, and $XV = V'X$.

The minimal isometric dilation $(\mathcal{K}_o, \mathcal{V}_o)$ of (K_o,V_o) is the unique (modulo the above coincidences) V_o - adequate isometry for which K_o is a *semi-invariant subspace* for $\mathcal{V}$, i.e.

$$V_o^n = P_{K_o}^{\mathcal{K}_o} \mathcal{V}_o^n | K_o. \tag{1.3}$$

In general, there are V_o - adequate isometries which do not coincide with $(\mathcal{K}_o,\mathcal{V}_o)$.

We shall recall here some parametrizations of the set of all V_o - adequate isometries which are particularizations to our situation of general results proved in [3], [2].

A sequence $\{(K_n,V_n)\}_{n\geq 1}$ will be called a *generating sequence* of a V_o-adequate isometry if it is defined recurrently by the formulas

$$K_n = K_{n-1} \oplus D_{V_{n-1}}$$

$$V_n = \begin{bmatrix} V_{n-1} & D_{V^*_{n-1}} C_{n-1} \\ D_{V_{n-1}} & -V^*_{n-1} C_{n-1} \end{bmatrix} \tag{1.4$_n$}$$

where, for each $n \geq 1$, C_{n-1} is a contraction from $D_{V_{n-1}}$ into $D_{V^*_{n-1}}$. Clearly the string $\{(K_n,V_n)\}_{n=1}^p$ is well determined by the string $\{C_n\}_{n=0}^{p-1}$. We shall refer at the sequence $\{C_n\}_{n\geq 0}$ as a generating sequence of a V_o-adequate isometry with the meaning that C_n appears in the form $(1.4)_{n+1}$ of (K_{n+1},V_{n+1}).

Let $\{(K_n,V_n)\}_{n\geq 1}$ be a generating sequence of a V_o-adequate isometry. Considering $K_{n-1} \subset K_n$ in a natural way it is easy to see that, for $n \geq 2$, $V_n|K_{n-2} = V_{n-1}|K_{n-2}$. It results that there exists the inductive limit $(K,V) = \varinjlim (K_n,V_n)$ and (K,V) is a V_o-adequate isometry.

With a natural notion of coincidence for the generating sequence of V_o - adequate isometries we have

PROPOSITION 1.1. *The map* $\{(K_n, V_n)\}_{n\geq 1} \to (K,V) = \varinjlim (K_n,V_n)$ *is a bijective*

correspondence between the set of all generating sequences of V_0 *- adequate isometries and the set of all* V_0 *- adequate isometries.*

If (K,V) is a V_0 - adequate isometry then setting $K_n = \bigvee_{j=0}^{n} V^j K_0$ and $V_n = P^K_{K_n} V|K_n$, it is easy to see that $\{(K_n,V_n)\}_{n\geq 1}$ is the generating sequence of V_0 - - adequate isometry, corresponding to (K,V).

A sequence $\{R_n\}_{n\geq 0}$ will be called a V_0 *- choice sequence* provided $R_0 = V_0$ and for $n\geq 1$ R_n is a contraction from $D_{R_{n-1}}$ into $D_{R^*_{n-1}}$.

For a V_0 - choice sequence $R = \{R_n\}_{n\geq 0}$ let us set

$$K = K(R) = K_0 \oplus D_{R_0} \oplus D_{R_1} \oplus \cdots \tag{1.5}$$

and let V be the operator on K having, with respect to the decomposition (1.5) of K, the matrix

$$V = V(R) = \begin{bmatrix} R_0 & D_{R_0^*}R_1 & D_{R_0^*}D_{R_1^*}R_2 & \cdots \\ D_{R_0} & -R_0^*R_1 & -R_0^*D_{R_1^*}R_2 & \cdots \\ 0 & D_{R_1} & -R_1^*R_2 & \cdots \\ \cdots & \cdots & \cdots & \cdots \end{bmatrix} \tag{1.6}$$

It is easy to see that (K, V) is a V_0 - adequate isometry.

PROPOSITION 1.2. *The map* $R \to (K(R),V(R))$ *is a bijective correspondence between the set of all* V_0 *- choice sequences and the set of all* V_0 *- adequate isometries.*

Let (K,V) be a V_0 - adequate isometry, $\{(K_n,V_n)\}_{n\geq 1}$ be its generating sequence and $\{R_n\}_{n\geq 0}$ be the V_0 - choice sequence of (K,V). We shall work freely with one of the three described above forms of (K_n,V_n):

I. The non specificated form

$$K_n = \bigvee_{j=0}^{n} V^j K_0$$

$$V_n = P^K_{K_n} V|K_0.$$

II. The recursive matricial form $(1.4)_n$.

III. The matricial form

$$K_n = K_o \oplus \mathcal{D}_{R_o} \oplus \cdots \oplus \mathcal{D}_{R_{n-1}}$$

$$V_n = \begin{bmatrix} R_o & D_{R_o^*}R_1 & D_{R_o^*}D_{R_1^*}R_2 & \cdots & D_{R_o^*}\cdots D_{R_{n-2}^*}R_{n-1} \\ D_{R_o} & -R_o^*R_1 & -R_o^*D_{R_1^*}R_2 & \cdots & -R_o^*D_{R_1^*}\cdots D_{R_{n-2}^*}R_{n-1} \\ 0 & D_{R_1} & -R_1^*R_2 & \cdots & -R_1^*D_{R_2^*}\cdots D_{R_{n-2}^*}R_{n-1} \\ \cdots & \cdots & \cdots & \cdots & \cdots \\ 0 & 0 & 0 & \cdots & -R_{n-2}^*R_{n-1} \end{bmatrix}.$$

The identification between forms I and II is made having in mind that

$$K_n = K_{n-1} \vee VK_{n-1} = K_{n-1} \oplus \overline{[V - V_{n-1}]K_{n-1}}$$

and $\|(V - V_{n-1})k_{n-1}\| = \|D_{V_{n-1}}k_{n-1}\|$. The identifications between forms II and III is made having in mind that

$$D_{V_{n-1}} \begin{bmatrix} k_{n-1} \\ D_{V_{n-2}} \begin{bmatrix} k_{n-2} \\ \ddots \begin{bmatrix} 0 \\ D_{V_o}k_o \end{bmatrix} \end{bmatrix} \end{bmatrix} = \|D_{R_{n-1}}\cdots D_{R_o}k_o\|.$$

Clearly (K, V) coincides to the minimal isometric dilation of V_o if and only if $C_n = 0$ for any $n \geq 0$, or, equivalently, $R_n = 0$ for any $n \geq 1$.

Let now $\hat{V}_o$ be a contraction on the Hilbert space $\hat{K}_o$ and let $K_o \subset \hat{K}_o$ be invariant to $\hat{V}_o$. Denote $V_o = \hat{V}_o | K_o$. Let $(\hat{K},\hat{V})$ be a $\hat{V}_o$ - adequate isometry. Denote

$$K = K_+(K_o) = \bigvee_{n \geq 0} \hat{V}^n K_o, \quad V = V_+(K_o) = \hat{V} | K.$$

Then clearly (K,V) is a V_o - adequate isometry. Let $\{(\hat{K}_n, \hat{V}_n)\}_{n \geq 1}$ and

$\{(K_n,V_n)\}_{n\geq 1}$ be the generating sequences of $(\hat{K},\hat{V})$ and (K,V) respectively. We have

$$K_n = \bigvee_{j=0}^{n} V^j K_o = \bigvee_{j=0}^{n} \hat{V}^j K_o \subset \bigvee_{j=0}^{n} \hat{V}^j \hat{K}_o = \hat{K}_n.$$

If we denote by Z_n the inclusion of K_n into $\hat{K}_n$ then, for $n\geq 1$, Z_n has the recursive matricial form

$$(1.7)_n \qquad Z_n = \begin{bmatrix} Z_{n-1} & a_{n-1} \\ 0 & b_{n-1} \end{bmatrix}$$

where a_{n-1} is the contraction from $D_{V_{n-1}}$ into $\hat{K}_{n-1}$ defined by

$$(1.8)_n \qquad a_{n-1}D_{V_{n-k}}k_{n-1} = (\hat{V}_{n-1}Z_{n-1} - Z_{n-1}V_{n-1})k_{n-1}, \quad k_{n-1} \in K_{n-1},$$

and b_{n-1} is the contraction from $D_{V_{n-1}}$ into $D_{\hat{V}_{n-1}}$ defined by

$$(1.9)_n \qquad b_{n-1}D_{V_{n-1}}k_{n-1} = D_{\hat{V}_{n-1}}Z_{n-1}k_{n-1}, \quad k_{n-1} \in K_{n-1}.$$

Let us also note that

$$(1.10)_n \qquad V_n = P^{\hat{K}_n}_{K_n}\hat{V}_n|K_n$$

or, in recursive matricial form

$$V_n = Z_n^*\hat{V}_nZ_n.$$

If we consider the matricial forms of (K_n,V_n) and $(\hat{K}_n,\hat{V}_n)$ then the matrix of Z_n has not a simple form. However we have the following

REMARK 1.1. Let $k_o \in K_o$ and for $n\geq 1$ let $k_n = (k_o)_n$ be the element of K_n which in the matricial form of (K_n,V_n) is given by $0\oplus\ldots\oplus 0\oplus D_{R_{n-1}}\ldots D_{R_o}k_o$. Then in the recursive matricial form of (K_n,V_n) we have

$$(k_o)_1 = \begin{bmatrix} 0 \\ D_{V_o}k_o \end{bmatrix}, \quad (k_o)_2 = \begin{bmatrix} 0 \\ D_{V_1}k_1 \end{bmatrix} \quad \ldots \quad (k_o)_n = \begin{bmatrix} 0 \\ D_{V_{n-1}}k_{n-1} \end{bmatrix}.$$

It results

$$D_{\hat{V}_n} Z_n k_n = D_{\hat{V}_n} Z_n \begin{bmatrix} 0 \\ D_{V_{n-1}} k_{n-1} \end{bmatrix} = \begin{bmatrix} D_{\hat{V}_n} & a_{n-1} D_{V_{n-1}} k_{n-1} \\ & b_{n-1} D_{V_{n-1}} k_{n-1} \end{bmatrix} =$$

$$= D_{\hat{V}_n} \begin{bmatrix} 0 \\ b_{n-1} D_{V_{n-1}} k_{n-1} \end{bmatrix} = \begin{bmatrix} D_{\hat{V}_n} & 0 \\ & D_{\hat{V}_{n-1}} Z_{n-1} k_{n-1} \end{bmatrix}.$$

This implies that in the matricial form of $(\hat{K}_{n+1}, \hat{V}_{n+1})$ the element $\begin{bmatrix} 0 \\ D_{\hat{V}_n} Z_n k_n \end{bmatrix}$ of $\hat{K}_{n+1}$ is given by $0 \oplus \ldots \oplus 0 \oplus D_{\hat{R}_n} \cdots D_{\hat{R}_o} k_o$.

Let us end this section with the following simple but useful

PROPOSITION 1.3. *Let (K,V) be a V_o-adequate isometry and let $\{(K_n, V_n)\}_{n \geq 1}$ be its generating sequence. Let H be a closed subspace of K_o. Then the following are equivalent*

(i) $V[K \ominus H] \subset K \ominus H$

(ii) $V_n[K_n \ominus H] \subset K_n \ominus H, \quad n \geq 0$

(iii) $\begin{cases} V_o[K_o \ominus H] \subset K_o \ominus H \\ C_n D_{V_n} \subset D_{V_n^*} \ominus \overline{D_{V_n^*} H}, \quad n \geq 0. \end{cases}$

PROOF. Since for $k_n \in K_n$, k_n is in $K_n \ominus H$ if and only if $k_n \in K \ominus H$ and because

$$V_{n+1} \begin{bmatrix} k_n \\ 0 \end{bmatrix} = V \begin{bmatrix} k_n \\ 0 \end{bmatrix},$$

we clearly have (i)$\Leftrightarrow$(ii).

Since $0 \oplus D_{V_n} k_n \in K_{n+1} \ominus H$ from (ii) we obtain $V_{n+1} \begin{bmatrix} 0 \\ D_{V_n} k_n \end{bmatrix} \in K_{n+1} \ominus H$.

But

$$V_{n+1} \begin{bmatrix} 0 \\ D_{V_n} k_n \end{bmatrix} = \begin{bmatrix} D_{V_n^*} C_n D_{V_n} k_n \\ - V_n^* C_n D_{V_n} k_n \end{bmatrix}$$

and (iii) clearly results.A simple induction shows that (iii)$\Leftrightarrow$(ii).

2. ANDO DILATIONS AS ADEQUATE ISOMETRIES

In what follows we shall denote by $(H, [T,S])$ a pair of commuting contractions T, S on the Hilbert space H.

We say that $(K,[U,V])$ is an *Ando dilation* of $(H,[T,S])$ provided K is a Hilbert space containing H as a closed subspace, U, V are two commuting isometries on K and

$$K = \bigvee_{n,m \geq 0} U^n V^m H \tag{2.1}$$

$$T^n S^m = P^K_H U^n V^m | H, \quad n,m \geq 0. \tag{2.2}$$

Two Ando dilations $(K,[U,V])$ and $(K',[U',V'])$ of $(H,[T,S])$ *coincide* if there exists a unitary operator X from K on K' such that $X|H = I_H$ and $XU = U'X$, $XV = V'X$.

Let us denote

$$K_o = \bigvee_{n \geq 0} U^n H, \quad U_o = U|K_o, \quad V_o = P^K_{K_o} V|K_o.$$

It is clear that (K_o, U_o) is an identification of the minimal isometric dilation of T. It is easy to see that K_o reduces U and V_o is a contractive intertwining dilation of the commutant S of T. That is

$$U_o V_o = V_o U_o, \quad SP^{K_o}_H = P^{K_o}_H V_o.$$

We say that the Ando dilation $(K,[U,V])$ of $(H,[T,S])$ *crosses through* $(K_o,[U_o,V_o])$ if $(K_o,[U_o,V_o])$ is attached to $(K,[U,V])$ as above.

Since

$$K = \bigvee_{n,m \geq 0} U^n V^m H = \bigvee_{m \geq 0} V^m K_o$$

and $V_o = P^K_{K_o} V|K_o$ we conclude that (K,V) is a V_o - adequate isometry.

For all what follows we shall fix $(K_o,[U_o,V_o])$ consisting from an identification (K_o,U_o) of the minimal isometric dilation of T and a contractive intertwining dilation V_o of the commutant S of T.

We say that the V_o - adequate isometry (K,V) *produces* an Ando dilation of $(H,[T,S])$ if there exists an isometry U on K such that $(K,[U,V])$ is an Ando dilation of $(H,[T,S])$ which crosses through $(K_o,[U_o,V_o])$.

The above considerations show that any Ando dilation of $(H,[T,S])$ which crosses through $(K_o,[U_o,V_o])$ is produced by a V_o-adequate isometry. In [4] it was proved, in a slightly different terminology the following

PROPOSITION 2.1. *Let (K,V) be a V_o- adequate isometry. Then (K,V) produces an Ando dilation of $(H,[T,S])$ if and only if the following conditions hold.*

(i) *If* $\{C_n\}_{n\geq 0}$ *is the generating sequence of* (K,V) *then*

$$C_n D_{V_n} \subset D_{V_n^*} \ominus \overline{D_{V_n^*} H}; \tag{2.3$_n$}$$

(ii) *The formulas*

$$\Gamma_n D_{V_n} = V_n U_n - U_n V_n, \quad n \geq 0 \tag{2.4$_n$}$$

$$Y_n D_{\Gamma_n} D_{V_n} = D_{V_n} U_n, \quad n \geq 0 \tag{2.5$_n$}$$

define a contraction Γ_n *from* D_{V_n} *into* **ker** U_n^* *and an isometry* Y_n *from* D_{Γ_n} *into* D_{V_n}. *The operator* U_n, $n \geq 1$, *which appears in* $(2.4)_n$, $(2.5)_n$ *is the isometry on* K_n *defined recursively, according to* $(2.4)_{n-1}$, $(2.5)_{n-1}$, *by the matrix*

$$U_n = \begin{bmatrix} U_{n-1} & \Gamma_{n-1} \\ 0 & Y_{n-1} D_{\Gamma_{n-1}} \end{bmatrix}. \tag{2.6$_n$}$$

REMARKS. Two V_0 - adequate isometries which produce coinciding Ando dilations of $(H,[T,S])$, coincide. If the V_0-adequate isometry (K,V) produces the Ando dilation $(K,[U,V])$ of $(H,[T,S])$ then U is uniquely determined by (K,V).

If $(\mathcal{K}_o,\mathcal{V}_o)$ is the minimal isometric dilation of V_0 then it produces an Ando dilation $(\mathcal{K}_o,[\mathcal{U}_o,\mathcal{V}_o])$ of $(H,[T,S])$. We shall call $(\mathcal{K}_o,[\mathcal{U}_o,\mathcal{V}_o])$ the *distinguished Ando dilation of* $(H,[T,S])$ which crosses through $(K_0,[U_0,V_0])$.

For an Ando dilation $(K,[U,V])$ of $(H,[T,S])$ let us denote by $(\hat{K},[\hat{U},\hat{V}])$ its *minimal unitary-isometric extension*. This means that $(\hat{K},\hat{U})$ is the minimal unitary extension of (K,U) and $\hat{V}$ is an isometry on $\hat{K}$ which commutes with $\hat{U}$ and extends V. Such an extension always exists and it is unique. Let $\hat{K}_0 = \bigvee_{n\geq 0} \hat{U}^{*n} K_0$ and $\hat{U}_0 = \hat{U}|\hat{K}_0$. Clearly $(\hat{K}_0,\hat{U}_0)$ is the minimal unitary extension of (K_0,U_0) and consequently the minimal unitary dilation of T. If $\hat{V}_0 = P^{\hat{K}}_{\hat{K}_o} \hat{V}|K_0$ then $\hat{V}_0$ is the unique extension of V_0 to a commutant of $\hat{U}_0$.

Hence $(\hat{K}_0,[\hat{U}_0,\hat{V}_0])$ is uniquely determined by $(K_0,[U_0,V_0])$. Recall also that denoting $K_{*0} = \bigvee_{n\geq 0} \hat{U}_0^{*n} H$ then (cf. [6])

$$\hat{K}_0 = [K_{*0} \ominus H] \oplus H \oplus [K_0 \ominus H]. \tag{2.7}$$

Clearly $(\hat{K}, \hat{V})$ is a $\hat{V}_0$-adequate isometry.

We say that a $\hat{V}_o$-adequate isometry $(\hat{K},\hat{V})$ *produces an Ando dilation* for $(H,[T,S])$ if there exists a unitary operator $\hat{U}$ on $\hat{K}$ which extends $\hat{U}_o$ and commutes with $\hat{V}$ such that

(2.8) $$T^n S^m = P^{\hat{K}}_H \hat{U}^n \hat{V}^m | H, \quad n, m \geq 0.$$

Setting $K = \bigvee_{n,m\geq 0} \hat{U}^n \hat{V}^m H$, $U = \hat{U}|K$, $V = \hat{V}|K$ then clearly $(K,[U,V])$ is an Ando dilation of $(H,[T,S])$ which crosses through $(K_o,[U_o,V_o])$ – the Ando dilation *produced* by $(\hat{K},\hat{V})$. Since $K = \bigvee_{n,m\geq 0} V^n U_o^m H = \bigvee_{n\geq 0} V^n K_o$, clearly the V_o-adequate isometry $(K,V) = (K_+(K_o),V_+(K_o))$ is attached to $(\hat{K},\hat{V})$ and K_o as in the preceding section and $(K,[U,V])$ is also produced by (K,V).

From

$$\bigvee_{n\geq 0} \hat{U}^{*n} K = \bigvee_{n\geq 0} \hat{U}^{*n} \bigvee_{m\geq 0} V^m K_o = \bigvee_{n\geq 0} V^n \bigvee_{m\geq 0} \hat{U}^{*m} K_o = \bigvee_{n\geq 0} V^n \hat{K}_o = \hat{K}$$

it results that $(\hat{K},\hat{U})$ is the minimal unitary extension of (K,U). Clearly then $(\hat{K},[\hat{U},\hat{V}])$ is the minimal unitary-isometric extension of $(K,[U,V])$.

We conclude that any Ando dilation of $(H,[T,S])$ which crosses through $(K_o,[U_o,V_o])$ is produced as above by a (uniquely determined) $\hat{V}_o$-adequate isometry $(\hat{K},\hat{V})$.

Suppose now that the $\hat{V}_o$-adequate isometry $(\hat{K},\hat{V})$ produces the Ando dilation $(K,[U,V])$. Let $(\hat{K}_n,\hat{V}_n)$ and (K_n,V_n) be the generating sequences of $(\hat{K},\hat{V})$ and (K,V) respectively. Then clearly $\hat{K}_n$ is a reducing subspace for $\hat{U}$ and K_n is invariant for $\hat{U}$. Let $\hat{U}_n = \hat{U}|\hat{K}_n$, $U_n = \hat{U}|K_n$. We shall call $(\hat{K}_n,[\hat{U}_n,\hat{V}_n])$ and $(K_n,[U_n,V_n])$ the *sequence of succesive dilations* of $(\hat{K},[\hat{U},\hat{V}])$ and $(K,[U,V])$ respectively. From

$$\bigvee_{j\geq 0} \hat{U}_n^{*j} K_n = \bigvee_{j\geq 0} \hat{U}^{*j} \bigvee_{p=0}^{n} \hat{V}^p K_o = \bigvee_{p=0}^{n} \hat{V}^p \bigvee_{j\geq 0} \hat{U}_o^{*} K_o = \bigvee_{p=0}^{n} \hat{V}^p \hat{K}_o = \hat{K}_n$$

it results that $(\hat{K}_n, \hat{U}_n)$ is the minimal unitary extension of (K_n,U_n). Since for any k_n, $g_n \in K_n$ we have

$$(U_n^* V_n U_n k_n , g_n) = (V_n U_n k_n , U_n g_n) = (\hat{V}\hat{U} k_n , \hat{U} g_n) = (\hat{V} k_n , g_n) = (V_n k_n , g_n)$$

it results

$(2.9)_n$ $$U_n^* V_n U_n = V_n.$$

That means that V_n is a U_n - *Toeplitz operator,* and because $P^{\hat{K}_n}_{K_n} \hat{V}_n | K_n = V_n$ we conclude that $\hat{V}_n$ is the *symbol* of the U_n - Toeplitz operator V_n. It is known then (cf. [5]) that we have

$$(2.10)_n \qquad \hat{V}_n = \text{s-}\lim_{p\to\infty} \hat{U}_n^{*p} V_n P_{\mathcal{K}_n}^{\hat{\mathcal{K}}_n} \hat{U}_n^p.$$

Note also that for $j \le n$, $\hat{\mathcal{K}}_j$ reduces $\hat{U}_n$ and in the matricial form of $(\hat{\mathcal{K}}_n, \hat{V}_n)$, $\hat{U}_n$ is given by the diagonal matrix

$$(2.11)_n \qquad \hat{U}_n = \begin{bmatrix} \hat{U}_o & 0 & \dots & 0 \\ 0 & \hat{Y}_o & \dots & 0 \\ & & & \\ \cdot & \cdot & \dots & \cdot \\ 0 & 0 & \dots & \hat{Y}_{n-1} \end{bmatrix}$$

where for $0 \le j \le n-1$, $\hat{Y}_j = \hat{U}_o | \mathcal{D}_{\hat{R}_j}$, and $\hat{U}$ itself has the corresponding infinite diagonal matrix. From Proposition 2.1 it results that for the recursive matricial form of $(\mathcal{K}_n, [U_n, V_n])$ the relations $(2.3)_n$ - $(2.5)_n$ hold and U_n is given by the matrix $(2.6)_n$. We shall prove now the main result of this section.

THEOREM 1.1. *Let* $(\hat{\mathcal{K}}, \hat{V})$ *be a* $\hat{V}_o$*-adequate isometry and* $\{\hat{R}_n\}_{n\ge0}$ *be its* $\hat{V}_o$*-choice sequence. Let* $(\mathcal{K}, V) = (\mathcal{K}_+(\mathcal{K}_o), V_+(\mathcal{K}_o))$ *be the corresponding* V_o*-adequate isometry and* $\{R_n\}_{n\ge0}$ *be its* V_o*-choice sequence. Then* $(\hat{\mathcal{K}}, \hat{V})$ *(and consequentely* $(\mathcal{K}, V)$*) produces an Ando dilation of* $(H, [T,S])$ *if and only if the following conditions hold:*

(i) *For any* $n \ge 1$ *we have*

$$(2.12)_n \qquad \hat{U}_o \hat{R}_n = \hat{R}_n \hat{U}_o | \mathcal{D}_{\hat{R}_{n-1}}.$$

(ii) *For any* $n \ge 1$, $k_o \in \mathcal{K}_o$, $k_{*o} \in \mathcal{K}_{*o}$ *we have*

$$(2.13)_n \qquad (\hat{R}_n D_{\hat{R}_{n-1}} \dots D_{\hat{R}_o} k_o, D_{\hat{R}^*_{n-1}} \dots D_{\hat{R}^*_o} k_{*o}) = -(\hat{V}^2_{n-1} S_{n-1} k_o, k_{*o})$$

where $S_o k_o = 0$ *and for* $n \ge 1$, S_n *is the operator from* $\mathcal{K}_o$ *into* $\hat{\mathcal{K}}_n$ *given by*

$$(2.14)_n \qquad S_n k_o = Z_n (0 \oplus \dots \oplus 0 \oplus D_{R_{n-1}} \dots D_{R_o} k_o).$$

PROOF. Suppose that $(\hat{\mathcal{K}}, \hat{V})$ produces an Ando dilation $(\mathcal{K}, [U,V])$ of $(H, [T,S])$. Writing $\hat{V}_n$ and $\hat{U}_n$ in the matricial form given by $(1.6)_n$ and $(2.11)_n$ respectively, and using $\hat{U}_n \hat{V}_n = \hat{V}_n \hat{U}_n$ we obtain $(2.12)_{n-1}$.

From $V_{n+1} = P^{\hat{K}_{n-1}}_{K_{n+1}} \hat{V}_{n+1}|K_{n+1}$ which in the recursive matricial form can be written as

$$V_{n+1} = Z^*_{n+1} \hat{V}_{n+1} Z_{n+1}$$

we obtain

$$(2.15)_n \qquad Z^*_n(\hat{V}_n a_n + D_{\hat{V}^*_n} \hat{C}_n b_n) = D_{V^*_n} C_n.$$

From Proposition 2.1 we have $D_{V^*_n} C_n D_{V_n} \subset K_n \ominus H$ and using $(2.15)_n$ we obtain

$$(2.16)_n \quad (D_{\hat{V}^*_n} \hat{C}_n D_{\hat{V}_n} Z_n k_n , h) = - (\hat{V}_n a_n D_{V_n} k_n , h) = - (\hat{V}^2_n Z_n k_n , h) + (V^2_n k_n , h).$$

For n = 0 it results

$$(2.16)_0 \qquad (D_{\hat{V}^*_0} \hat{C}_0 D_{V_0} k_0 , h) = - (\hat{V}^2_o k_0 , h) + (V^2_o k_0 , h) = 0$$

which in the matricial form gives

$$(D_{\hat{R}^*_0} \hat{R}_1 D_{\hat{R}_0} k_0 , h) = 0.$$

Since

$$(D_{\hat{R}^*_0} \hat{R}_1 D_{\hat{R}_0} k_0 , \hat{U}^{*p}_o h) = (\hat{U}^p_o D_{\hat{R}^*_0} \hat{R}_1 D_{\hat{R}_0} k_0 , h) = (D_{\hat{R}^*_0} \hat{R}_1 D_{\hat{R}_0} U^p_o k_0 , h) = 0$$

for any $p \geq 0$, we obtain $(2.13)_1$.

Suppose now $n \geq 1$. Since $V_n(K_n \ominus H) \subset K_n \ominus H$ (see Proposition 1.3) from $(2.16)_n$ it results that for any $k_n \in K_n \ominus H$ we have

$$(D_{\hat{V}^*_n} \hat{C}_n D_{\hat{V}_n} Z_n k_n , h) = - (\hat{V}^2_n Z_n k_n , h).$$

Since $U_n[K_n \ominus H] \subset K_n \ominus H$ we obtain

$$(D_{\hat{V}^*_n} \hat{C}_n D_{\hat{V}_n} Z_n k_n , \hat{U}^{*p}_o h) = (\hat{U}^p_n D_{\hat{V}^*_n} \hat{C}_n D_{\hat{V}_n} Z_n k_n , h) =$$

$$= (D_{\hat{V}^*_n} \hat{C}_n \hat{U}^p_n Z_n k_n , h) = (D_{\hat{V}^*_n} \hat{C}_n Z_n U^p_n k_n , h) =$$

$$= (\hat{V}^2_n Z_n U^p_n k_n , h) = (\hat{V}^2_n \hat{U}^p_n Z_n k_n , h) = - (\hat{V}^2_n Z_n k_n , \hat{U}^{*p}_o h).$$

Hence

$$(2.17)_n \quad (D_{\hat{V}^*_n} \hat{C}_n D_{\hat{V}_n} Z_n k_n , k_{*o}) = - (\hat{V}^2_n Z_n k_n , k_{*o}), \quad k_n \in K_n \ominus H, \ k_{*o} \in K_{*o}.$$

Let now $k_0 \in K_0$ and denote by k_n the element of $K_n \ominus H$ which in the matricial form of (K_n, V_n) is given by $k_n = 0 \oplus \ldots \oplus 0 \oplus D_{R_{n-1}} \ldots D_{R_0} k_0$. We have

$$(\hat{V}_{n+1}\begin{bmatrix}0\\ D_{\hat{V}_n} Z_n k_n\end{bmatrix}, k_{*o}) = (D_{\hat{V}_n^*}\hat{C}_n D_{\hat{V}_n} Z_n k_n, k_{*o}) = -(\hat{V}_n^2 Z_n k_n, k_{*o}).$$

But according to Remark 1.1, in the **matricial** form of $(\hat{K}_{n+1}, \hat{V}_{n+1})$ the element $0\oplus D_{\hat{V}_n} Z_n k_n$ of $\hat{K}_{n+1}$ is given by $0\oplus\ldots\oplus 0\oplus D_{\hat{R}_n}\ldots D_{\hat{R}_o} k_o$. It results

$$(D_{\hat{R}_o^*}\ldots D_{\hat{R}_n^*}\hat{R}_{n+1} D_{\hat{R}_n}\ldots D_{\hat{R}_o} k_o, k_{*o}) = (\hat{V}_{n+1}\begin{bmatrix}0\\ D_{V_n} Z_n k_n\end{bmatrix}, k_{*o}) = -(\hat{V}_n^2 Z_n k_n, k_{*o})$$

which is $(2.13)_{n+1}$.

We proved that $(2.12)_n$ and $(2.13)_n$, $n\geq 1$ are necessary conditions in order that $(\hat{K}, \hat{V})$ produces an Ando dilation.

Suppose now that for any $n\geq 1$, $(2.12)_n$ and $(2.13)_n$ hold. From $(2.12)_n$ it results that $(2.11)_n$ defines a unitary operator $\hat{U}_n$ on $\hat{K}_n$ which extends $\hat{U}_o$ and commutes with $\hat{V}_n$ and consequently we can construct the unitary extension $\hat{U}$ of $\hat{U}_o$ to $\hat{K}$ which commutes with $\hat{V}$. If

$$P_H^K V = P_H^K V P_H^K \tag{2.18}$$

holds then

$$P_H^{\hat{K}}\hat{V}^n\hat{U}^m h = P_H^K V^n U_o^m h = P_H^K V P_H^K V^{n-1} U_o^m h =$$

$$= P_H^{K_o} P_{K_o}^K V P_H^K V^{n-1} U_o^m h = P_H^{K_o} V_o P_H^K V^{n-1} U_o^m h = \ldots =$$

$$= P_H^{K_o} V_o^n U_o^m h = S^n T^m h.$$

So if (2.18) holds then $(\hat{K}, \hat{V})$ produces an Ando dilation of $(H,[T,S])$.

But (2.18) is equivalent to $V[K\ominus H]\subset K\ominus H$ and, according to Proposition 1.3 this is equivalent to $V_n(K_n\ominus H)\subset K_n\ominus H$, for any $n\geq 0$ and also equivalent to $D_{V_n^*} C_n D_{V_n}\subset$ $\subset K_n\ominus H$ for any $n\geq 0$. For $n = 0$ we have

$$(D_{V_o^*} C_o D_{V_o} k_o, h) = (V_1\begin{bmatrix}0\\ D_{V_o} k_o\end{bmatrix}, h) = (Z_1^*\hat{V}_1 Z_1\begin{bmatrix}0\\ D_{V_o} k_o\end{bmatrix}, h) =$$

$$= (\hat{V}_1\begin{bmatrix}0\\ D_{\hat{V}_o} k_o\end{bmatrix}, h) = (D_{\hat{R}_o^*}\hat{R}_1 D_{\hat{R}_o} k_o, h) = 0$$

because of $(2.13)_1$.

Suppose now, by induction, that for some $n\geq 1$ we have $V_n(K_n\ominus H)\subset K_n\ominus H$.

Then for any $k_n \in K_n \ominus H$ we have

$$(D_{V_n^*} C_n D_{V_n} k_n, h) = (V_{n+1} \begin{bmatrix} 0 \\ D_{V_n} k_n \end{bmatrix}, h) = (Z^*_{n+1} \hat{V}_{n+1} Z_{n+1} \begin{bmatrix} 0 \\ D_{V_n} k_n \end{bmatrix}, h) =$$

$$= (\hat{V}_{n+1} \begin{bmatrix} a_n D_{V_n} k_n \\ b_n D_{V_n} k_n \end{bmatrix}, h) = (\hat{V}_n a_n D_{V_n} k_n, h) + (\hat{V}_{n+1} \begin{bmatrix} 0 \\ D_{\hat{V}_n} Z_n k_n \end{bmatrix}, h) =$$

$$= (\hat{V}_n^2 Z_n k_n, h) - (V_n^2 k_n, h) + (\hat{V}_{n+1} \begin{bmatrix} 0 \\ D_{\hat{V}_n} Z_n k_n \end{bmatrix}, h) =$$

$$= (\hat{V}_n^2 Z_n k_n, h) + (\hat{V}_{n+1} \begin{bmatrix} 0 \\ D_{V_n} Z_n k_n \end{bmatrix}, h).$$

If k_n is the element of $K_n \ominus H$ which in the matricial form of (K_n, V_n) is given by $0 \oplus \ldots \oplus 0 \oplus D_{R_{n-1}} \ldots D_{R_o} k_o$ then using $(2.13)_n$ we obtain

$$(D_{V_n^*} C_n D_{V_n} k_n, h) = (\hat{V}_n^2 Z_n k, h) + (D_{\hat{R}_o^*} \ldots D_{\hat{R}_{n-1}^*} \hat{R}_n D_{\hat{R}_{n-1}} \ldots D_{\hat{R}_o} k_o, h) = 0.$$

Since for $n \geq 1$ $D_{V_n} K_n = D_{V_n}(0 \oplus \ldots \oplus 0 \oplus D_{R_{n-1}})$ we obtain

$$D_{V_n^*} C_n D_{V_n} K_n \subset K_n \ominus H$$

which clearly implies $V_{n+1}(K_{n+1} \ominus H) \subset K_{n+1} \ominus H$ and the induction is complete.

This finishes the proof of the theorem.

The conditions $(2.12)_n$ and $(2.13)_n$, $n \geq 1$, describe all the obstructions on the $\hat{V}_o$ - choice sequence $\{\hat{R}_n\}_{n \geq 0}$ in order that the corresponding $\hat{V}_o$ - adequate isometry $(\hat{K}, \hat{V})$ produces an Ando dilation $(K, [U, V])$ of $(H, [T, S])$. They also suggest a recursive construction of $\{\hat{R}_n\}_{n \geq 0}$. Indeed if we denote

$$W_n = \hat{U}_o | \overline{D_{\hat{R}_n} \ldots D_{\hat{R}_o} K_o}$$

$$W_{*n} = \hat{U}_o^* | \overline{D_{\hat{R}_n^*} \ldots D_{\hat{R}_o^*} K_{*o}}$$

then W_n and W_{*n} are isometries. Let Q_n be the operator from $\overline{D_{\hat{R}_n} \ldots D_{\hat{R}_o} K_o}$ into $\overline{D_{\hat{R}_n^*} \ldots D_{\hat{R}_o^*} K_{*o}}$ defined by

$$(2.19)_n \qquad (Q_n D_{\hat{R}_n} \cdots D_{\hat{R}_o} k_o , D_{\hat{R}^*_n} \cdots D_{\hat{R}^*_o} k_{*o}) = -(\hat{V}_n^2 S_n k_o , k_{*o}).$$

According to $(2.13)_{n+1}$, Q_n is a contraction. We also have

$$(2.20)_n \qquad Q_n W_n = W^*_{*n} Q_n.$$

So $\hat{R}_{n+1}$ is a contractive intertwining dilation of the contraction Q_n which intertwines W_n and W^*_{*n} and can be constructed following the methods from intertwining dilation theory (cf. [2]). Unfortunately the choice of $\hat{R}_{n+1}$ as a contractive intertwining dilation of Q_n is not free. It has to be chosen such that Q_{n+1} defined by $(2.19)_{n+1}$ be a contraction. It can happen that $(2.13)_j$ holds for any $1 \leq j \leq n$ but the choice of $\hat{R}_{n+1}$ which satisfies $(2.13)_{n+1}$ be impossible. An example in this sense is given in [4;III].

In [4;I] we indirectly proved that if we start with $\hat{R}_1$ verifying $(2.13)_1$ and having sufficiently small norm we can produce a sequence $\{\hat{R}_n\}_{n\geq 1}$ which verifies $(2.13)_n$ for any $n \geq 1$.

In the next section we shall study a class of Ando dilations of $(H,[T,S])$ which cross through $(K_o,[U_o,V_o])$ for which we can exhibit a system of free parameters which describe it.

3. U-DIAGONAL ANDO DILATION

Let $(K,[U,V])$ be an Ando dilation of $(H,[T,S])$ and let $(K_n,[U_n,V_n])_{n\geq 0}$ be its sequence of succesive dilations. We say that $(K,[U,V])$ is U - *diagonal* provided $U_n V_n = V_n U_n$ for any $n \geq 0$.

Since $\Gamma_n D_{V_n} = V_n U_n - U_n V_n$ it results from Proposition 2.1 that $(K,[U,V])$ is U-diagonal if and only if for each $n \geq 1$ U_n has the diagonal form

$$U_n = \begin{bmatrix} U_{n-1} & 0 \\ 0 & Y_{n-1} \end{bmatrix}$$

where $Y_{n-1} D_{V_{n-1}} = D_{V_{n-1}} U_{n-1}$. Clearly then U itself has a diagonal matricial form.

PROPOSITION 3.1. *Let $(K,[U,V])$ be an Ando dilation of $(H,[T,S])$, $(\hat{K},[\hat{U},\hat{V}])$ be the minimal unitary-isometric extension of $(K,[U,V])$ and $\{\hat{K}_n,[\hat{U}_n,\hat{V}_n]\}_{n\geq 0}$, $\{K_n,[U_n,V_n]\}_{n\geq 0}$ be the corresponding sequences of succesive dilations. The following assertions are equivalent:*

(i) $(K,[U,V])$ *is* U - *diagonal*.

(ii) *For each* $n \geq 0$, $\hat{V}_n | K_n = V_n$.

(iii) *The recursive matricial form of the embedding* Z_n *of* K_n *into* $\hat{K}_n$ *is diagonal, i.e.*

$$Z_n = \begin{bmatrix} Z_{n-1} & 0 \\ 0 & b_{n-1} \end{bmatrix}$$

with $b_{n-1} D_{V_{n-1}} = D_{\hat{V}_{n-1}} Z_{n-1}$.

PROOF. Since according to $(2.10)_n$, for any $n \geq 0$ and $k_n \in K_n$ we have

$$\hat{V}_n k_n = s - \lim_{p \to \infty} \hat{U}_n^{*p} V_n U_n^p k_n$$

clearly (i)$\Longrightarrow$(ii). If (ii) holds then for any $k_n \in K_n$ we have

$$(V_n U_n - U_n V_n) k_n = \hat{V}_n \hat{U}_n k_n - \hat{U}_n \hat{V}_n k_n = 0$$

hence (ii)$\Longrightarrow$(i).

Since in general

$$Z_n = \begin{bmatrix} Z_{n-1} & a_{n-1} \\ 0 & b_{n-1} \end{bmatrix}$$

with $a_{n-1} D_{V_{n-1}} k_{n-1} = (\hat{V}_{n-1} Z_{n-1} - Z_{n-1} V_{n-1}) k_{n-1}$ clearly (ii)$\Longleftrightarrow$(iii).

Let us fix again $(K_o, [U_o, V_o])$ and consequently $(\hat{K}_o, [\hat{U}_o, \hat{V}_o])$ with the semnification given in the preceding section.

THEOREM 3.1. *Let* $(\hat{K}, \hat{V})$ *be a* $\hat{V}_o$ *- adequate isometry and let* $\{R_n\}_{n \geq 0}$ *be its* $\hat{V}_o$ *- choice sequence. Then* $(\hat{K}, \hat{V})$ *produces a* U*-diagonal Ando dilation of* $(H, [T, S])$ *if and only if for any* $n \geq 1$ *we have*

$(3.1)_n$ $$\hat{U}_o \hat{R}_n = \hat{R}_n \hat{U}_o | D_{R_{n-1}}$$

$(3.2)_{n,0}$ $$D_{\hat{R}^*_{n-1}} \cdots D_{\hat{R}^*_o} \hat{R}_n D_{\hat{R}_{n-1}} \cdots D_{\hat{R}_o} K_o \subset K_o \ominus H$$

$(3.2)_{n,j}$ $$\hat{R}^*_{j-1} D_{\hat{R}^*_j} \cdots D_{\hat{R}^*_{n-1}} \hat{R}_n D_{\hat{R}_{n-1}} \cdots D_{\hat{R}_o} K_o \subset \overline{D_{\hat{R}_{n-1}} \cdots D_{\hat{R}_o} K_o}, \quad 1 \leq j \leq n-1,$$

$(3.2)_{n,n}$ $$\hat{R}^*_{n-1} \hat{R}_n D_{\hat{R}_{n-1}} \cdots D_{\hat{R}_o} K_o \subset \overline{D_{\hat{R}_{n-1}} \cdots D_{\hat{R}_o} K_o}.$$

PROOF. Suppose that $(\hat{K},\hat{V})$ produces the U-diagonal Ando dilation $(K,[U,V])$ of $(H,[T,S])$ and let $(\hat{K},[\hat{U},\hat{V}])$ be the minimal unitary - isometric extension of $(K,[U,V])$. From Proposition 3.1 it results that $\hat{V}_n|K_n = V_n$ and the embedding Z_n of K_n in $\hat{K}_n$ has a diagonal recursive matricial form. It results that if we write $\hat{K}_n$ as

$$\hat{K}_n = \hat{K}_0 \oplus D_{\hat{R}_0} \oplus \dots \oplus D_{\hat{R}_{n-1}}$$

then the subspace K_n of $\hat{K}_n$ is

$$K_n = K_0 \oplus \overline{D_{\hat{R}_0} K_0} \oplus \overline{D_{\hat{R}_1} D_{\hat{R}_0} K_0} \oplus \dots \oplus \overline{D_{\hat{R}_{n-1}} \dots D_{\hat{R}_0} K_0}.$$

From

$$\hat{V}_n \begin{bmatrix} 0 \\ 0 \\ \cdot \\ \cdot \\ \cdot \\ 0 \\ D_{R_{n-1}} \dots D_{R_0} k_0 \end{bmatrix} = \begin{bmatrix} D_{\hat{R}_0^*} \dots D_{\hat{R}_{n-1}^*} \hat{R}_n D_{\hat{R}_{n-1}} \dots D_{\hat{R}_0} k_0 \\ -\hat{R}_0^* D_{\hat{R}_1^*} \dots D_{\hat{R}_{n-1}^*} \hat{R}_n D_{\hat{R}_{n-1}} \dots D_{\hat{R}_0} k_0 \\ \cdot \\ \cdot \\ \cdot \\ -\hat{R}_{n-2}^* D_{\hat{R}_{n-1}^*} \hat{R}_n D_{\hat{R}_{n-1}} \dots D_{\hat{R}_0} k_0 \\ \hat{R}_{n-1}^* \hat{R}_n D_{\hat{R}_{n-1}} \dots D_{\hat{R}_0} k_0 \end{bmatrix}$$

and

$$\hat{V}_n \begin{bmatrix} 0 \\ 0 \\ \cdot \\ \cdot \\ \cdot \\ 0 \\ D_{\hat{R}_{n-1}} \dots D_{\hat{R}_0} k_0 \end{bmatrix} = V_n \begin{bmatrix} 0 \\ 0 \\ \cdot \\ \cdot \\ \cdot \\ 0 \\ D_{\hat{R}_{n-1}} \dots D_{\hat{R}_0} k_0 \end{bmatrix} \in K_n \ominus H.$$

we obtain $(3.2)_{n,j}$ for $0 \leq j \leq n$. The conditions $(3.1)_n$ result as in the proof of Theorem 2.1.

Suppose now that for any $n \geq 1$, $(3.1)_n$ and $(3.2)_{n,j}$, $0 \leq j \leq n$, hold. Using $(3.1)_n$ we shall construct as in the proof of Theorem 2.1 the unitary operator $\hat{U}$ on $\hat{K}$ which extends $\hat{U}_o$ and commutes with $\hat{V}$. Let $(K,V) = (K_+(K_o), V_+(K_o))$ be the V_o - adequate isometry attached to $(\hat{K},\hat{V})$, K_o as in the first section and (K_n,V_n) be its generating sequence.

We shall prove first that

$$(3.3)_n \qquad K_n = K_o \oplus \overline{D_{\hat{R}_o} K_o} \oplus \dots \oplus \overline{D_{\hat{R}_{n-1}} \dots D_{\hat{R}_o} K_o}.$$

Since $K_{n+1} = K_n \vee VK_n$ proceeding by induction, it is sufficient to prove that for any $k_o \in K_o$ we have

$$V \begin{bmatrix} 0 \\ 0 \\ \cdot \\ \cdot \\ \cdot \\ 0 \\ D_{\hat{R}_{n-1}} \dots D_{\hat{R}_o} k_o \end{bmatrix} \in K_o \oplus \overline{D_{\hat{R}_o} K_o} \oplus \dots \oplus \overline{D_{\hat{R}_{n-1}} \dots D_{\hat{R}_o} K_o}.$$

But from $(3.2)_{n,j}$, $0 \leq j \leq n$ we have

$$V \begin{bmatrix} 0 \\ 0 \\ \cdot \\ \cdot \\ \cdot \\ 0 \\ D_{\hat{R}_{n-1}} \dots D_{\hat{R}_o} k_o \end{bmatrix} = \hat{V} \begin{bmatrix} 0 \\ 0 \\ \cdot \\ \cdot \\ \cdot \\ 0 \\ D_{\hat{R}_{n-1}} \dots D_{\hat{R}_o} k_o \end{bmatrix} =$$

$$= \begin{bmatrix} D_{\hat{R}_o^*} \dots D_{\hat{R}_{n-1}^*} \hat{R}_n D_{\hat{R}_{n-1}} \dots D_{\hat{R}_o} k_o \\ -\hat{R}_o^* D_{\hat{R}_1^*} \dots D_{\hat{R}_{n-1}^*} \hat{R}_n D_{\hat{R}_{n-1}} \dots D_{\hat{R}_o} k_o \\ \cdot \\ \cdot \\ \cdot \\ -\hat{R}_{n-1}^* \hat{R}_n D_{\hat{R}_{n-1}} \dots D_{\hat{R}_o} k_o \\ D_{\hat{R}_n} D_{\hat{R}_{n-1}} \dots D_{\hat{R}_o} k_o \end{bmatrix} \in K_o \oplus \overline{D_{\hat{R}_o} K_o} \oplus \dots \oplus \overline{D_{\hat{R}_n} \dots D_{\hat{R}_o} K_o}$$

and $(3.3)_n$ is proved for any $n \geq 1$.

It results

$$(3.4) \qquad K = K_o \oplus \overline{D_{\hat{R}_o} K_o} \oplus \overline{D_{\hat{R}_1} D_{\hat{R}_o} K_o} \oplus \dots .$$

From (3.4) and $(3.2)_{n,0}$ we obtain $V[K \ominus H] \subset K \ominus H$ which together with the existence of $\hat{U}$ shows that $(\hat{K},\hat{V})$ produces an Ando dilation $(K,[U,V])$ of $(H,[T,S])$. Using again $(3.3)_n$ and $(3.2)_{n,j}$, $0 \leq j \leq n$ we obtain $\hat{V}_n | K_n = V_n$ and by Proposition 3.1 we conclude that $(K,[U,V])$ is U - diagonal.

The proof of the theorem is complete.

Based on Theorem 3.1 we shall give now a labelling by a system of free parameters of the set of all U-diagonal Ando dilations of $(H, [T,S])$ which cross through a fixed $(K_o,[U_o,V_o])$.

A sequence $\Theta = \{[L_{n-1},F_{n-1},\Theta_n(\lambda)]\}_{n \geq 1}$ will be called $(K_o,[U_o,V_o])$ - *adequate sequence of analytic functions* if its terms $[L_{n-1},F_{n-1},\Theta_n(\lambda)]$ are analytic functions defined in the open disc **D** of the complex plane, taking values contractions from the Hilbert space L_{n-1} into the Hilbert space F_{n-1} which are recurrently constructed following the procedure presented below:

Let us denote

$$(3.5)_o \qquad \hat{R}_o = \hat{V}_o$$

and

$$(3.6) \qquad \begin{aligned} L_o &= \overline{D_{\hat{R}_o} K_o} \ominus \hat{U}_o \overline{D_{\hat{R}_o} K_o}; \qquad R_o = \bigcap_{p \geq 0} \hat{U}_o^p \overline{D_{\hat{R}_o} K_o} \\ L_{*o} &= \overline{D_{\hat{R}_o^*} K_{*o}} \ominus U_o^* \overline{D_{\hat{R}_o^*} K_{*o}}; \qquad R_{*o} = \bigcap_{p \geq 0} \hat{U}_o^{*p} \overline{D_{\hat{R}_o^*} K_{*o}} . \end{aligned}$$

Then L_o and L_{*o} are wandering subspaces for $\hat{U}_o$ and

$$(3.8)_o \qquad \begin{aligned} \overline{D_{\hat{R}_o} K_o} &= M_+(L_o) \oplus R_o \\ \overline{D_{R_o^*} K_{*o}} &= M_-(L_{*o}) \oplus R_{*o} \end{aligned}$$

when, for the wandering subspace L of the unitary operator U we denoted

$$M(L) = \bigoplus_{n \in \mathbf{Z}} U^n L, \quad M_+(L) = \bigoplus_{n \geq 0} U^n L, \quad M_-(L) = \bigoplus_{n \geq 0} U^{*n} L.$$

Let us denote

$$(3.9)_o \quad \begin{aligned} D_o &= \bigoplus_{p\geq 1} \hat{U}_o^{*p} L_o = M(L_o) \ominus M_+(L_o) = D_{\hat{R}_o} \ominus \overline{D_{\hat{R}_o} K_o} \\ D_{*o} &= \bigoplus_{p\geq 1} \hat{U}_o^{p} L_{*o} = M(L_{*o}) \ominus M_-(L_{*o}) = D_{\hat{R}_o^*} \ominus \overline{D_{\hat{R}_o^*} K_{*o}} \end{aligned}$$

and let $H_{o,o}$ be the contraction from D_{*o} into D_o defined by

$$(3.10)_{o,o} \qquad H_{o,o} = P_{D_o} \hat{R}_o^* | D_{*o}.$$

Since we have

$$(3.11)_{o,o} \qquad H_{o,o} \hat{U}_o | D_{*o} = P_{D_o} \hat{U}_o H_{o,o}$$

it results that the subspace $M_o = \mathbf{ker}\, H_{o,o}$ of D_{*o} is invariant to $\hat{U}_o$. Since $\hat{U}_o | D_{*o}$ is a unilateral shift, $\hat{U}_o | M_o$ will be a unilateral shift too. Hence we have

$$(3.12)_o \qquad M_o = \mathbf{ker}\, H_{o,o} = M_+(F_o)$$

with

$$(3.13) \qquad F_o = M_o \ominus \hat{U}_o M_o.$$

We shall choose as the first term in our sequence *an arbitrary contractive analytic function* $[L_o, F_o, \Theta_1(\lambda)]$. Let us remark that what is imposed in this choice are the subspace L_o, F_o (in fact only their dimensions) the choice of the parameter $\Theta_1(\lambda)$ as a contractive analytic function from **D** into $L(L_o, F_o)$ being totally free.

Let us denote by $\Phi_{U,L} = \Phi_L$ the Fourier representation of the bilateral shift U on $M(L)$ as multiplication by coordinate function e^{it} on $L^2(L)$, i.e.

$$\Phi_L \sum_{n\in\mathbf{Z}} U^n l_n = \sum_{n\in\mathbf{Z}} e^{int} l_n, \qquad l_n \in L, \ \sum_{n\in\mathbf{Z}} \|l_n\|^2 < \infty .$$

Using $(3.8)_o$ we shall define the contraction $\hat{R}_1$ from $D_{\hat{R}_o}$ into $D_{\hat{R}_o^*}$ by

$$(3.5)_1 \qquad \hat{R}_1 = \Phi_{F_o}^* \hat{\Theta}_1 \Phi_{L_o} \oplus 0_{R_o}$$

where $\hat{\Theta}_1$ we denoted the contraction from $L^2(L_o)$ into $L^2(F_o)$ given by the pointwise multiplication with the boundary values $\Theta_1(e^{it})$ of $\Theta_1(\lambda)$.

Let us mention the following properties of $\hat{R}_1$. From the definition $(3.5)_1$ it is clear that

$$(3.14)_1 \qquad \hat{R}_1 \hat{U}_o | D_{\hat{R}_o} = \hat{U}_o \hat{R}_1.$$

Using

$$\hat{R}_1 \overline{D_{\hat{R}_o} K_o} \subset M_+(F_o) \subset D_{*o} = D_{\hat{R}_o^*} \ominus \overline{D_{\hat{R}_o^*} K_{*o}}$$

we obtain

$$(3.15)_{1,0} \qquad (\hat{R}_1 D_{\hat{R}_o} k_o\,,\, D_{\hat{R}_o^*} k_{*o}) = 0,\quad k_o \in K_o,\ k_{*o} \in K_{*o}.$$

Note also that from $H_{o,o}\hat{R}_1 D_{\hat{R}_o} K_o = \{0\}$ it results

$$(3.15)_{1,1} \qquad \hat{R}_o^* \hat{R}_1 D_{\hat{R}_o} K_o \subset \overline{D_{\hat{R}_o} K_o}.$$

Let now $n \geq 1$. Suppose that, after the nth step of our construction for any p, $0 \leq p \leq n-1$ we produced L_p, L_{*p} by $(3.6)_p$, D_p, D_{*p} by $(3.9)_p$, $H_{p,j}$, $0 \leq j \leq p$ by $(3.10)_{p,j}$, M_p by $(3.12)_p$ and F_p by $(3.13)_p$. Suppose that $[L_p, F_p, \Theta_{p+1}]$ was chosen as an arbitrary contractive analytic function on D and $\hat{R}_{p+1}$ was constructed by $(3.5)_{p+1}$. Suppose also the relations $(3.14)_{p+1}$ and $(3.15)_{p+1,j}$, $0 \leq j \leq p+1$ hold.

We shall describe now the $(n+1)^{th}$-step of our construction.

Define

$$(3.6)_n \qquad \begin{aligned} L_n &= \overline{D_{\hat{R}_n} \cdots D_{\hat{R}_o} K_o} \ominus \hat{U}_o \overline{D_{\hat{R}_n} \cdots D_{\hat{R}_o} K_o}; \quad R_n = \bigcap_{p \geq 0} \hat{U}_o^p \overline{D_{\hat{R}_n} \cdots D_{\hat{R}_o} K_o} \\ L_{*n} &= \overline{D_{\hat{R}_n^*} \cdots D_{\hat{R}_o^*} K_{*o}} \ominus \hat{U}_o^* \overline{D_{\hat{R}_n^*} \cdots D_{\hat{R}_o^*} K_{*o}}\ ; \quad R_{*n} = \bigcap_{p \geq 0} \hat{U}_o^{*p} \overline{D_{\hat{R}_n^*} \cdots D_{\hat{R}_o^*} K_{*o}}. \end{aligned}$$

We have

$$(3.7)_n \qquad \begin{aligned} \overline{D_{\hat{R}_n} \cdots D_{\hat{R}_o} K_o} &= M_+(L_n) \oplus R_n \\ \overline{D_{\hat{R}_n^*} \cdots D_{\hat{R}_o^*} K_{*o}} &= M_-(L_{*n}) \oplus R_{*n} \end{aligned}$$

and

$$(3.8)_n \qquad \begin{aligned} D_{\hat{R}_n} &= M(L_n) \oplus R_n \\ D_{\hat{R}_n^*} &= M(L_{*n}) \oplus R_{*n}\,. \end{aligned}$$

Denote

$$(3.9)_n \qquad \begin{aligned} D_n &= \bigoplus_{p \geq 1} \hat{U}_o^{*p} L_n = M(L_n) \ominus M_+(L_n) = D_{\hat{R}_n} \ominus \overline{D_{\hat{R}_n} \cdots D_{\hat{R}_o} K_o} \\ D_{*n} &= \bigoplus_{p \geq 1} \hat{U}_o^p L_{*n} = M(L_{*n}) \ominus M_-(L_{*n}) = D_{\hat{R}_n^*} \ominus \overline{D_{\hat{R}_n^*} \cdots D_{\hat{R}_o^*} K_{*o}} \end{aligned}$$

and let $H_{n,j}$, $0 \leq j \leq n$ be the contraction from D_{*n} to D_j defined by

$$(3.10)_{n,j} \qquad H_{n,j} = \begin{cases} P_{D_j} \hat{R}_j^* D_{\hat{R}_{j+1}^*} \cdots D_{\hat{R}_n^*} | D_{*n}\,, & 0 \leq j \leq n-1 \\ P_{D_n} \hat{R}_n^* | D_{*n}\,, & j = n. \end{cases}$$

Since for each j, $1\le j\le n$, $\hat{R}_j$ is a contraction from $D_{\hat{R}_{j-1}}$ into $D_{\hat{R}^*_{j-1}}$ we have $D_{\hat{R}_j}\subset D_{\hat{R}_{j-1}}$, $D_{\hat{R}^*_j}\subset D_{\hat{R}^*_{j-1}}$. Also $D_j\subset D_{\hat{R}_j}$, $D_{*j}\subset D_{\hat{R}^*_j}$. Hence $(3.10)_{n,j}$ has sense for any j, $0\le j\le n$, and defines a contraction $H_{n,j}$ from D_{*n} to D_j. Moreover it is easy to see that

$$(3.11)_{n,j} \qquad H_{n,j}\hat{U}_0|D_{*n} = P_{D_j}\hat{U}_0H_{n,j} \quad 0\le j\le n.$$

It results that the subspace $M_n = \bigcap_{j=0}^{n}\ker H_{n,j}$ of D_{*n} is invariant to $\hat{U}_0$ and $\hat{U}_0|D_{*n}$ being a unilateral shift, $\hat{U}_0|M_n$ will be a unilateral shift too. Hence we have

$$(3.12)_n \qquad M_n =: \bigcap_{j=0}^{n}\ker H_{n,j} = M_+(F_n)$$

with

$$(3.13)_n \qquad F_n = M_n\ominus\hat{U}_0M_n.$$

We shall choose as the $(n+1)^{th}$-term of our sequence an *arbitrary* contractive analytic function $[L_n,F_n,\Theta_{n+1}(\lambda)]$.

Using $(3.8)_n$ we shall define the contraction $\hat{R}_{n+1}$ from $D_{\hat{R}_n}$ into $D_{\hat{R}^*_n}$ by

$$(3.5)_{n+1} \qquad \hat{R}_{n+1} = \Phi^*_{F_n}\hat{\Theta}_{n+1}\Phi_{L_n}\oplus 0_{R_n}.$$

Clearly $\hat{R}_{n+1}$ verifies

$$(3.14)_{n+1} \qquad \hat{R}_{n+1}\hat{U}_0|D_{\hat{R}_n} = \hat{U}_0\hat{R}_{n+1}.$$

Using

$$\hat{R}_{n+1}\overline{D_{\hat{R}_n}\cdots D_{\hat{R}_0}K_0}\subset M_+(F_n)\subset D_{*n} = D_{\hat{R}^*_0}\ominus\overline{D_{\hat{R}^*_n}\cdots D_{\hat{R}^*_0}K_{*0}}$$

we obtain

$$(3.15)_{n+1,0} \qquad (\hat{R}_{n+1}D_{\hat{R}_n}\cdots D_{\hat{R}_0}k_0,\ D_{\hat{R}^*_n}\cdots D_{\hat{R}^*_o}k_{*o}) = 0,\quad k_o\in K_0,\ k_{*0}\in K_{*o}.$$

From $H_{n,j}\hat{R}_{n+1}D_{\hat{R}_n}\cdots D_{\hat{R}_o}k_o = 0$, $k_o\in K_o$, $0\le j\le n$, we obtain

$$(3.15)_{n+1,j+1} \qquad \begin{aligned} &\hat{R}^*_jD_{\hat{R}^*_{j+1}}\cdots D_{\hat{R}^*_n}\hat{R}_{n+1}D_{\hat{R}_n}\cdots D_{\hat{R}_0}K_0\subset\overline{D_{\hat{R}_j}\cdots D_{\hat{R}_o}K_o}, \quad 0\le j\le n-1\\ &\hat{R}^*_n\hat{R}_{n+1}\overline{D_{\hat{R}_n}\cdots D_{\hat{R}_0}K_0}\subset\overline{D_{\hat{R}_n}\cdots D_{\hat{R}_0}K_0}, \quad j = n.\end{aligned}$$

In this way we described completely the recursive construction of a

$(K_o,[U_o,V_o])$-choice sequence of analytic functions $\Theta = \{[L_{n-1},F_{n-1},\Theta_n(\lambda)\}_{n\geq 1}$. This construction also produces the V_o - choice sequence $\{\hat{R}_n\}_{n\geq 0} = \{\hat{R}_n(\Theta)\}_{n\geq 0}$. We shall call $\{\hat{R}_n(\Theta)\}_{n\geq 0}$ the $\hat{V}_o$-choice sequence *canonically attached to* the $(K_o,[U_o,V_o])$-choice sequence of analytic functions $\{[L_{n-1},F_{n-1},\Theta_n(\lambda)]\}_{n\geq 1}$.

We can state now the main result of this section.

THEOREM 3.2. *Let* $\Theta = \{[L_{n-1},F_{n-1},\Theta_n(\lambda)]\}_{n\geq 1}$ *be a* $(K_o,[U_o,V_o])$ *- choice sequence of analytic function and let* $\{\hat{R}_n\}_{n\geq 0} = \{\hat{R}_n(\Theta)\}_{n\geq 0}$ *be the* $\hat{V}_o$ *- choice sequence cannonically attached to* Θ. *Then the* $\hat{V}_o$ *- adequate isometry* $(\hat{K},\hat{V}) = (\hat{K}(\Theta),\hat{V}(\Theta))$ *corresponding to* $\{\hat{R}_n\}_{n\geq 0} = \{\hat{R}_n(\Theta)\}_{n\geq 0}$ *produces a* U *- diagonal Ando dilation* $(K,[U,V]) = (K(\Theta),[U(\Theta),V(\Theta)])$ *of* $(H,[T,S])$ *which crosses through* $(K_o,[U_o,V_o])$. *Moreover, the map* $\Theta \to (K(\Theta),[U(\Theta),V(\Theta)])$ *is a one-to-one correspondence between the set of all* $(K_o,[U_o,V_o])$ *- adequate sequences of analytic functions and the set of all* U *- diagonal Ando dilations of* $(H,[T,S])$ *which cross through* $(K_o,[U_o,V_o])$.

PROOF. Let $\Theta = \{[L_{n-1},F_{n-1},\Theta_n(\lambda)]\}_{n\geq 1}$ be a $(K_o,[U_o,V_o])$ - choice sequence of analytic functions and $\{\hat{R}_n\}_{n\geq 0} = \{\hat{R}_n(\Theta)\}_{n\geq 0}$. Then the conditions $(3.14)_n$ and $(3.15)_{n,j}$, $0\leq j\leq n$ hold for any $n\geq 1$. But $(3.14)_n$ is the same as $(3.1)_n$ and $(3.15)_{n,j}$ is the same as $(3.2)_{n,j}$, $0\leq j\leq n$. From Theorem 3.1 it results that $(\hat{K}(\Theta),\hat{V}(\Theta))$ produces a U-diagonal Ando dilation.

So the map $\Theta \to (K(\Theta),[U(\Theta),V(\Theta)])$ is well defined and we can show easily that it is an injective map.

In order to prove that it is also a surjective map let $(K,[U,V])$ be a U-diagonal Ando dilation of $(H,[T,S])$ which crosses through $(K_o,[U_o,V_o])$ and let $(\hat{K},[\hat{U},\hat{V}])$ be its unitary - isometric extension. If $\{\hat{R}_n\}_{n\geq 0}$ is the $\hat{V}_o$ - choice sequence of $(\hat{K},\hat{V})$ then from Theorem 3.1 it results that for any $n\geq 1$ the relations $(3.1)_n$ and $(3.2)_{n,j}$, $0\leq j\leq n$, hold.

As in the recursive construction of $(K_o,[U_o,V_o])$ - choice sequences of analytic functions we shall produce all the elements defined by the formulas $(3.6)_n$ - $(3.13)_n$. The relations $(3.2)_{n,j}$, $0\leq j\leq n$ imply

$$(3.16)_n \qquad \hat{R}_n\overline{D_{\hat{R}_{n-1}}\cdots D_{\hat{R}_o}K_o} \subset M_+(F_{n-1}).$$

Since R_{n-1} is reducing for $\hat{U}_o$ so will be $\hat{R}_nR_{n-1}$ which together with $(3.16)_n$ implies $\hat{R}_n|R_{n-1} = 0$. It results that there exists a contractive analytic function

$[L_{n-1}, F_{n-1}, \Theta_n(\lambda)]$ such that

$$\hat{R}_n = \Phi^*_{F_{n-1}} \hat{\Theta}_n \Phi_{L_{n-1}} \oplus 0_{R_{n-1}}.$$

The resulting $\Theta = \{[L_{n-1}, F_{n-1}, \Theta_n(\lambda)]\}_{n \geq 1}$ is a $(K_o,[U_o,V_o])$ - choice sequence of analytic functions and $\hat{R}_n = \hat{R}_n(\Theta)$. Hence $(K,[U,V]) = (K(\Theta),[U(\Theta,V(\Theta)])$.

The proof of the theorem is complete.

COROLLARY 3.1. *The distinguished Ando dilation* $(K_o,[U_o,V_o])$ *of* $(H,[T,S])$ *is the only* U *- diagonal Ando dilation of* $(H,[T,S])$ *which crosses through* $(K_o,[U_o,V_o])$ *if and only if either* $L_o = \{0\}$ *or* **ker** $H_{o,o} = \{0\}$.

REFERENCES

1. **Ando, T.** : On a pair of commutative contractions, *Acta Sci. Math. (Szeged)* **24**(1963), 88-90.
2. **Ceauşescu, Zoia** : *Operatorial extrapolations* (Romanian), Thesis, Bucharest, 1980.
3. **Ceauşescu, Zoia ; Foiaş, C.** : On intertwining dilations. VI, *Rev. Roumaine Math. Pures Appl.* **23**(1978), 1471-1482.
4. **Ceauşescu, Zoia ; Suciu, I.** : Isometric dilations of commuting contractions.I, *J. Operator Theory* **12**(1984), 65-88; II, in *Dilation theory, Toeplitz operators and other topics*, Birkhäuser Verlag, 1983, pp.55-80; III, in *Spectral theory of linear operators and related topics*, Birkhäuser Verlag, 1984, pp.47-59.
5. **Douglas, R.G.** : On the operator equations $S^* X = X$ and related topics, *Acta Sci. Math. (Szeged)* **30**(1969), 19-32.
6. **Sz. - Nagy, B. ; Foiaş, C.** : *Harmonic analysis of operators on Hilbert space*, North-Holland – Akádemiai Kiadó, Amsterdam – Budapest, 1970.

Zoia Ceauşescu and Ion Suciu

Department of Mathematics, INCREST
Bdul Păcii 220, 79622 Bucharest
Romania.

Operator Theory:
Advances and Applications, Vol.17

RECENT RESULTS ON REFLEXIVITY OF OPERATORS AND ALGEBRAS OF OPERATORS

Bernard Chevreau

INTRODUCTION

Let H be a complex Hilbert space and let $L(H)$ denote the algebra of all bounded linear operators on H. For any subset S of $L(H)$ let, as usual, **Lat**(S) be the set of (closed) subspaces M of H invariant under any element of S (i.e. $TM \subset M$ for any $T \in S$) and **Alg Lat**(S) the subalgebra of $L(H)$ consisting of those operators T such that **Lat**(T)$\supset$ $\supset$**Lat**(S). Of course **Alg Lat**(S) is closed in the weak operator topology of $L(H)$ (**WOT** for short). A (necessarily **WOT**-closed) subalgebra A of $L(H)$ is *reflexive* if A = **Alg Lat**(A). An operator T is reflexive if W_T (the unital **WOT**-closed algebra generated by T) is reflexive. Thus roughly speaking an operator is reflexive if its lattice of invariant subspaces is rich enough so as to determine the **WOT**-closed subalgebra it generates. This is indirectly confirmed by the fact that a unicellular operator (i.e. an operator T such that **Lat**(T) is linearly ordered) cannot be reflexive.

It is interesting to note that reflexivity is a relatively recent area of investigation. In [9] Sarason gave the first examples of reflexive operators by showing that normal operators and analytic Toeplitz operators are reflexive. Later Deddens [5] extended this result to isometries. Both results are particular cases of the more recent theorem of Olin and Thomson [8] which says that all subnormal operators are reflexive. It is worth mentioning that the finite dimensional case was settled only recently in [6]. Results for C_0-operators, in some sense generalizing the finite dimensional case, appear in [4].

In [2] it was proved that any (BCP) operator (that is a completely non unitary contraction whose essential spectrum $\sigma_e(T)$ intersects the open unit disc **D** in a dominating set ∂**D**) is reflexive. (Recall that a subset E of **D** is dominating for ∂**D** if,

$$\|h\|_\infty = \sup_{\lambda \in E} |h(\lambda)|$$

for any $h \in H^\infty$, the algebra of bounded analytic functions in the open unit disc.) In [1], in collaboration with H. Bercovici, C. Foiaş and C. Pearcy, we proved that this result extends to the larger class $\mathbf{A}_{\aleph_0}$ defined below):

THEOREM A (cf. [1, Theorem 1.7]). *Every operator in* $\mathbf{A}_{\aleph_0}$ *is reflexive.*

The main purpose of this note is to present a detailed and nearly self-contained outline of the proof of Theorem A, illustrating at the same time some of the structure theory of the predual of the dual algebra generated by T in $\mathbf{A}_{\aleph_0}$ (see Section 2 for the terminology). It goes along the ideas of [1] and [2], presented here may be in a somewhat less technical form.

2. PRELIMINARIES

For T in $L(H)$, A_T will denote the unital ultraweakly closed algebra generated by T (also called the dual algebra generated by T). The algebra A_T can be identified with the dual space of the quotient space $Q_T = (\tau c)/{}^{\perp}A_T$ (of the Banach space (τc) of trace-class operators on H, by the preannihilator ${}^{\perp}A_T$ of A_T in (τc)) under the pairing

$$\langle A , [L]\rangle = \mathrm{tr}(AL), \qquad A \in A_T, \quad [L] \in Q_T.$$

The class $\mathbf{A}$ ($= \mathbf{A}(H)$) consists of all those absolutely continuous contractions T on H (i.e. those contractions T whose unitary part is either trivial or has spectral measure absolutely continuous with respect to Lebesgue measure on the unit circle $\mathbf{T}$) such that the Sz.-Nagy–Foiaş functional calculus $\Phi_T : H^\infty \to A_T$ is an isometry. For T in $\mathbf{A}(H)$ then Φ_T turns out to be onto and is the adjoint of an isometry ϕ_T of Q_T onto the predual L^1/H^1_0 of H^∞. In particular for $\lambda \in \mathbf{D}$ we will denote as usual by $[C_\lambda]$ the image under ϕ_T^{-1} of the weak* continuous linear functional on H^∞ defined by evaluation at λ. In order words $[C_\lambda]$ is the element in Q_T such that

$$\langle h(T) , [C_\lambda]\rangle = h(\lambda) \qquad h \in H^\infty, \quad \lambda \in \mathbf{D}.$$

If $x,y \in H$ we write $x\otimes y$ for the rank one operator defined by

$$x\otimes y(u) = (u,y)x \qquad u \in H,$$

and by $[x\otimes y]$ its image in Q_T. If n is any cardinal number satisfying $1 \leq n \leq \aleph_0$, $\mathbf{A}_n$ ($= \mathbf{A}_n(H)$) will denote the set of all those T in $\mathbf{A}$ for which every system of simultaneous equations

$$[x_i\otimes y_j] = [L_{ij}], \qquad 1 \leq i, j \leq n$$

has a solution $(x_i)_{1\le i\le n}$, $(y_j)_{1\le j\le n}$ for any $n\times n$ array of elements $[L_{ij}]$ in Q_T.

We include in that section a result on compressions which has now become a standard device in the theory of the classes $\mathbf{A}_n$. If $T\in L(H)$ and E is a subspace of H we denote T_E the compression of T to E, that is the operator on E defined by $T_E(x) = P_E Tx$, $x\in E$, where P_E is the orthogonal projection on E.

This notion is particularly interesting when E is a semiinvariant subspace for T (i.e. $E = M\ominus N$ with $M,N\in \mathbf{Lat}\,T$, $N\subset M$) because then $(T_E)^n = (T^n)_E$ for any integer n and consequently for $T\in \mathbf{A}$ and $h\in H^\infty$, $(h(T))_E = h(T_E)$. It will be convenient to use the following notation: if V is any set of vectors in H, then $\bigvee^T V$ will denote the smallest subspace invariant for T containing V.

PROPOSITION 1. *Let F be any finite subset in* $\mathbf{D}$, *p the monic polynomial whose zeros are exactly the elements of F,* $\{u_i, v_j\}$ $1\le i,j\le n$ *vectors in H such that* $[u_i\otimes v_j]\in \operatorname{span}_{\lambda\in F}\{[C_\lambda]\}$ $1\le i,j\le n$. *Then* $p(T_{M\ominus N}) = 0$ *where* $M = \bigvee^T\{u_1,\dots,u_n\}$, $M_* = \bigvee^{T^*}\{v_1,\dots,v_n\}$ *and* $N = M\cap M_*^\perp$.

PROOF. As mentioned above $p(T_{M\ominus N}) = p(T)_{M\ominus N}$ and thus we have to show that for any $x\in M\ominus N$, $p(T)x\in N$ that is $p(T)x\perp M_*$.

Now for any polynomials $q_1,\dots,q_n$, $r_1,\dots,r_n$ we have

$$(p(T)\sum_{i=1}^{n} q_i(T)u_i\,,\ \sum_{j=1}^{n} r_j(T^*)v_j) = \sum_{1\le i,j\le n} \langle [u_i\otimes v_j]\,,\ p(T)q_i(T)\tilde{r}_j(T)\rangle$$

(where we have set $\tilde{r}_j(z) = \overline{r_j(\bar{z})}$), and since, for $1\le i,j\le n$, $[u_i\otimes v_j]$ is a finite combination of the $[C_\lambda]$, $\lambda\in F$ and the polynomial $pq_i\tilde{r}_j$ vanishes at all λ in F, all terms in the second member of the above equality vanish. Since vectors of the form $\sum_{i=1}^{n} q_i(T)u_i$ (resp. $\sum_{j=1}^{n} r_j(T^*)v_j$) form a dense subset of M (resp. of M_*) we have for every $x\in M$ (and in particular for every x in $M\ominus N$) $p(T)x\perp M_*$. This completes the proof.

In connection with compressions we will need the following elementary but useful fact.

LEMMA 2. (cf. [2, Lemma 3]). *Let* $T\in L(H)$, M, $N\in \mathbf{Lat}\,T$ *with* $N\subset M$ *and let* $S\in \mathbf{Alg\,Lat}\,T$; *then*

$$S_{M\ominus N}\in \mathbf{Alg\,Lat}\,T_{M\ominus N}\,.$$

We state also for future reference the following elementary result.

LEMMA 3. (cf. [2, Lemma 2]). *If* B *is an operator acting on any complex Hilbert space and there exists a nonzero polynomial* p *with simple zeros annihilating* B, *then* B *is reflexive and* W_B *coincides with the algebra of all polynomials in* B.

An important ingredient in the proof of the reflexivity of (BCP)-operators was not only the possibility of solving finite equations in the predual of $\mathbf{A}_T$ but of doing it with reasonable estimates on the distance from the initial data to the solution. The following proposition says the same result holds for T in $\mathbf{A}_{\aleph_0}$.

PROPOSITION 4. (cf. [1, Theorem 1.5]). *Suppose* $T \in \mathbf{A}_{\aleph_0}$, $\{[L_{ij}]\}_{1\leq i,j\leq n}$ *is a finite array of elements from* Q_T, $u_1,\dots,u_n$ *and* $v_1,\dots,v_n$ *are sequences from H such that*

$$\|[L_{ij}] - [u_i \otimes v_j]\| < \delta \qquad \text{for } 1 \leq i, j \leq n,$$

for some $\delta > 0$. *Then there exist sequences* $\{u'_1,\dots,u'_n\}$ *and* $\{v'_1,\dots,v'_n\}$ *from H such that:*

(1) $[L_{ij}] = [u'_i \otimes v'_j] \quad 1 \leq i, j \leq n$ *and*

(2) $\|u'_i - u_i\| < n\delta^{\frac{1}{2}}, \ \|v'_i - v_i\| < n\delta^{\frac{1}{2}} \quad 1 \leq i \leq n.$

This result can be obtained as a straightforward consequence of [3, Proposition 6.1 and Theorem 3.4]. We proved it in [1] "reducing it" so to speak to the already known similar result for (BCP) operators via the following proposition which seems of interest in its own right.

PROPOSITION 5. (cf. [1, Proposition 1.3]). *Suppose* $T \in \mathbf{A}_{\aleph_0}$, $\{u_1,\dots,u_n\}$ *is any finite subset of H, and* $\varepsilon > 0$. *Then there exists* $M \in \mathbf{Lat}\, T$ *such that*

(1) *both* $T|M$ *and* $T_{H \ominus M}$ *are (BCP) - operators, and*

(2) $\|P_M u_i\| < \varepsilon \quad$ *for* $i = 1,\dots,n$.

Finally we recall the well-known elementary criteria for membership in W_T.

LEMMA 6. (cf. [2, Lemma 1]). *Suppose* S *and* T *belong to* $L(H)$. *If for every positive integer* n *and every pair of finite sequences* $\{u_1,\dots,u_n\}$ *and* $\{v_1,\dots,v_n\}$ *of vectors in H the equality* $\sum_{i=1}^{n}[u_i \otimes v_i] = 0$ *implies* $\sum_{i=1}^{n}(Au_i, v_i) = 0$, *then* $A \in W_T$.

3. PROOF OF THEOREM A

Let S belong to **Alg Lat** T (T being in $\mathbf{A}_{\aleph_0}$) and let $\{u_1,\dots,u_n\}$ and $\{v_1,\dots,v_n\}$ sequences of vectors from H satisfying (*) $\sum_{i=1}^{n}[u_i\otimes v_i] = 0$. In view of Lemma 6 we have to show that $\sum_{i=1}^{n}(Su_i\,,\,v_i) = 0$.

If n = 1 (*) says that $\bigvee T/u_1$ is orthogonal to v_1. Since $\bigvee T/u_1$ must belong to **Lat** S we have also $(Su_1\,,\,v_1) = 0$. We now consider the case $n > 1$. We claim first that the desired conclusion holds under the hypothesis (in addition to (*)) that there exists F finite subset of **D** such that

$$[u_i\otimes v_j] \in \operatorname*{span}_{\lambda \in F}\,[C_\lambda] \qquad 1 \leq i, j \leq n.$$

Indeed in that case by Proposition 1, if we set $M = \bigvee T/\{u_1,\dots,u_n\}$ $M_* = \bigvee T^*\{v_1,\dots,v_n\}$, $N = M \cap M_*^{\perp}$ then $p(T_{M\ominus N}) = 0$ where p is the monic polynomial whose zeros are exactly the elements of F. From Lemmas 2 and 3 we obtain that $S_{M\ominus N} = q(T_{M\ominus N})$ for some polynomial q. Consequently (setting $E = M\ominus N$ for convenience) we have

$$\sum_{i=1}^{n}(Su_i\,,\,v_i) = \sum_{i=1}^{n}(S_E u_i\,,\,v_i) = \sum_{i=1}^{n}(q(T_E u_i)\,,\,v_i) = \sum_{i=1}^{n}(q(T)u_i\,,\,v_i) = \langle q(T)\,,\,\sum_{i=1}^{n}[u_i\otimes v_i]\rangle = 0$$

as claimed.

For the general case we use a kind of density argument based on Proposition 4. Given any $\varepsilon > 0$ there exists F finite set in **D** such that $\max_{1\leq i,j\leq n} \operatorname{dist}\{[u_i\otimes v_j]\,,\,\operatorname*{span}_{\lambda\in F}[C_\lambda]\} < \varepsilon$ (this follows from the fact – valid for every T in **A** – that the absolutely closed convex hull of the $[C_\lambda]$'s, $\lambda \in \mathbf{D}$, is the closed unit ball of Q_T). Choose $[L_{ij}]$ in $\operatorname*{span}_{\lambda\in F}[C_\lambda]$ such that $\|[u_i\otimes v_j] - [L_{ij}]\| < \varepsilon$ for $1 \leq i, j \leq n$ $(i,j) \neq (n,n)$ and $\sum_{i=1}^{n}[L_{ii}] = 0$.

Note that this together with $\sum_{i=1}^{n}[u_i\otimes v_i] = 0$ implies

$$\|[L_{n,n}] - [u_n\otimes v_n]\| < (n-1)\varepsilon.$$

Thus by Proposition 4 there exist $h_1,\dots,h_n$, $k_1,\dots,k_n$ satisfying

$$[L_{ij}] = [h_i\otimes k_j],$$

$$\|u_i - h_i\| < \sqrt{\varepsilon}\,n(n-1)$$

$$\|v_i - k_i\| < \sqrt{\varepsilon}\,n(n-1) \qquad 1 \leq i, j \leq n.$$

Since $\sum_{i=1}^{n} [h_i \otimes h_i] = 0$ and the $[h_i \otimes k_j]$'s belong to $\underset{\lambda \in F}{\text{span}} [C_\lambda]$ we have by what has just been proved $\sum_{i=1}^{n} (Sh_i , h_i) = 0$. Now from

$$|(Sh_i , k_i) - (Su_i , v_i)| \leq |(S(h_i - u_i) , k_i)| + |(Su_i , k_i - v_i)|$$

we get

$$|(Sh_i , k_i) - (Su_i , v_i)| \leq C\, n(n - 1)\sqrt{\varepsilon},$$

with $C = \|S\| \max_{i,j}\{\|u_i\| , \|k_j\|\}$. Therefore

$$|\sum_{i=1}^{n} (Su_i , v_i)| \leq Cn^2(n - 1)\sqrt{\varepsilon}$$

and since ε is arbitrary, the proof is complete.

To conclude we mention that by considering the problem of solving equations in the predual of a **(WOT)**-closed algebra A the authors of [3] (cf. Proposition 9.16 there ; an idea already implicit in [7]) have managed to show that such an algebra is reflexive provided that 2×3 systems of equations in the predual of A (as opposed to $\aleph_0 \times \aleph_0$ systems in the predual of A_T for T in $\mathbf{A}_{\aleph_0}$) can be solved with estimates - a priori looser than the ones offered by Proposition 4 above - on the distance from the initial data to the solution (cf. [3, Theorem 9.20]).

REFERENCES

1. **Bercovici, H. ; Chevreau, B. ; Foiaş, C. ; Pearcy, C. :** Dilation theory and systems of simultaneous equations in the predual of an operator algebra. II, *Math. Z.* **187**(1984), 97-103.
2. **Bercovici, H. ; Foiaş, C. ; Langsam, J. ; Pearcy, C. :** (BCP)-operators are reflexive, *Michigan Math. J.* **29**(1982), 371-379.
3. **Bercovici, H. ; Foiaş, C. ; Pearcy, C. :** *Dual algebras with applications to invariant subspaces and dilation theory*, CBMS-NSF Lecture Notes nr.56, A.M.S., Providence, 1985.
4. **Bercovici, H. ; Foiaş, C. ; Sz.-Nagy, B. :** Reflexive and hyper-reflexive operators of class C_0, *Acta Sci. Math. (Szeged)* **43**(1981), 5-13.
5. **Deddens, J. A. :** Every isometry is reflexive, *Proc. Amer. Math. Soc.* **28**(1971), 509-512.
6. **Deddens, J. A. ; Fillmore, P. A. :** Reflexive linear transformations, *Linear Algebra Appl.* **10**(1975), 89-93.
7. **Larson, D. :** Annihilators of operator algebras, in *Invariant subspaces and other topics*, Birkhäuser Verlag, 1982, pp.119-130.

8. **Olin, R. ; Thomson, J.** : Algebras of subnormal operators, *J. Functional Analysis* **37**(1980), 271-301.

9. **Sarason, D.** : Invariant subspaces and unstarred operator algebras, *Pacific J. Math.* **17**(1966), 511-517.

Bernard Chevreau

UÉR de Mathématiques et Informatique
Université de Bordeaux I
351, Cours de la Libération, 33405 Talence
France.

Operator Theory:
Advances and Applications, Vol.17

SCHUR ANALYSIS FOR MATRICES WITH A FINITE NUMBER OF NEGATIVE SQUARES

Tiberiu Constantinescu

Writing the form associated with a positive matrix as a sum of positive squares, there appears a sequence of complex numbers determining the given matrix. In a certain sense, these calculations are equivalent to the classical algorithm of I.Schur [24] which associates a similar sequence of parameters to an analytic contractive function on the unit disc (such sort of parameters are known as Schur-Szegö parameters). The formalism using these Schur-Szegö parameters (we call it Schur analysis) can be extended to a general framework (operators on Hilbert spaces – see [5], where the operatorial version of the Schur-Szegö parameters is called choice sequence) and can be used to solve some extension problems (as Carathéodory-Fejér problem, Nehari problem and so on – see [1], [5]) and to describe the Kolmogorov decomposition of positive-definite kernels on the set of integers – see [13].

On the other hand, a Schur analysis can be given for a larger class of functions. Thus, in [10], it is showed that with a meromorphic function f on the unit disc, having a finite number of poles different from zero and $|f(z)| \leq 1$ for $|z| = 1$ one associates a sequence of complex numbers with the property that from a certain rank it becomes a sequence of Schur-Szegö parameters.

This phenomenon proves to be generic for the generalization of the classical extension problems as stated e.g. in [26], [21].

The aim of this note is to adapt the Schur analysis of positive matrices given in [13] for a matrix having a finite number κ of negative squares (Section 2); this will also give the possibility of determining the signature of a matrix from the associated parameters. Then, we shall develop some elements of orthogonal polynomials in the Pontrjagin space associated with an ascendent sequence of matrices having κ negative squares (Section 4). In the last section we will deal with two applications. The first one refers to the asymptotic properties of the Toeplitz determinants (having κ negative squares) including variants for the two classical Szegö limit theorems. The second one is about an extremum problem inspired from a paper of Arov and Krein [4].

The history of these problems is a very long one. Here are some samples. In

Section 3 we will deal with the indefinite trigonometric moment problem which was first considered in [16], [17]. Several sorts of orthogonal polynomials in spaces with indefinite metric appear in [20], [21], [19]. The extension problems which are connected with the trigonometric problem are treated in [2] and in the vast program developed in the series of papers [21] (see also [8]).

1. In this section we introduce some objects we will need and discuss some of their properties. Let H and H' be (complex) Hilbert spaces; for a selfadjoint operator A in $L(H)$ we denote by E_A its spectral measure. Define

$$J_A = \mathbf{sgn}\, A = \int_{-\infty}^{\infty} \mathbf{sgn}\, t \, dE_A(t),$$

(here $\mathbf{sgn}\, t = \begin{cases} 1 & t \geq 0 \\ -1 & t < 0 \end{cases}$),

$$Q_A = |A|^{\frac{1}{2}} = \int_{-\infty}^{\infty} |t|^{\frac{1}{2}} dE_A(t);$$

the following relations hold: $J_A = J_A^*$, $J_A^2 = I$, $Q_A \geq 0$, $A = Q_A^2 J_A$.

It will be convenient to say that A has κ ($<\infty$) negative squares if $\mathbf{dim}\, E_A(-\infty, 0) = \kappa$ and our first task is to describe the structure of the 2×2 matrices having κ negative squares.

We have the following basic result for our analysis in Section 2 (see [22]). As usual, **Ran** T will denote the range of the operator T.

1.1. PROPOSITION. *Let* A *and* C *be selfadjoint operators in* $L(H)$ *having* κ *negative squares and* $H = \begin{bmatrix} A & B \\ B^* & C \end{bmatrix}$. *Then* H *has* κ *negative squares if and only if there is an operator* G *in* $L(\overline{\mathbf{Ran}\, Q_C}, \overline{\mathbf{Ran}\, Q_A})$ *such that* $B = Q_A G Q_C$ *and* $J_C - G^* J_A G \geq 0$.

PROOF. We first show that if A and C are invertible operators and A has κ_0 negative squares, then H has $\kappa \geq \kappa_0$ negative squares if and only if there is a unique operator G in $L(H)$ such that $B = Q_A G Q_C$ and $J_C - G^* J_A G$ has $\kappa - \kappa_0$ negative squares. To show this we have only to write the equality:

$$H = \begin{bmatrix} I & 0 \\ B^*A^{-1} & I \end{bmatrix} \begin{bmatrix} A & 0 \\ 0 & C-B^*A^{-1}B \end{bmatrix} \begin{bmatrix} I & A^{-1}B \\ 0 & I \end{bmatrix}$$

and, as $\begin{bmatrix} I & A^{-1}B \\ 0 & I \end{bmatrix}$ is invertible, H has κ negative squares if and only if $C - B^*A^{-1}B$ has $\kappa - \kappa_0$ negative squares. Now, we obtain that $Q_C(J_C - Q_C^{-1}B^*Q_A^{-1}J_AQ_A^{-1}BQ_C^{-1})Q_C$ has $\kappa - \kappa_0$ negative squares and we define $G = Q_C^{-1}BQ_A^{-1}$.

For the general situation, supposing that B has the representation from the statement, it follows that

$$H = \begin{bmatrix} Q_A & 0 \\ 0 & Q_C \end{bmatrix} \begin{bmatrix} J_A & G \\ G^* & J_C \end{bmatrix} \begin{bmatrix} Q_A & 0 \\ 0 & Q_C \end{bmatrix}$$

and from the properties of G, $\begin{bmatrix} J_A & G \\ G^* & J_C \end{bmatrix}$ has κ negative squares. Therefore, H has κ negative squares.

Conversely, we can suppose that we are in the following situation: for a sufficiently large N, $H + 1/n$ has κ negative squares, $A + 1/n$ and $C + 1/n$ have κ negative squares and are invertible operators, $J_{A+1/n} = J_A$, $J_{C+1/n} = J_C$ for every $n \geq N$. It results that:

$$J_C - Q_{C+1/n}^{-1}B^*Q_{A+1/n}^{-1}J_AQ_{A+1/n}^{-1}BQ_{C+1/n}^{-1} \geq 0$$

and let us define $G_n = Q_{A+1/n}^{-1}BQ_{C+1/n}^{-1}$, $n > N$.

If $Q_{C+1/n} = \begin{bmatrix} (Q_{C+1/n})_- & 0 \\ 0 & (Q_{C+1/n})_+ \end{bmatrix}$ and $Q_{A+1/n} = \begin{bmatrix} (Q_{A+1/n})_- & 0 \\ 0 & (Q_{A+1/n})_+ \end{bmatrix}$ are the decompositions with respect to the negative and positive spaces then $G_n = \begin{bmatrix} G_n^1 & G_n^2 \\ G_n^3 & G_n^4 \end{bmatrix}$ is the matrix of G_n with respect to these decompositions. As $(Q_{A+1/n})_-^{-1}$ and $(Q_{C+1/n})_-^{-1}$ converge to bounded operators, G_n^1 also converge to a bounded operator G^1.

Further on, as $J_C - G_n^*J_AG_n \geq 0$ for $n > N$, we have:

$$-I + (G_n^1)^*(G_n^1) - (G_n^3)^*(G_n^3) \geq 0$$

$$I + (G_n^2)^*(G_n^2) - (G_n^4)^*(G_n^4) \geq 0.$$

Using the remark at the beginning of the proof, $J_A - G_nJ_CG_n^* \geq 0$, then

$$-I + (G_n^1)(G_n^1)^* - (G_n^2)(G_n^2)^* \geq 0.$$

From these inequalities, the sequences $\{G_n^2\}_{n>N}, \{G_n^3\}_{n>N}, \{G_n^4\}_{n>N}$ are bounded and let G^2, G^3, G^4 be limit points of them in the corresponding w-topologies. Defining $G = \begin{bmatrix} G^1 & G^2 \\ G^3 & G^4 \end{bmatrix}$, G satisfies the conditions of the proposition and, moreover, it is uniquely determined as an operator acting from $\overline{\text{Ran}\, Q_C}$ into $\overline{\text{Ran}\, Q_A}$. ■

1.2. REMARK. If we note $D = J_C - G^* J_A G$, then, in the conditions of Proposition 1.1, the following factorization holds:

$$H = \begin{bmatrix} Q_A & 0 \\ 0 & Q_C \end{bmatrix} \begin{bmatrix} I & 0 \\ G^* J_A & D^{\frac{1}{2}} \end{bmatrix} \begin{bmatrix} J_A & 0 \\ 0 & I \end{bmatrix} \begin{bmatrix} I & J_A G \\ 0 & D^{\frac{1}{2}} \end{bmatrix} \begin{bmatrix} Q_A & 0 \\ 0 & Q_C \end{bmatrix}. \tag{1.1}$$

We next describe the form of the row-operators

$$X_n = (G_1, K_2, \ldots, K_n) : H_n\, (= \underbrace{H \oplus \ldots \oplus H}_{n}) \to H$$

having the property:

$$\begin{bmatrix} J_{S_o} & 0 \\ 0 & I_{n-1} \end{bmatrix} - X_n^* J_{S_o} X_n \geq 0 \tag{1.2}$$

where $S_o \in L(H)$ has κ negative squares, $G_1 \in L(H)$ and $J_{S_o} - G_1^* J_{S_o} G_1 \geq 0$.

We can obtain more general variants, but we are concerned here exactly with what will be necessary in the next section. First, let us introduce some more notation; for a contraction $T \in L(H)$ (i.e. $\|T\| \leq 1$), we define $D_T = (I - T^*T)^{\frac{1}{2}}$ and $\mathcal{D}_T = \overline{\text{Ran}\, D_T}$ (see [25]). Then, we define $D_1 = (J_{S_o} - G_1^* J_{S_o} G_1)^{\frac{1}{2}}$, $\mathcal{D}_1 = \overline{\text{Ran}\, D_1}$; as also $J_{S_o} - G_1 J_{S_o} G_1^* \geq 0$ (see [18], Chapter III), we define $D_{*1} = (J_{S_o} - G_1 J_{S_o} G_1^*)^{\frac{1}{2}}$, $\mathcal{D}_{*1} = \overline{\text{Ran}\, D_{*1}}$.

Moreover, with an abuse of notation, for an operator G such that $J_1 - G^* J_2 G \geq 0$ we note $D_G = (J_1 - G^* J_2 G)^{\frac{1}{2}}$ and $\mathcal{D}_G = \overline{\text{Ran}\, D_G}$; when we will use this notation it will be clear whether G is a contraction or a (J_1,J_2)-contraction. It is clear that $D_1 = D_{G_1}$, but, as G_1 will have a special role in the sequel, we prefer a particular notation.

1.3. LEMMA. *In the above conditions on S_o and G_1, the equation*

$$D_1 Z_1 = G_1^* J_{S_o} D_{*1} \tag{1.3}$$

has a unique solution as a bounded operator from D_{*1} *into* D_1.

PROOF. By direct computations,

$$D_1^2 - G_1^* J_{S_o} D_{*1}^2 J_{S_o} G_1 = D_1^2 J_{S_o} D_1^2$$

and as $J_{S_o} + I \geq 0$ we see that $D_1^2 J_{S_o} D_1^2 \geq -D_1^4$; consequently, $G_1^* J_{S_o} D_{*1}^2 J_{S_o} G_1 \leq D_1^2 + D_1^4 =$ $= D_1^2(1 + D_1^2) \leq kD_1^2$, k being a constant such that $I + D_1^2 \leq kI$. From the obtained inequality the existence of Z_1 readily follows. ■

1.4. REMARK. If S_o is a contraction, then G_1 is a contraction and the equation (1.3) becomes:

$$D_{G_1} Z_1 = G_1^* D_{G_1^*}$$

consequently, $Z_1 = G_1^* : D_{G_1^*} \to D_{G_1}$ (see [25]). ■

1.5. PROPOSITION. *With* S_o *and* G_1 *as above, we have:*

(1) *The row-operator* X_n *satisfies* (1.2) *if and only if there exist a contraction* $G_2 \in L(H, D_{*1})$ *and contractions* $G_p \in L(H, D_{G_{p-1}^*})$, $p > 2$, *such that*

$$K_p = D_{*1} D_{G_2^*} \cdots D_{G_{p-1}^*} G_p, \quad 2 \leq p \leq n. \tag{1.4}$$

(2) *The following factorizations hold:*

$$D_{X_n^*}^2 = D_{*1} D_{G_2^*} \cdots D_{G_n^*}^2 \cdots D_{*1} \tag{1.5}$$

$$D_{X_n}^2 = E_n^* E_n \tag{1.6}$$

where $E_1 = D_1$, *and for* $n > 1$, $E_n = \begin{bmatrix} E_{n-1} & -Z_{n-1} G_n \\ 0 & D_{G_n} \end{bmatrix}$; Z_1 *is the solution of the equation* (1.3) *and for* $1 < k \leq n$, $Z_k = (Z_{k-1} D_{G_k^*}, G_k^*)^t$ *("t" standing for the matrix transpose).*

PROOF. This is similar to the proof of Lemma 1.2 in [12]. ■

Let us define $d_{*n} = D_{G_2^*} \cdots D_{G_n^*}$, $d_n = D_{G_n} \cdots D_{G_2}$, $n \geq 2$ and $X_n^{(2)} =$ $= (G_2, d_{*,2} G_3, \ldots, d_{*,n-1} G_n)$, $n \geq 2$.

By Lemma 1.2 of [12], $X_n^{(2)}$ is a contraction and we remark that $X_n = (G_1, D_{*1}X_n^{(2)})$. Let us consider now the case of an infinite row operator.

1.6. PROPOSITION. *Suppose* $\{X_n\}_{n\geq 1}$ *are as in Proposition 1.4. Then the sequence* $\{X_n P_{H_n}^{\ell^2(\mathbf{N},H)}\}_{n\geq 1}$ *converges in the* s*-topology to an operator* X_∞ *which satisfies*

$$\begin{bmatrix} J_{S_o} & 0 \\ 0 & I \end{bmatrix} - X_\infty^* J_{S_o} X_\infty \geq 0$$

($P_{H'}^H$ *denotes the orthogonal projection of* H *onto* $H'\subset H$ *and* H_n *is viewed as a subspace of* $\ell^2(\mathbf{N},H)$).

PROOF. By Lemma 1.3 of [12], $\{X_n^{(2)} P_{H_{n-1}}^{\ell^2(\mathbf{N},H)}\}_{n=2}^{\infty}$ converges in the s-topology to a contraction $X_\infty^{(2)}$, consequently, the sequence $\{X_n P_{H_n}^{\ell^2(\mathbf{N},H)}\}_{n\geq 1}$ converges in the same topology to the operator $X_\infty = (G_1, D_{*1}X_\infty^{(2)})$ which satisfies:

$$D_{X_\infty^*}^2 = D_{*1}(I - X_\infty^{(2)} X_\infty^{(2)*})D_{*1} \geq 0.$$

Using Propositions 1.4 and 1.6 in [12], $\{E_n\}_{n=1}^{\infty}$ converges in the s-topology to

$$E_\infty = \begin{bmatrix} D_1 & -Z_1 X_\infty^{(2)} \\ 0 & D_\infty^{(2)} \end{bmatrix}$$

(where $D_\infty^{(2)} = \text{s-}\lim\limits_{n\to\infty} D_n^{(2)} P_{H_n}^{\ell^2(\mathbf{N},H)}$ and $D_2^{(2)} = D_{G_2}$, $D_n^{(2)} = \begin{bmatrix} D_{n-1}^{(2)} & -Z_{n-1}^{(2)} G_n \\ 0 & D_{G_n} \end{bmatrix}$, $Z_2^{(2)} = -G_2^*$, $Z_n^{(2)} = (Z_{n-1}^{(2)} D_{G_n^*}, G_n^*)^t$) and we conclude that

$$D_{X_\infty}^2 = E_\infty^* E_\infty$$

$$D_{X_\infty^*}^2 = \text{s-}\lim_{n\to\infty} (D_{*1} d_{*n} d_{*n}^* D_{*1}). \qquad \blacksquare$$

Let us define for further use the unitary operators:

$$\begin{cases} \alpha_\infty : \mathcal{D}_{X_\infty} \to \mathcal{D}_1 \oplus \mathcal{D}_{G_2} \oplus \mathcal{D}_{G_3} \oplus \cdots \\ \alpha_\infty D_{X_\infty} = E_\infty \end{cases}$$

$$\begin{cases} \tilde{\alpha}_\infty : \mathcal{D}_{X^*_\infty} \to \mathcal{D}_* \\ \tilde{\alpha}_\infty D_{X^*_\infty} = (\text{s-lim}_{n\to\infty} (D_{*1} d_{*n} d^*_{*n} D_{*1}))^{\frac{1}{2}} \end{cases}$$

where $\mathcal{D}_* = \overline{\text{Ran}(\text{s-lim}_{n\to\infty} (D_{*1} d_{*n} d^*_{*n} D_{*1}))}$.

Similar considerations can be made for column-operators $\tilde{X}_n$, $\tilde{X}_\infty$, ...

1.7. REMARK. Let us sketch another application of Proposition 1.1. Let $T_1 \in L(H)$ be such that $I - T_1^* T_1$ has κ negative squares. First, we establish the form of the operators $L = (T_1, T_2)$ such that $I - L^* L$ has κ negative squares. We obtain that $T_2 = Q_{I - T_1 T_1^*} G_2$, $I - G_2^* J_{I - T_1 T_1^*} G_2 \geq 0$ and G_2 is uniquely determined by L as an operator acting from H into $\overline{\text{Ran}\, Q_{I - T_1 T_1^*}}$. In a similar way we obtain that $R = (T_1, T_3)^t$ is such that $I - RR^*$ has κ negative squares if and only if $T_3 = G_3 Q_{I - T_1^* T_1}$, $I - G_3 J_{I - T_1^* T_1} G_3^* \geq 0$ and G_3 is uniquely determined as an operator from $\overline{\text{Ran}\, Q_{I - T_1^* T_1}}$ into H.

Finally, we obtain that $T = \begin{bmatrix} T_1 & T_2 \\ T_3 & X \end{bmatrix}$ is such that $I - TT^*$ has κ negative squares if and only if

$$X = -G_3 J_{I - T_1^* T_1} T_1^* G_2 + Q_{I - G_3 J_{I - T_1^* T_1} G_3^*} G Q_{I - G_2^* J_{I - T_1 T_1^*} G_2},$$

where G is a uniquely determined contraction acting between the obvious spaces.

We note that the case $\kappa = 0$ is used in [3] in order to give a solution to the Nehari problem and that the general form for $\kappa = 0$ appears in [7]; for other aspects we mention [23]. ■

2. This section is devoted to the main result of the paper.

We begin by remarking that the algorithm (the scalar case) in [13] can be formally applied to any matrix in order to obtain a uniquely determined sequence of complex numbers $\{g_n\}_{n=1}^\infty$ with the only precaution to replace $(1 - |g_n|^2)^{\frac{1}{2}}$ by $(\text{sgn}(1 - |g_n|^2)(1 - |g_n|^2))^{\frac{1}{2}}$; the demand of such procedure should be to determine the signature of the matrix from the associated parameters. In this section we will show that this is indeed the case. We shall work out only the Toeplitz case, and, actually, even a more particular case, but illustrating the difference from the positive case. To derive the general case is then a simple matter of notation.

To begin with, let S_o have κ negative squares. Our goal is to determine the form of the operators $S_n \in L(H)$ such that the Toeplitz matrices

$$T_n = \begin{bmatrix} S_o & S_1 & \dots & S_n \\ & & \ddots & \\ S_n^* & S_{n-1}^* & \dots & S_o \end{bmatrix}$$

have κ negative squares for $n \geq 0$. We need some more notation. For a contraction $T \in L(H,H')$ the unitary operator

$$\begin{cases} R(T) : H \oplus D_{T^*} \to H' \oplus D_T \\ R(T) = \begin{bmatrix} T & D_{T^*} \\ D_T & -T^* \end{bmatrix} \end{cases}$$

plays a central role in the dilation theory. For $G_1 \in L(H)$ such that $J_{S_o} - G_1^* J_{S_o} G_1 \geq 0$ a substitute of the above R is:

$$\begin{cases} R(G_1) : H \oplus D_{*1} \to H \oplus D_1 \\ R(G_1) = \begin{bmatrix} J_{S_o} G_1 J_{S_o} & J_{S_o} D_{*1} \\ D_1 J_{S_o} & -Z_1 \end{bmatrix} \end{cases}$$

where Z_1 is the solution of the equation (1.3). The operator $R(G_1)$ is $\begin{bmatrix} J_{S_o} & 0 \\ 0 & I \end{bmatrix}$-unitary. A κ - choice sequence is a sequence $\{G_n\}_{n=o}^{\infty}$ of operators such that $G_o = S_o$, $J_{S_o} - G_1^* J_{S_o} G_1 \geq 0$, $G_2 \in L(D_1, D_{*1})$ is a contraction and $G_n \in L(D_{G_{n-1}}, D_{G^*_{n-1}})$ are contractions for $n > 2$. The notion of 0 - choice sequence coincides with the usual choice sequence in [5].

For a κ - choice sequence we define $V_o = J_{S_o}$, $U_o = J_{S_o}$, $V_1 = R(G_1)$, $U_1 = V_1$, and

$$V_n = \begin{bmatrix} R(G_1) & 0 \\ 0 & I_{n-1} \end{bmatrix} \begin{bmatrix} I & 0 & 0 \\ 0 & R(G_2) & 0 \\ 0 & 0 & I_{n-1} \end{bmatrix} \dots \begin{bmatrix} I_{n-1} & 0 \\ 0 & R(G_n) \end{bmatrix}, \quad n \geq 2$$

$$U_n = V_n \begin{bmatrix} J_{S_o} & 0 \\ 0 & I_n \end{bmatrix} \begin{bmatrix} U_{n-1} & 0 \\ 0 & I \end{bmatrix}, \quad n \geq 2$$

$$F_o = I, \; F_1 = \begin{bmatrix} I & J_{S_o} G_1 \\ 0 & D_1 \end{bmatrix}, \; F_n = \begin{bmatrix} F_{n-1} & U_{n-1}\tilde{X}_n \\ 0 & d_n D_1 \end{bmatrix}, \quad n \geq 2.$$

First we need some preliminaries on F_n.

2.1. LEMMA. *For* $n \geq 1$,

$$F_n = V_n \begin{bmatrix} X_n^* & \begin{bmatrix} J_{S_o} & 0 \\ 0 & I \end{bmatrix} F_{n-1} \\ d_{*n}^* D_{*1} & 0 \end{bmatrix}. \tag{2.1}$$

PROOF. By induction we have

$$V_n = \begin{bmatrix} J_{S_o} & 0 \\ 0 & I \end{bmatrix} \begin{bmatrix} X_n & D_{*1} d_{*n} \\ E_n & -Z_n \end{bmatrix} \begin{bmatrix} J_{S_o} & 0 \\ 0 & I \end{bmatrix}. \tag{2.2}$$

Then,

$$U_{n-1}\tilde{X}_n = \begin{bmatrix} J_{S_o} X_{n-1} U_{n-2} \tilde{X}_{n-1} + J_{S_o} D_{*1} d_{*(n-1)} G_n d_{n-1} D_1 \\ E_{n-1} U_{n-2} \tilde{X}_{n-1} - Z_{n-1} G_n d_{n-1} D_1 \end{bmatrix}.$$

Using this equality, we obtain by induction that

$$F_n = \begin{bmatrix} I & J_{S_o} X_n F_{n-1} \\ 0 & E_n F_{n-1} \end{bmatrix} \tag{2.3}$$

and now, by a direct computation using (2.3), we conclude the proof. ∎

Having these preliminaries, we can derive the main result.

2.2. THEOREM. *There exists a one-to-one correspondence between the set of the sequences* $\{T_n\}_{n \geq 0}$, T_n *having* κ *negative squares for every* $n \in \mathbf{N}$ *and the set of the* κ*-choice sequences, given by the formulae :*

$$S_1 = Q_{S_o} G_1 Q_{S_o}, \quad S_2 = Q_{S_o}(G_1 J_{S_o} G_1 + D_{*1} G_2 D_1) Q_{S_o},$$

$$S_n = Q_{S_o}(X_{n-1}U_{n-2}\tilde{X}_{n-1} + D_{*1}d_{*(n-1)}G_n d_{n-1}D_1)Q_{S_o}, \quad n \geq 3.$$

PROOF. Having the model of the proof of Theorem 1.2 [11], one can prove by induction for $n \geq 3$:

$(2.4)_n$ $$(S_1,\ldots,S_n) = Q_{S_o}X_n F_{n-1}Q_{S_o}^{(n)}, \quad Q_{S_o}^{(n)} = \overbrace{Q_{S_o}\oplus\cdots\oplus Q_{S_o}}^{n},$$

$(2.5)_n$ $$(S_n,\ldots,S_1)^t = Q_{S_o}^{(n)}F_{n-1}^* \begin{bmatrix} J_{S_o} & 0 \\ 0 & I_{n-1} \end{bmatrix} U_{n-1}\tilde{X}_n,$$

$(2.6)_n$ $$T_n = Q_{S_o}^{(n)} F_n^* \begin{bmatrix} J_{S_o} & 0 \\ 0 & I_{n-1} \end{bmatrix} F_n Q_{S_o}^{(n)}.$$

$(2.7)_n$ there exists a contraction $G_{n+1} \in L(D_{G_n}, D_{G_n^*})$ such that

$$S_{n+1} = Q_{S_o}(X_n U_{n-1}\tilde{X}_n + D_{*1}d_{*n}G_n d_n D_1)Q_{S_o}.$$

The first two steps can be obtained by a direct computation. ■

2.3. REMARK. Let $\{T_n\}_{n\geq 0}$ be a sequence of Toeplitz matrices such that T_p has κ_p negative squares for $p \geq 0$. Then the sequence of parameters produced by the corresponding variant of the algorithm in Theorem 2.2 is: $G_o = S_o$, $J_{S_o} - G_1^* J_{S_o} G_1$ has $\kappa_1 - \kappa_o$ negative squares; $J_{J_{S_o} - G_1^* J_{S_o} G_1} - G_2^* J_{J_{S_o} - G_1 J_{S_o} G_1^*} G_2$ has $\kappa_2 - \kappa_1$ negative squares, and so on.

The case $\kappa_p = 0$, $p \geq 0$ reduces to the usual notion of choice sequence [5]. ■

2.4. REMARK. In the scalar case, we can use the parameters $\{g_n\}_{n\geq 0}$ in order to determine the signature of the matrices T_n. Thus, if we consider (for simplicity of notation),

(2.8) $$T_n = \begin{bmatrix} -1 & s_1 & \cdots & s_n \\ \bar{s}_1 & -1 & \cdots & s_{n-1} \\ & & \cdots & \\ \bar{s}_n & & \cdots & -1 \end{bmatrix}$$

then:

$$T_1 \text{ has } \begin{cases} 1 \text{ negative square iff } |g_1| \geq 1 \\ 2 \text{ negative squares iff } |g_1| < 1. \end{cases}$$

$$T_2 \text{ has } \begin{cases} 1 \text{ negative square iff } |g_1|>1,\ |g_2|\leq 1 \\ 2 \text{ negative squares iff } |g_1|>1,\ |g_2|>1 \text{ or } |g_1|<1,\ |g_2|\geq 1 \\ 3 \text{ negative squares iff } |g_1|<1,\ |g_2|<1, \end{cases}$$

and so on. ■

We end this section with a formula for computing the determinants of the matrices T_n.

2.5. PROPOSITION. *In the conditions of Theorem* 2.2, *with* $\dim H < \infty$,

$$\det T_n = (-1)^{\kappa} \Big(\prod_{m=2}^{n} (\det D_{G_m})^{2(n-m+1)} \Big) \det D_1^{2n} \det Q_{S_0}^{2(n+1)}.$$

PROOF. Is a direct consequence of $(2.6)_n$. ■

3. In this section we indicate the Naimark dilation in the setting developed in Section 2. Thus, we shall obtain a Pontrjagin space K containing H and a unitary operator W on this space such that

$$(W^n h, h)_K = (Q_{S_0}^{-1} S_n Q_{S_0}^{-1} h, h), \quad h \in H.$$

We will describe the Naimark construction (given for the indefinite trigonometric moment problem in [17]) in terms of the κ-choice sequence associated with $\{S_n\}_{n=0}^{\infty}$.

From the very beginning, we can suppose Q_{S_0} to be invertible (and even $Q_{S_0} = I$) and we define: $\hat{W}_1 = G_1$,

$$\hat{W}_n = \begin{bmatrix} J_{S_0} & 0 \\ 0 & I_{n-1} \end{bmatrix} V_{n-1} \begin{bmatrix} J_{S_0} & 0 \\ 0 & I_{n-1} \end{bmatrix} \begin{bmatrix} I_{n-1} & 0 \\ 0 & G_n \end{bmatrix}.$$

We have

$$\hat{W}_n = \begin{bmatrix} G_1 & D_{*1} X_n^{(2)} \\ \begin{bmatrix} D_1 \\ 0_{n-1} \end{bmatrix} & \begin{bmatrix} -Z_1 & 0 \\ 0 & I_{n-2} \end{bmatrix} W_n^{(2)} \end{bmatrix},$$

where $W_n^{(2)}$ is defined by the same formula as above, but for the choice sequence $\{G_n\}_{n=2}^{\infty}$ (see (2.5) in [12]). We also define the spaces:

$$K_n = H \oplus D_1 \oplus D_{G_2} \oplus \cdots \oplus D_{G_{n-1}},$$

$$K_+ = H \oplus D_1 \oplus \bigoplus_{n=2}^{\infty} D_{G_n}$$

and using Lemma 2.2 in [12] for $W_n^{(2)}$ it results that there exists $\hat{W}_+ : K_+ \to K_+$, $\hat{W}_+ = \text{s-}\lim\limits_{n\to\infty} \hat{W}_n P_{K_n}^{K_+}$; let us define $J_+ = J_{S_o} \oplus I_{D_1} \oplus \bigoplus_{n=2}^{\infty} D_{G_n}$.

3.1. PROPOSITION. *For* $n \geq 1$, $P_H^{K_+} \hat{W}_+ J_+ \hat{W}_+ \cdots J_+ \hat{W}_+ | H = Q_{S_o}^{-1} S_n Q_{S_o}^{-1}$.

PROOF. By the definition of $\hat{W}_+$,

$$P_H^{K_+} \hat{W}_+ J_+ \hat{W}_+ \cdots J_+ \hat{W}_+ P_H^{K_+} = P_H^{K_n} \hat{W}_n \begin{bmatrix} J_{S_o} & 0 \\ 0 & I_{n-1} \end{bmatrix} \hat{W}_n \cdots \hat{W}_n P_H^{K_n}.$$

Using the formula (2.2),

$$\hat{W}_n = \begin{bmatrix} X_{n-1} & D_{*1} d_{*(n-1)} G_n \\ E_{n-1} & -Z_{n-1} G_n \end{bmatrix} \tag{3.2}$$

which shows that

$$(I, 0_{n-1}) \hat{W}_n = X_n. \tag{3.3}$$

Then, by induction,

$$\left(\begin{bmatrix} J_{S_o} & 0 \\ 0 & I_{n-1} \end{bmatrix} \hat{W}_n \right)^{n-1} \begin{bmatrix} I \\ 0_{n-1} \end{bmatrix} = \begin{bmatrix} U_{n-2} \tilde{X}_{n-1} \\ d_{n-1} D_1 \end{bmatrix}. \tag{3.4}$$

Now, (3.3), (3.4) and Theorem 2.2 conclude the proof. ∎

On the space K_+ we introduce the indefinite scalar product

$$(x, y)_{J_+} = (J_+ x, y)$$

and $(K_+, (\cdot,\cdot)_{J_+})$ is a π_κ - space (see [18]). We define $W_+ = J_+ \hat{W}_+$, then

$$(W_+^n h, h)_{J_+} = (Q_{S_o}^{-1} S_n Q_{S_o}^{-1} h, h), \quad h \in H. \tag{3.5}$$

As $\hat{W}_n^* \begin{bmatrix} J_{S_o} & 0 \\ 0 & I_{n-1} \end{bmatrix} \hat{W}_n = \begin{bmatrix} J_{S_o} & 0 & 0 \\ 0 & I_{n-2} & 0 \\ 0 & 0 & G_n^* G_n \end{bmatrix}$, it results that $\hat{W}_+^* J_+ \hat{W}_+ = J_+$ and $W_+^* J_+ W_+ = J_+$. Finally, we can deduce the minimality condition,

$$K_+ = \bigvee_{n=0}^{\infty} W_+^n H. \tag{3.6}$$

Actually, we can continue as in [12] in order to obtain the unitary extension of W_+. Thus, we define

$$\hat{W}_{red} = \begin{bmatrix} I & 0 \\ 0 & \alpha_\infty \end{bmatrix} R(X_\infty) \begin{bmatrix} 0 & I \\ \tilde{\alpha}_\infty^* & 0 \end{bmatrix}$$

and

$$K = \ldots \oplus D_* \oplus H \oplus D_1 \oplus \bigoplus_{n=2}^{\infty} D_{G_n},$$

$$\hat{W} = I \oplus \hat{W}_{red} \quad \text{on } K.$$

Then, $\hat{W} = \begin{bmatrix} * & 0 \\ * & \hat{W}_+ \end{bmatrix}$ and $\hat{W}^* J \hat{W} = J$, where $J = I \oplus J_+$ on K ; defining $W = J\hat{W}$ and introducing on the space K the indefinite scalar product $(x , y)_J = (Jx , y)$ then W is unitary on K and

$$(W^n h , h)_J = (Q_{S_o}^{-1} S_n Q_{S_o}^{-1} h , h), \quad h \in H$$

$$\bigvee_{h=-\infty}^{\infty} W^n H = K.$$

4. In this section we make some connections with certain results in [21]. Thus, we first consider orthogonal polynomials in the Pontrjagin space K defined in Section 2.

Then we shall show the relevance of the κ-choice sequence in a representation result obtained in [21]. Let us consider the polynomials:

$$\phi_n(z) = L_{n,n} z^n + L_{n,n-1} z^{n-1} + \ldots + L_{n,o}, \quad L_{n,k} \in L(H),\ L_{n,n} \geq 0$$

having the properties:

$$\phi_o(z) S_o \phi_o^*(z) = J_{S_o} \tag{4.1}$$

(4.2) $$\phi_n(z)\tilde{T}_n\phi_k^*(z) = 0, \quad 0 < k < n$$

(4.3) $$\phi_n(z)\tilde{T}_n\phi_n^*(z) = I, \quad n > 0$$

where

$$\tilde{T}_n = \begin{bmatrix} S_o & S_1^* & S_n^* \\ S_1 & S_o & S_{n-1}^* \\ & \cdots & \\ S_n & \cdots & S_o \end{bmatrix}.$$

It is plain that $\phi_o(z) = Q_{S_o}^{-1}$ and we remark that (4.2) and (4.3) are equivalent to the following system:

(4.4) $$(L_{n,o}, L_{n,1}, \ldots, L_{n,n})T_n \begin{bmatrix} I_n & 0 \\ 0 & L_{n,n} \end{bmatrix} = (0_n, I)$$

4.1. PROPOSITION. *If* D_1 *and* $\{D_{G_n}\}_{n\geq 2}$ *are invertible operators, then*

$$L_{n,n} = (Q_{S_o}D_1 d_n^* d_n D_1 Q_{S_o})^{-\frac{1}{2}}$$

and

$$(L_{n,o}, \ldots, L_{n,n-1}) = -L_{n,n}Q_{S_o}\tilde{X}_n^* U_{n-1}^*(F_{n-1}^*)^{-1}(Q_{S_o}^{(n)})^{-1}.$$

PROOF. We consider the system

$$(\mathcal{L}_{n,o}, \ldots, \mathcal{L}_{n,n})T_n = (0_n, I)$$

and using (2.6) and the triangular form of F_n we obtain a recurrence formula for F_n^{-1}. From this formula we easily conclude the proof. ■

The following two results state the recurrence formulas for the polynomials ϕ_n and the Christoffel-Darboux formulas for these polynomials. In order to write recurrence formulas in the operatorial case we also need polynomials "to the right".

Thus, we consider the polynomials:

$$\tilde{\phi}_n(z) = R_{n,n}z^n + R_{n,n-1}z^{n-1} + \ldots + R_{n,o}, \quad R_{n,k} \in L(H), \; R_{n,n} \geq 0$$

having the properties

(4.5) $$\tilde{\phi}_o^*(z)S_o\tilde{\phi}_o(z) = J_{S_o}$$

(4.6)
$$\tilde{\phi}_n^*(z)T_n\tilde{\phi}_k(z) = 0, \quad 0 < k < n$$

(4.7)
$$\tilde{\phi}_n^*(z)T_n\tilde{\phi}_n(z) = I, \quad n > 0.$$

It is easy to see that $R_{n,k} = \tilde{L}_{n,k}$ (here we define for a formula depending on $G_1,\dots,G_n$, $\widetilde{\text{formula}}(G_1,\dots,G_n) = (\text{formula}(G_1^*,\dots,G_n^*))^*$ and the above relation between $R_{n,k}$ and $L_{n,k}$ justifies the notation $\tilde{\phi}_n$ for the right polynomials).

Now, we define: $\Phi_o(z) = I$, $\Phi_1(z) = z - G_1^*J_{S_o}$, $\Phi_n(z) = Q_{S_o}^{-1}L_{n,n}^{-1}\phi_n(z)Q_{S_o}$; $\tilde{\Phi}_o(z) = I$, $\tilde{\Phi}_n(z) = Q_{S_o}\tilde{\phi}_n(z)R_{n,n}^{-1}Q_{S_o}^{-1}$. For a polynomial p of degree n we use the notation: $p^{\circledast}(z) = z^n\bar{p}(1/z)$, $\bar{p}(z) = \overline{p(\bar{z})}$.

4.2. THEOREM. *The following formulas hold:*

$$\Phi_o(z) = I, \ \Phi_1(z) = z - G_1^*J_{S_o}, \ \tilde{\Phi}_o(z) = I, \ \tilde{\Phi}_1(z) = z - J_{S_o}G_1^*,$$

(4.8)
$$\Phi_{n+1}(z) = z\Phi_n(z) - D_1d_n^*G_{n+1}^*d_{*n}^{-1}D_{*1}^{-1}\tilde{\Phi}_n^{\circledast}(z), \quad n \geq 1.$$

(4.9)
$$\tilde{\Phi}_{n+1}(z) = z\tilde{\Phi}_n(z) - \Phi_n^{\circledast}(z)D_1^{-1}d_n^{-1}G_{n+1}^*d_{*n}D_{*1}, \quad n \geq 1.$$

PROOF. We write $\Phi_n(z) = C_{n,o}z^n + \dots + C_{n,n}$ and $\tilde{\Phi}_n(z) = \tilde{C}_{n,o}z^n + \dots + \tilde{C}_{n,n}$. Using the results in Sections 2 and 3, we have

(4.10)
$$F_{n-1}^{-1}U_{n-1}\tilde{X}_n = \begin{bmatrix} J_{S_o}(D_{*1}d_{*(n-1)} - X_{n-1}E_{n-1}^{-1}Z_{n-1})G_nd_{n-1}D_1 \\ F_{n-2}^{-1}U_{n-2}\tilde{X}_{n-1} + F_{n-2}^{-1}E_{n-1}^{-1}Z_{n-1}G_nd_{n-1}D_1 \end{bmatrix}.$$

By induction we show that for $n \geq 1$,

(4.11)
$$D_{*1}d_{*(n-1)} - X_{n-1}E_{n-1}^{-1}Z_{n-1} = J_{S_o}D_{*1}^{-1}d_{*(n-1)}^{-1},$$

consequently, by Proposition 4.1, it results:

$$C_{n,n} = -D_1d_{n-1}^*G_n^*d_{*(n-1)}^{-1}D_{*1}^{-1}.$$

Then, again by induction

(4.12)
$$U_{n-1}F_{n-1}^* \begin{bmatrix} 0_{n-k} \\ I \\ 0_{k-1} \end{bmatrix} = F_{n-1} \begin{bmatrix} 0_{k-1} \\ I \\ 0_{n-k} \end{bmatrix}.$$

This equality implies by a direct computation that

$$C_{n+1,k} = C_{n,k} - D_1 d_n^* G_{n+1}^* d_{*n}^{-1} D_{*1}^{-1} \tilde{C}_{n,n-k+1}^*$$

which are exactly the desired recurrence formulas. ■

Having the recurrence formulas for ϕ_n (and $\tilde{\phi}_n$), we can state the analogous of the Christoffel-Darboux formulas.

4.3. THEOREM. *The following formulas hold:*

$$(4.13)\quad (1 - \bar{z}\zeta)(\phi_0^*(z)J_{S_0}\phi_0(\zeta) + \sum_{k=1}^{n} \phi_k^*(z)\phi_k(\zeta)) = \tilde{\phi}_{n+1}^{\circledast *}(z)\tilde{\phi}_{n+1}^{\circledast}(\zeta) - \phi_{n+1}^*(z)\phi_{n+1}(\zeta)$$

$$(4.14)\quad (1 - z\bar{\zeta})(\tilde{\phi}_0(z)J_{S_0}\tilde{\phi}_0^*(\zeta) + \sum_{k=1}^{n} \tilde{\phi}_k(z)\tilde{\phi}_k^*(\zeta)) = \phi_{n+1}^{\circledast}(z)\phi_{n+1}^{\circledast *}(\zeta) - \tilde{\phi}_{n+1}(z)\tilde{\phi}_{n+1}^*(\zeta).$$

PROOF. By a direct computation

$$(1 - \bar{z}\zeta)\phi_0^*(z)J_{S_0}\phi_0(\zeta) = \tilde{\phi}_1^{\circledast *}(z)\tilde{\phi}_1^{\circledast}(\zeta) - \phi_1^*(z)\phi_1(\zeta).$$

The rest is as in the positive-definite case [14]. ■

From now on, we restrict ourselves to the scalar case. Thus, we consider the Toeplitz matrices (2.8) such that T_k has κ_k negative squares for $k < r$ and T_k has κ negative squares for $k \geq r$.

According to Theorem 2.2, we associate to the family $\{T_n\}_{n\geq 0}$ a sequence of complex numbers $\{g_n\}_{n=1}^{\infty}$ such that $|g_k| < 1$ for $k \geq r+2$. By Theorem 4.2 it results: $\Phi_1(z) = z + \bar{g}_1$, $\Phi_n(z) = z\Phi_{n-1}(z) - \bar{g}_n\Phi_{n-1}^{\circledast}(z)$.

The next result establishes the explicit connection between the orthogonal polynomials and the dilation (for the definite case see [11], [19], [12]).

4.4. PROPOSITION. *For* $n \geq 1$, $\Phi_n(z) = \mathbf{det}(z - W_n^*)$, *where* $W_n = P_{\mathbf{C}^n}^{\ell^2(\mathbf{N})} W P_{\mathbf{C}^n}^{\ell^2(\mathbf{N})}$.

PROOF. As in the definite case, by a direct computation. ■

Let J_n be the signature of T_n, then, for a sufficiently large n,

$$W_n^* \begin{bmatrix} J_\kappa & 0 \\ 0 & I \end{bmatrix} W_n = \begin{bmatrix} J_\kappa & 0 \\ 0 & |g_n|^2 \end{bmatrix} \leq \begin{bmatrix} J_\kappa & 0 \\ 0 & 1 \end{bmatrix}$$

and using Proposition 4.4 we can obtain some information on the zeros of the polynomials ϕ_n (see [20]).

Let us consider the function

$$F(z) = 1 - 2 \sum_{n=1}^{\infty} s_n z^n$$

then (Satz 6.2 in [21, I]) F is in the class C_κ and is holomorphic at 0 (the class C_κ is defined in [21, I] as the set of the meromorphic functions in the unit disc for which the kernel $C_F(z,\zeta) = (F(z)-\overline{F(\zeta)})/(1 - z\overline{\zeta})$ has κ negative squares).

Now let us consider the positive measure on the unit circle with $\mu(1) = 1$ associated with the choice sequence $\{g_k\}_{k \geq r+2}$ and the function

$$F_\mu(z) = \frac{1}{2\pi} \int_0^{2\pi} [(e^{it} + z)/(e^{it} - z)] d\mu(t),$$

which is in C_0.

Let $\{a_k\}$, $\{b_k\}$ be the orthogonal polynomials of first and second kind (see[14]) associated with μ and $\psi_n(z) = \phi_n(z\,; -g_1, -g_2, \ldots, -g_n)$. It easy to determine the connection between $\{a_k\}$, $\{b_k\}$ and $\{\phi_n\}$, $\{\psi_n\}$.

4.5. PROPOSITON. $\psi_{r+k+1} = \frac{1}{2}(P_1^{\circledast} b_k + P_2^{\circledast} a_k)$, $\phi_{r+k+1} = \frac{1}{2}(Q_1^{\circledast} b_k + Q_2^{\circledast} a_k)$ *where*

$$P_1 = \psi_{r+1} + \psi_{r+1}^{\circledast}, \quad P_2 = \phi_{r+1}^{\circledast} - \phi_{r+1}$$

$$Q_1 = \psi_{r+1}^{\circledast} - \psi_{r+1}, \quad Q_2 = \phi_{r+1} + \phi_{r+1}^{\circledast}. \qquad \blacksquare$$

PROOF. By induction.

Now, the connection between F and F_μ is almost clear.

4.6. THEOREM.

(4.15) $$F = (P_1 F_\mu + P_2)/(Q_1 F_\mu + Q_2).$$

PROOF. The Schur analysis shows that the computations are as in the definite case. Comparing the developments around 0 we obtain the desired formules. $\blacksquare$

In [21, I], it is showed that the structure of the functions in the class C_κ is given exactly by the transformation (4.15). So, we pointed out the operation to be made at the choice sequence level in order to obtain such a structure (see also [10]).

Further on, we suppose that $\log(d\mu/dt) \in L^1(\mathbf{T})$ and let g be the outer function factorizing $d\mu/dt$ and g^- the outer function factorizing $d\mu^-/dt$ (μ^- is the positive measure on the unit circle associated with the choice sequence $\{-g_k\}_{k\geq r+2}$). Then, by several classical results (see [14]), $\{a_k^{\circledast}\}$ converges to $1/g$ uniformly on the compact sets in the unit disc and $\{b_k^{\circledast}\}$ converges to $1/g^-$. Consequently, $\{\phi_n^{\circledast}\}$ converges to $\frac{1}{2}(Q_1/g^- + Q_2/g)$ and $\{\psi_n^{\circledast}\}$ to $\frac{1}{2}(P_1/g^- + P_2/g)$. Let us define the meromorphic functions:

$$G = 2[1/(Q_1/g^- + Q_2/g)] \tag{4.16}$$

$$G^- = 2[1/(P_1/g^- + P_2/g)] \tag{4.17}$$

then, by Theorem 4.6,

$$F = G(1/G^-). \tag{4.18}$$

One more remark on the Christoffel-Darboux formulas. By Theorem 4.3,

$$(1 - \bar{z}\zeta) \sum_{k=0}^{n} \overline{\phi_k(z)}\varepsilon_k\phi_k(\zeta) = \overline{\phi^{\circledast}_{n+1}(z)}\phi^{\circledast}_{n+1}(\zeta) - \overline{\phi_{n+1}(z)}\phi_{n+1}(\zeta)$$

(here we denoted $J_n = (\varepsilon_k)$), then, for $|z| < 1$, $|\zeta| < 1$ different from the poles and zeros of G,

$$\sum_{n=0}^{\infty} \overline{\phi_n(z)}\varepsilon_n\phi_n(\zeta) = [1/(1-\bar{z}\zeta)][1/(\overline{G(z)}G(\zeta))]. \tag{4.19}$$

5. In this section we indicate two applications.

A. The first application refers to the asymptotic properties of the Toeplitz determinants $\det T_n$. For the definite case we have the classical theorems of Szegö [15]. For the indefinite case, by Proposition 2.5

$$(\det T_{n+1})/(\det T_n) = \det{}^2 D_{S_o} \prod_{k=1}^{n+1} \det{}^2 D_{G_k}$$

and there exists the limit:

$$\lim_{n\to\infty} (\det T_{n+1})/(\det T_n) = \det{}^2 D_{S_o} \prod_{n=1}^{\infty} \det{}^2 D_{G_n}$$

because, from a certain rank, G_n become contractions.

We note the above limit by $G(F)$ ($= G(0)$); then there exists the limit

$$\lim_{n\to\infty} (\det T_n)/G(F)^{n+1} = (-1)^{\kappa}(\prod_{n=1}^{\infty} \det {}^{2n}D_{G_n})^{-1}.$$

We also remark that we have at hand all devices in order to take again the considerations in [6] for the present setting.

B. The last application is connected with a paper of Arov and Krein [4]. For the sake of simplicity we treat only the scalar case.

Let μ be a positive measure on the unit circle, $\mu(1) = 1$ and we define the "entropy" of μ by the formula:

$$h(\mu,z) = -\frac{1}{4\pi}\int_0^{2\pi} \mathbf{Re}[(e^{it}+z)/(e^{it}-z)]\mathbf{ln}[(d\mu/dt)(t)]dt.$$

We denote by $\{s_n(\mu)\}$ the Fourier coefficients of μ. Now, we consider the following problem: let there be given the complex numbers $s_1, \ldots, s_n$ such that

$$T_n = \begin{bmatrix} 1 & s_1 & s_2 & \cdots & s_n \\ \bar{s}_1 & 1 & s_1 & \cdots & s_{n-1} \\ & \cdots & & & \\ \bar{s}_n & \bar{s}_{n-1} & & \cdots & 1 \end{bmatrix}$$

is positive and the set $B_n = \{\mu$ positive measure on the unit circle with $\mu(1) = 1$ and $s_k(\mu) = s_k$, $k = 1, \ldots, n\}$. Our purpose is to determine $\min_{\mu\in B_n} h(\mu,z)$.

We obtain a simple solution in the following way: let g be the outer function factorizing $d\mu/dt$, then $h(\mu,z) = -\mathbf{ln}\, g(z)$.

Let $\mu \in B_n$, then the first n terms in the associated choice sequence coincide with the choice sequence associated with T_n. Let $\{\phi_n\}_{n=0}^{\infty}$ be the orthogonal polynomials of μ, and using the Christoffel-Darboux formula:

$$h(\mu,z) = \mathbf{ln}(1 - |z|^2)^{\frac{1}{2}} + \tfrac{1}{2}\,\mathbf{ln}(\sum_{k=0}^{\infty} |\phi_k(z)|^2) \geq$$

$$\geq \mathbf{ln}(1 - |z|^2)^{\frac{1}{2}} + \tfrac{1}{2}\mathbf{ln}(\sum_{k=0}^{n} |\phi_k(z)|^2)$$

and $\{\phi_k\}_{k=0}^{n}$ depend only on the choice sequence $\{g_k\}_{k=1}^{n}$ associated with T_n (therefore, on $s_1, \ldots, s_n$). Now, we have only to determine a choice sequence $\{g_1, \ldots, g_n, g_{n+1}(z), \ldots\}$ such that for every fixed z, $\phi_{n+1}(z) = 0$, $\phi_{n+2}(z) = 0, \ldots$. From the recurrence formulas, it is easy to see that the only posibility is $g_{n+1}(z) = \overline{z\phi_n(z)/\phi_n^{*}(z)}$, $g_{n+k}(z) = 0$, for $k \geq 2$. An immediate consequence of the Christoffel-Darboux formulas is that $|\phi_n(z)/\phi_n^{*}(z)| \leq 1$, then $\{g_1, \ldots, g_n, \overline{z\phi_n(z)/\phi_n^{*}(z)}, 0, 0, \ldots\}$ is indeed a choice sequence for every $|z| < 1$.

The above problem can be seen as a generalization of the Jensen inequality, and is a particular case of [4]. The paper [4] is more elaborate and is connected with the paper [9], where the case $z = 0$ is considered. For this situation, we also have

$$h(\mu,0) = -\tfrac{1}{2}\ln\det(I - B_{1,1})$$

where $B_{1,1} = P^{K}_{W*H} P^{K}_{K_+} P^{K}_{W*H}$ (see [6]); the notation $B_{1,1}$ is exactly the one used in Section 3, but for the positive definite case. This last formula provides an operatorial variant (without determinants) for such sort of considerations.

Now, we pass on to the indefinite case. Let $s_1, \ldots, s_n$ be fixed complex numbers such that

$$\overset{o}{T}_n = \begin{bmatrix} -1 & s_1 & s_2 & \cdots & s_n \\ \bar{s}_1 & -1 & s_1 & \cdots & s_{n-1} \\ & \ddots & & & \\ \bar{s}_n & \bar{s}_{n-1} & & \cdots & -1 \end{bmatrix}$$

has κ negative squares and consider the set $B_{n,\kappa} = \{T = \{T_k\}_{k=0}^{\infty}, T_k$ Toeplitz matrices having κ negative squares and $T_n = \overset{o}{T}_n\}$. The problem to be considered is the following one: $\max\limits_{T \in B_{n,\kappa}} |G_T(z)|^2$, where G_T is associated with T by the formula (4.16). Using (4.19),

$$|G(z)|^2 = 1/[(1 - |z|^2) \sum_{k=0}^{\infty} \varepsilon_k |\phi_k(z)|^2]$$

then, if n is sufficiently large with respect to κ (this means that at the step n the choice sequence is stabilized: $|g_n| < 1$, $|g_{n+1}| < 1, \ldots$) we have

$$|G(z)|^2 \leq 1/[(1 - |z|^2) \sum_{k=0}^{n} \varepsilon_k |\phi_k(z)|^2]$$

with the same choice for the solution attaining the maximum.

For illustrating the rest, let us consider the following example: fix $|g_1| < 1$, $|g_2| < 1$ (T_1 has 2 negative squares, T_2 has 3 negative squares) and let $T \in B_{2,3}$, then $|g_3| > 1$, $|g_4| < 1, \ldots$ and $G(0) = (1 - |g_1|^2)(1 - |g_2|^2)(|g_3|^2 - 1) \prod\limits_{k \geq 4} (1 - |g_k|^2)$; if $g_3 \to \infty$, $G(0) \to \infty$.

REFERENCES

1. **Adamjan, V.M.; Arov, D.Z.; Krein, M.G. :** Infinite Hankel matrices and generalized Carathéodory-Fejér and Schur problems (Russian), *Funck. Analiz. Prilozen.* **2**:4(1968), 1-17.

2. **Adamjan,V.M.; Arov,D.Z.; Krein, M.G. :** Analytic properties of Schmidt pairs for a Hankel operator and the generalized Schur-Takagi problem (Russian), *Math. Sb.* **86**(128) (1971), 34-75.

3. **Adamjan, V.M.; Arov, D.Z. ; Krein, M.G. :** Infinite Hankel block matrices and related extension problems (Russian), *Izv. Akad. Nauk. Armjan. SSR. Ser. Mat.* **6**(1971), 87-112.

4. **Arov, D.Z. ; Krein, M.G. :** Problem of search of the minimum of entropy in indeterminate extension problems (Russian), *Funck. Analiz. Prilozen.* **15**(1981), 61-64.

5. **Arsene, Gr. ; Ceaușescu, Z. ; Foiaș, C. :** On intertwining dilations. VIII, *J. Operator Theory* **4**(1980), 55-91.

6. **Arsene, Gr. ; Constantinescu, T. :** The structure of the Naimark dilation and Gaussian stationary processes, *Integral Equations Operator Theory*, **8**(1985), 181-204.

7. **Arsene, Gr. ; Gheondea, A. :** Completing matrix contractions, *J. Operator Theory*, **7**(1982), 179-189.

8. **Ball, J.A. ; Helton, J.W. :** A Beurling-Lax theorem for the Lie group U(m,n) which contains most classical interpolation theory, *J. Operator Theory* **9**(1983), 107-142.

9. **Burg, J. P. :** *Maximal entropy spectral analysis*, Ph.D. dissertation, Geophysics Dept., Stanford University, Stanford, CA, May 1975.

10. **Chamfy, C. :** Functions méromorphes dans le cercle-unité et leurs series de Taylor, *Ann. Inst. Fourier* **8**(1958), 211-251.

11. **Constantinescu, T. :** On the structure of positive Toeplitz forms, in *Dilation theory, Toeplitz operators and other topics*, Birkhäuser Verlag, 1983, pp.127-149.

12. **Constantinescu, T. :** On the structure of the Naimark dilation, *J. Operator Theory*, **12**(1984), 159-175.

13. **Constantinescu, T. :** The structure of n×n positive operator matrices, INCREST preprint, no.14/1984, 54/1984.

14. **Geronimus, Ja. L. :** *Orthogonal polynomials*, New York, Consultants Bureau, 1961.

15. **Grenander, U. ; Szegö, G. :** *Toeplitz forms and their applications*, University of California Press, 1950.

16. **Iohvidov, I. S. ;** On the theory of indefinite Toeplitz forms (Russian), *Dokl. Akad. Nauk. SSSR* **101**:2(1955), 213-216.

17. **Iohvidov, I. S. ; Krein, M. G. :** Spectral theory of operators in spaces with an indefinite metric. II (Russian), *Trudy Moskov. Mat. Obsc.* **8**(1959), 413-496.

18. **Iohvidov, I. S. ; Krein, M.G. ; Langer, H.:** *Introduction to the spectral theory of operators in spaces with an indefinite metric*, Akademie-Verlag, Berlin, 1982.

19. **Kailath, T. ; Porat, B.** : State-space generators for orthogonal polynomials, in *Prediction theory and harmonic analysis*, North-Holland, 1983.

20. **Krein, M.G.** : On the location of the roots of polynomials which are orthogonal on the unit circle with respect to an indefinite weight (Russian), *Teor. Funkcii, Funck. Analiz. Prilozen* 2(1966), 131-137.

21. **Krein, M. G. ; Langer, H.** : Über einige Fortsetzungsprobleme, die eng mit der Theorie hermitescher Operatoren im Raume π_κ zusammenhangen. Teil I., *Math. Nachr.* **77**(1977), 187-236; Teil II, *J. Functional Analysis* **30**(1978), 390-447; III Part(1), *Beiträge zur Analysis* **14**(1981), 25-40; III Part(2), *Beiträge zur Analysis* **15**(1981), 27-45.

22. **Potapov, V.P.** : The multiplicative structure of J-contractive matrix-functions (Russian), *Trudy Moskov. Mat. Obsc.* **4**(1955), 125-236.

23. **Redheffer, R. M.** : Inequalities for a matrix Ricatti equation, *J. Math. Mech.* **8**(1959), 349-377.

24. **Schur, I.** : Über Potenzreihen, die im Innern des Einheitskreises beschränkt sind, *J.Reine Angew. Math.* **148**(1918), 122-145.

25. **Sz.-Nagy, B. ; Foiaş, C.** : *Harmonic analysis of operators on Hilbert spaces*, Amsterdam-Budapest, 1970.

26. **Takagi, T.** : On an algebraic problem related to an analytic theorem of Carathéodory and Fejér, *Japan. J. Math.* **1**(1924), 83-93.

T.Constantinescu

Department of Mathematics, INCREST
Bdul Păcii 220, 79622 Bucharest
Romania.

Operator Theory:
Advances and Applications, Vol.17

UNIFORM ALGEBRAS, HANKEL OPERATORS AND INVARIANT SUBSPACES

Raul E. Curto*, Paul S. Muhly*, Takahiko Nakazi†

§1. HANKEL OPERATORS AND SEMIGROUPS

Let $\mathbf{T}$ be the unit circle and let $H^2(\mathbf{T})$ ($\subset L^2(\mathbf{T})$) be the classical Hardy space. Let $P: L^2(\mathbf{T}) \to H^2(\mathbf{T})$ be the orthogonal projection and let $\phi \in L^\infty(\mathbf{T})$. The *Hankel operator* H_ϕ associated with ϕ is given by $H_\phi f = (I - P)\phi f$, $f \in H^2(\mathbf{T})$. Clearly $H_\phi = 0$ if and only if $\phi \in H^\infty(\mathbf{T}) := L^\infty(\mathbf{T}) \cap H^2(\mathbf{T})$; moreover, H_ϕ is compact if and only if $\phi \in H^\infty(\mathbf{T}) + C(\mathbf{T})$ ([3]).

The classical Hankel operators can be generalized in several ways. For instance, let Γ be a discrete abelian group, let $G := \hat{\Gamma}$ be its dual group, and let Γ_+ be a subsemigroup of Γ such that $\Gamma_+ - \Gamma_+ = \Gamma$. A subset $E \subset F$ is said to be a Γ_+*-module* if $E + \Gamma_+ \subset$ $\subset E$. The Hardy space associated with E is defined by $H^2(E) := \{\xi \in L^2(G,dt) : \hat{\xi}(\gamma) = 0$ for all $\gamma \notin E\}$ (here dt is the Haar measure for G). For $\phi \in L^\infty(G)$, the *Toeplitz operator* associated with ϕ is $T_\phi := PM_\phi | H^2(E)$, where $P: L^2(G) \to H^2(E)$ is the orthogonal projection and M_ϕ is the multiplication operator on $L^2(G)$ associated with ϕ. The two generalizations of the classical Hankel operators that we consider in this section are defined as follows.

DEFINITION 1.1. a) $H_\phi^{(1)} = (I-P)M_\phi | H^2(E)$,

b) $H_\phi^{(2)} = PJM_\phi | H^2(E)$,

where $J: L^2(G) \to L^2(G)$ is induced by the map $t \mapsto -t$, i.e., $(J\xi)(t) = \xi(-t)$, $\xi \in L^2(G)$.

EXAMPLE 1.2. In his Ph.D. dissertation, H. Salas studied C^*-algebras of Toeplitz operators associated with $\mathbf{Z}_+^2$-modules ([7]). Whether those C^*-algebras contain the compact operators or not is related to the compactness of the self-commutators of the Toeplitz operators, which in turn is closely tied to the problem of determining those functions ϕ for which the Hankel operator H_ϕ is compact. In the context studied by Salas, $\Gamma = \mathbf{Z}^2$, $\Gamma_+ = \mathbf{Z}_+^2$, $G = \mathbf{T}^2$, and E is determined by its boundary ∂E which, loosely

---

* Research partially supported by a grant from the National Science Foundation (U.S.A.).

†Research partially supported by KAKENHI (Japan).

speaking, can be thought of as a staircase descending from left to right. It then follows easily that $H_\phi^{(1)}$ and $H_\phi^{(2)}$ are in general distinct, but more important, one may be compact while the other is not. If E is such that $|E\cap(-E)| < \infty$ and $E\cup(-E) = \Gamma$, however, $H_\phi^{(1)}$ and $H_\phi^{(2)}$ are "approximately" the same.

LEMMA 1.3. *Let* E *be a* Γ_+*-module such that* $E\cup(-E) = \Gamma$. *If* $|E\cap(-E)| < \infty$ *then the compactness of* $H_\phi^{(1)}$ *is equivalent to that of* $H_\phi^{(2)}$.

PROOF. Observe that $P^\perp := I - P$ projects onto the subspace $\{\xi \in L^2(G) : \hat{\xi}(\gamma) = 0$ for all $\gamma \in E\}$, and that JPJ projects onto the subspace of L^2-functions whose Fourier transforms are supported in (-E). Since $E\cup(-E) = \Gamma$, $JPJ - P^\perp$ is a projection, and it has finite rank if and only if $|E\cap(-E)| < \infty$. Therefore, when $|E\cap(-E)| < \infty$, $H_\phi^{(1)} - JH_\phi^{(2)}$ is a finite rank operator ($\phi \in L^\infty(G)$).

The converse of Lemma 1.3 is true, and we state it in Corollary 1.6 below.

DEFINITION 1.4. For i = 1, 2, $B^{(i)} := \{\phi \in L^\infty(G) : H_\phi^{(i)}$ is a compact operator $\}$.

The previous lemma says that if $|E\cap(-E)| < \infty$ (and $\Gamma = E\cup(-E)$) then $B^{(1)} = B^{(2)}$. In what follows we shall investigate the relation of $B^{(1)}$ and $B^{(2)}$ to the subalgebra of $L^\infty(G)$ consisting of "analytic functions". We shall let $H^\infty(G) := L^\infty(G)\cap$ $\cap H^2(\Gamma_+)$.

LEMMA 1.5. a) $B^{(1)}$ *and* $B^{(2)}$ *are norm-closed subspaces of* $L^\infty(G)$;

b) $H^\infty(G)\subset B^{(1)}$ *and* $B^{(1)}$ *is a subalgebra of* $L^\infty(G)$;

c) $H^\infty(G)\subset B^{(2)} \Leftrightarrow |E\cap(-E)| < \infty$.

PROOF. a) Obvious.

b) Let $\phi \in H^\infty(G)$ and $\xi \in H^2(E)$. Then

$$\phi\xi = \sum_{\substack{\alpha\in E \\ \beta\in\Gamma^+}} \hat{\xi}(\alpha)\hat{\phi}(\beta)e_{\alpha+\beta}$$

(e_γ denotes the character associated with $\gamma \in \Gamma$). Thus, $H_\phi^{(1)}\xi = P^\perp\Big(\sum_{\substack{\alpha\in E \\ \beta\in\Gamma^+}} \hat{\xi}(\alpha)\hat{\phi}(\beta)e_{\alpha+\beta}\Big) = 0$,

since $E + \Gamma_+\subset E$. Moreover, the formula $H_{\phi\psi}^{(1)} = H_\phi^{(1)}T_\psi + U_\phi H_\psi^{(1)}$ ($\phi, \psi \in L^\infty(G)$), where $T_\psi := PM_\psi|H^2(E)$ and $U_\phi := P^\perp M_\psi|H^2(E)^\perp$, shows that $B^{(1)}$ is an algebra.

c) As in b), let $\phi \in H^{\infty}(G)$ and $\xi \in H^2(E)$. Then

$$JH_{\phi}^{(2)}\xi = JPJ(\phi\xi) = JPJ\Big(\sum_{\substack{\alpha\in E\\ \beta\in\Gamma_+}} \xi(\alpha)\phi(\beta)e_{\alpha+\beta}\Big) = \sum_{\substack{\alpha\in E,\beta\in\Gamma_+\\ \alpha+\beta\in(-E)}} \hat{\xi}(\alpha)\hat{\phi}(\beta)e_{\alpha+\beta},$$

because JPJ projects $L^2(G)$ onto the subspace of L^2-functions supported in (-E). Since $E + \Gamma_+ \subset E$, it follows that if $|E\cap(-E)| < \infty$ then $H^{\infty}(G)\subset B^{(2)}$. Conversely, suppose $H^{(2)}_{e_{\gamma_0}}$ is compact for every $\gamma_0 \in \Gamma_+$.The above computation then shows that $|E\cap(-\gamma_0 - E)|$ must be finite for every $\gamma_0 \in \Gamma_+$, and, in particular, $|E\cap(-E)|$ must be finite.

COROLLARY 1.6. *Assume* $\Gamma = E\cup(-E)$. *Then* $B^{(1)} = B^{(2)}$ *if and only if* $|E\cap(-E)| < \infty$.

PROOF. Only the necessity must be established. If $B^{(1)} = B^{(2)}$ then $B^{(2)}\supset H^{\infty}(G)$, by 1.5 b), and therefore $|E\cap(-E)| < \infty$ by 1.5 c).

EXAMPLE 1.7. Let $\Gamma = \mathbf{Z}^2$, $\Gamma_+ = \mathbf{Z}_+^2$ and $E = \mathbf{Z}_+\times\mathbf{Z}$. Then $E\cup(-E) = \Gamma$ and $|E\cap(-E)| = \infty$. One ϕ which belongs to $H^{\infty}(G)$ but does not belong to $B^{(2)}$ is $\phi(s,t) \equiv 1$, $s,t \in \mathbf{T}^2$.

Our next goal is to show that $B^{(1)}$ and $B^{(2)}$ are both translation invariant.

DEFINITION 1.8. For $t \in G$ let $U_t : L^2(G) \to L^2(G)$ be given by $(U_t\xi)(s) = \xi(s - t)$ $(\xi \in L^2(G),\ s \in G)$.

REMARK 1.9. Let $\xi \in H^2(E)$ and $t \in G$. Since $U_t e_\gamma = \langle\gamma,t\rangle^{-1}e_\gamma$ for all $\gamma \in \Gamma$, it follows that $\widehat{U_t\xi}(\gamma_0) = \langle\gamma_0,t\rangle^{-1}\hat{\xi}(\gamma_0)$ for all $\gamma_0 \in \Gamma$, so that U_t leaves $H^2(E)$ invariant, and therefore $U_tH^2(E) = H^2(E)$. In fact, if M is a (closed) subspace of $L^2(G)$ such that $U_tM = M$ (for all $t \in G$), then M is of the form $\{\xi \in L^2(G) : \hat{\xi}(\gamma) = 0 \text{ for all } \gamma \notin E\}$, for some set $E\subset\Gamma$ (E need not be a module). We shall write U_t for any of its restrictions to reducing subspaces.

LEMMA 1.10. *For* $t \in G$ *and* $\phi \in L^{\infty}(G)$, *the following hold:*

a) $U_tH_{\phi}^{(1)}U_t^* = H_{\phi_t}^{(1)}$, *where* $\phi_t(s) = \phi(s - t)$, $s \in G$;

b) $U_tJH_{\phi}^{(2)}U_t^* = JH_{\phi_t}^{(2)}$.

PROOF. a) $\quad U_tH_{\phi}^{(1)}U_t^*|H^2(E) = U_tP^{\perp} M_{\phi}U_{-t}|H^2(E) =$

$$= P^{\perp} U_t M_\phi U_{-t} | H^2(E) = P^{\perp} M_{\phi_t} | H^2(E) = H^{(1)}_{\phi_t}.$$

b) $$U_t J H^{(2)}_\phi U_t^* | H^2(E) = U_t (JPJ) M_\phi U_{-t} | H^2(E) =$$

$$= (JPJ) U_t M_\phi U_{-t} | H^2(E) = (JPJ) M_{\phi_t} | H^2(E) = J H^{(2)}_{\phi_t}$$

(recall that JPJ projects onto the subspace of $L^2(G)$-functions supported in (-E) and it therefore commutes with each U_t).

COROLLARY 1.11. *For* $i = 1,2$, $B^{(i)}$ *is* ***translation invariant****, i.e., if* $\phi \in B^{(i)}$ *and* $t \in G$, *then* $\phi_t \in B^{(i)}$. *Moreover, for* $\phi \in B^{(i)}$ *fixed, the map* $\sigma_\phi : t \to H^{(i)}_{\phi_t}$ *is norm-continuous.*

PROOF. The fisrt assertion follows from the previous lemma. As for the second, we first recall that map $t \mapsto U_t$ is strongly continuous.Fix $t_0 \in G$; for arbitrary $t \in G$,

$$\| H^{(1)}_{\phi_t} - H^{(1)}_{\phi_{t_0}} \| = \| U_t H^{(1)}_\phi U_t^* - U_{t_0} H^{(1)}_\phi U_{t_0}^* \| \leq$$

$$\leq \| U_t H^{(1)}_\phi U_t^* - U_{t_0} H^{(1)}_\phi U_t^* \| + \| U_{t_0} H^{(1)}_\phi U_t^* - U_{t_0} H^{(1)}_\phi U_{t_0}^* \| \leq$$

$$\leq \| (U_t - U_{t_0}) H^{(1)}_\phi \| + \| H^{(1)}_\phi (U_t^* - U_{t_0}^*) \| .$$

Since $H^{(1)}_\phi$ is compact and $U_t - U_{t_0} \to 0$ in the strong operator topology, we see that $H^{(1)}_{\phi_t} - H^{(1)}_{\phi_{t_0}}$ in the norm topology. A similar computation works for i=2.

COROLLARY 1.12. *Let* $f \in L^1(G)$ *and* $\phi \in B^{(1)}$ *(resp.* $\phi \in B^{(2)}$*). Then*

$$H_\phi * f := \int_G H_{\phi_t}(\cdot) f(t) dt$$

is compact, and

$$H_\phi * f = H_{\phi * f}.$$

PROOF. First, we observe that the above integral converges in light of the norm continuity of $t \mapsto H_{\phi_t}$. Moreover, since $\phi \in B^{(i)}$, $H^{(i)}_\phi$ is compact and, by Lemma 1.10, so is $H^{(i)}_{\phi_t}$ (all $t \in G$). It readily follows that $H_\phi * f$ is compact. To verify that $H_\phi * f = H_{\phi * f}$, consider first the case when $f = e_{\gamma_0}$ for some $\gamma_0 \in \Gamma$: if $\alpha \in E$ we have

$$(H_\phi^{(1)} * e_{\gamma_0})e_\alpha = \int_G U_t H_\phi^{(1)} U_t^* e_\alpha \langle\gamma_0, t\rangle dt = P^\perp\left(\int_G \langle\gamma_0,t\rangle\langle\alpha,t\rangle U_t(\phi e_\alpha)dt\right) =$$

$$= P^\perp\left(\int_G \langle\gamma_0,t\rangle\langle\alpha,t\rangle\left(\sum_{\gamma\in\Gamma} \hat\phi(\gamma)\langle-\alpha-\gamma,t\rangle e_{\alpha+\gamma}\right)dt\right) = P^\perp\left(\sum_{\gamma\in\Gamma}\hat\phi(\gamma)\left(\int_G \langle\gamma_0,t\rangle\overline{\langle\gamma,t\rangle}dt\right)e_{\alpha+\gamma}\right) =$$

$$= P^\perp\,(\hat\phi(\gamma_0)e_{\alpha+\gamma_0}) = (\hat\phi(\gamma_0)H^{(1)}_{e_{\gamma_0}})e_\alpha = H^{(1)}_{\phi * e_{\gamma_0}} e_\alpha .$$

Therefore, $H_\phi^{(1)} * e_{\gamma_0} = H^{(1)}_{\phi * e_{\gamma_0}}$ $(\phi \in B^{(1)}, \gamma_0 \in \Gamma)$. Let $f \in C(G)$ and let $\{p_i\}_{i\in I}$ be a net of polynomials in the characters converging uniformly to f. For each $i \in I$, $H_\phi^{(1)} * p_i = H^{(1)}_{\phi * p_i}$, and since $H_\phi^{(1)} * p_i \to H_\phi^{(1)} * f$ and $\phi * p_i \to \phi * f$ (both in norm), we get $H_\phi^{(1)} * f = H^{(1)}_{\phi * f}$. Similarly, since C(G) is dense in $L^1(G)$ (in the L^1-norm), and $\| H_{\phi_t} \| \leq \| H_\phi \|$ (all $t \in G$), we see that $H_\phi^{(1)} * f = H^{(1)}_{\phi * f}$ holds for any $\phi \in B^{(1)}$ and $f \in L^1(G)$. A similar argument works for i = 2.

As a consequence of the previous result we get the following analogue of Hartman's Theorem.

COROLLARY 1.13. *Let* $i = 1$ *or* 2 *and* $\phi \in B^{(i)}$. *Then there exists* $\psi \in C(G)$ *such that* $H_\phi^{(i)} = H_\psi^{(i)}$.

PROOF. We shall make use of E. Hewitt's generalization of P. Cohen's factorization theorem ([5, (32.22)]). By the preceding corollary, the family $\{H_\phi^{(i)} : \phi \in B^{(i)}\}$ is an *essential module* over $L^1(G)$, and therefore $\{H_\phi^{(i)} : \phi \in B^{(i)}\} * L^1(G) = \{H_\phi^{(i)} : \phi \in B^{(i)}\}$. Thus, for a given $\phi \in B^{(i)}$ there exist $\tilde\psi \in B^{(i)}$ and $f \in L^1(G)$ such that $H_\phi^{(i)} = H_{\tilde\psi}^{(i)} * f = H^{(i)}_{\tilde\psi * f}$. Now let $\psi = \tilde\psi * f$ and we are done.

We are now ready to establish criteria for $B^{(i)}$ to equal $H^\infty(G)$.

THEOREM 1.14. (a) $B^{(1)} = H^\infty(G)$ *if and only if for every* $\gamma \notin \Gamma_+$, $|E \cap (\Gamma \setminus (-\gamma + E))| = \infty$.

(b) $B^{(2)} \subset H^\infty(G)$ *if and only if for every* $\gamma \notin \Gamma_+$, $|(-\gamma - E) \cap E| = \infty$.

(c) $B^{(2)} = H^\infty(G)$ *if and only if for every* $\gamma \notin \Gamma_+$, $|(-\gamma - E) \cap E| = \infty$.

PROOF. (a) For $\gamma \in \Gamma$, $H^{(1)}_{e_\gamma} = P^\perp M_{e_\gamma} | H^2(E) = M_{e_\gamma}(M^*_{e_\gamma} P^\perp M_{e_\gamma}) | H^2(E)$. Let $F : L^2(G) \to l^2(\Gamma)$ be the Fourier transform. Then $F(M^*_{e_\gamma} P^\perp M_{e_\gamma})F^{-1}$ is the projection onto $l^2(\Gamma \setminus (E - \gamma))$, so that $H^{(1)}_{e_\gamma}$ can be thought of as $M_{e_\gamma} Q$, where Q is the inverse Fourier transform of the projection of $l^2(\Gamma)$ onto $l^2(E \cap (\Gamma \setminus (-\gamma + E)))$. Now, if $|(\Gamma \setminus (-\gamma_0 + E)) \cap E| < \infty$ for some $\gamma_0 \notin \Gamma_+$ then e_{γ_0} is a function not in $H^\infty(G)$ which produces a compact Hankel operator $H^{(1)}_{e_{\gamma_0}}$, so that $H^\infty(G) \subsetneq B^{(1)}$. Conversely, if there exists $\phi \in B^{(1)} \setminus H^\infty(G)$, and $\gamma \in \Gamma$, then $H^{(1)}_\phi * e_\gamma \in B^{(1)}$, so that $\hat{\phi}(\gamma) H^{(1)}_{e_\gamma} \in B^{(1)}$ for all $\gamma \in \Gamma$. Therefore, $H^{(1)}_{e_{\gamma_0}} \in B^{(1)}$ for some $\gamma_0 \notin \Gamma_+$ (otherwise ϕ would be in $H^\infty(G)$). This of course forces $E \cap (\Gamma \setminus (-\gamma_0 + E))$ to be finite.

(b) As above, let $\phi \in B^{(2)} \setminus H^\infty(G)$. Then for some $\gamma_0 \notin \Gamma_+$, $H^{(2)}_{e_{\gamma_0}}$ is compact. Since

$$JH^{(2)}_{e_{\gamma_0}} = M_{e_{\gamma_0}} (M^*_{e_{\gamma_0}} JPJM_{e_{\gamma_0}}) | H^2(E),$$

and $M^*_{e_{\gamma_0}} JPJM_{e_{\gamma_0}}$ is associated with $l^2(-\gamma_0 - E)$, we see that $|(-\gamma_0 - E) \cap E| < \infty$. Conversely, if for some $\gamma_0 \notin \Gamma_+$, $|(-\gamma_0 - E) \cap E| < \infty$, then $H^{(2)}_{e_{\gamma_0}} \in B^{(2)} \setminus H^\infty(G)$. Thus $B^{(2)} \not\subset H^\infty(G)$.

(c) Use (b) above and Lemma 1.5 b).

REMARK 1.15. In Corollary 1.6, the assumption $\Gamma = E \cup (-E)$ cannot be removed. For, if $\Gamma = \mathbf{Z}^2$, $\Gamma_+ = \mathbf{Z}^2_+$ and $E = \Gamma_+$ then $B^{(1)} = H^\infty(G)$ and $B^{(2)} \supsetneq H^\infty(G)$. $E \cap (-E)$ is, however, finite.

COROLLARY 1.16. *Let Γ be isomorphic to a subgroup of the reals* $\mathbf{R}$, *and* $E = \Gamma_+$. *Then there exist nonzero compact Hankel operators if and only if Γ is isomorphic to the integers* $\mathbf{Z}$.

PROOF. By Corollary 1.6, $B^{(1)} = B^{(2)}$, so that only (a) or (c) in Theorem 1.14 must be checked. In either case, it follows that $H^\infty(G) = B^{(1)} = B^{(2)}$ if and only if Γ is dense, as desired.

COROLLARY 1.17. *If Γ is isomorphic to $\mathbf{Z}^2$ with the colexicographic order (so that Γ_+ can be identified with $\{(k, l) \in \mathbf{Z}^2 : l > 0$ or $(l = 0$ and $k \geq 0)\}$), and $E = \Gamma_+$, then there exist nonzero compact Hankel operators.*

PROOF. Since $\Gamma = E \cup (-E)$ and $E \cap (-E) = \{(0,0)\}$, we again see that $B^{(1)} = B^{(2)}$. Now, for $\gamma = (-1,0)$, $(-\gamma - E) \cap E = \{(0,0), (1,0)\}$, so that by Theorem 1.14 (c), $B^{(2)} \supsetneq H^\infty(G)$.

We have said in our opening remarks that classical Hankel operators can be generalized in different ways. We shall see in Section 2 some results on Hankel operators associated with weak-$*$ Dirichlet algebras. Our main theorem there relates the existence of nonzero compact Hankel operators to the structure of the Gleason part through the multiplicative measure that "comes" with the algebra. Corollaries 1.16 and 1.17 above can be interpreted as suggesting such a result. For instance, the case when $E = \Gamma_+ = \{(m,n) \in \mathbf{Z}^2 : n - \alpha m \geq 0\}$ for an irrational number α corresponds to a trivial Gleason part, while the case exemplified in 1.17 corresponds to a nontrivial Gleason part.

§ 2. HANKEL OPERATORS AND UNIFORM ALGEBRAS

Let $(X,\mathcal{A},m)$ be a fixed nontrivial probability measure space and let A be a subalgebra of L^∞ $(= L^\infty(m))$ containing the constants and satisfying the conditions

(i) $$\int_X fg\,dm = \left(\int_X f\,dm\right)\left(\int_X g\,dm\right), \qquad f,g \in A;$$

and

(ii) $$[A + \bar{A}]_* = L^\infty \qquad ([\ \ ]_* \text{ denotes weak-}*\text{ closure}).$$

These algebras, called *weak-$*$ Dirichlet algebras*, were introduced and studied by Srinivasan and Wang in [8].

The abstract Hardy spaces $H^p = H^p(m)$, $0 < p \leq \infty$ are defined as follows: $H^p = [A]_p$ $(0 < p < \infty)$, where $[\ \ ]_p$ denotes closure in L^p, and $H^\infty = [A]_*$. The measure m is multiplicative on H^∞ and therefore it determines a point τ_0 in the maximal ideal space of H^∞. The Gleason part determined by τ_0, denoted $G(\tau_0)$, is $\{\tau \in M_{H^\infty} : \|\tau - \tau_0\| < 2\}$. It is well known that either $G(\tau_0) = \{\tau_0\}$ or $G(\tau_0)$ has an analytic structure holomorphically equivalent to the unit disc.

The projection from L^2 to H^2 will be denoted by P. If $\phi \in L^\infty$ we let M_ϕ denote the operator on L^2 of multiplication by ϕ, i.e., $M_\phi f = \phi f$, $f \in L^2$. The *Hankel operator* determined by $\phi \in L^\infty$ is defined by the equation $H_\phi f = (I - P)M_\phi f$, $f \in H^2$. In the decomposition $L^2 = H^2 \oplus (H^2)^\perp$, M_ϕ is given by

$$M_\phi = \begin{bmatrix} T_\phi & * \\ H_\phi & * \end{bmatrix},$$

where T_ϕ is the Toeplitz operator with symbol ϕ.

Our main result of this section is the following theorem.

THEOREM 2.1.([1]). $G(\tau_0) = \{\tau_0\}$ *if and only if each* **compact** *Hankel operator is zero.*

A key lemma in our proof is stated below.

LEMMA 2.2. *Let* $h \in H_0^1$ *and let* $\varepsilon > 0$. *Then there exist* $f \in H_0^2$ *and* $g \in H^2$ *such that* $\|f\|_2^2 \leq \|g\|_1$, $\|g\|_2^2 \leq \|h\|_1 + \varepsilon$, *and* $h = fg$.

COROLLARY 2.3. *Let* $\phi \in L^\infty$. *Then* $\|H_\phi\| = \|\phi + H^\infty\|$, *where* $\|\phi + H^\infty\| = \inf\{\|\phi + g\|_\infty : g \in H^\infty\}$.

We shall now sketch a proof of Theorem 2.1. The methods are rather different from those employed in the previous section. Let $B = \{\phi \in L^\infty : H_\phi$ is compact $\}$. Observe that H^∞ B.

Sufficiency. If $G(\tau_0) \neq \{\tau_0\}$ then there exists a nonconstant inner function Z (called *Wermer's embedding function*) such that $ZH^\infty = H_0^\infty = \{f \in H^\infty : \int_X f\, dm = 0\}$. One can then show that for every polynomial p, $p(\bar{Z})$ produces a Hankel operator of rank $\leq$ degree p. Therefore, $[H^\infty, \bar{Z}] \subset B$. This inclusion is actually an equality, as follows from the following result.

LEMMA 2.4. *Let* q *be a nonconstant inner function. Then* $B \subset [H^\infty, \bar{q}]$.

Necessity. We consider three cases:

I) H^∞ is a *maximal* weak-$*$ closed subalgebra of L^∞.

II) $H^\infty = \bigcap\{D : D \supset H^\infty$, D weak-$*$ closed subalgebra of $L^\infty\}$.

III) $H^\infty \subset D = \bigcap\{\tilde{D} : \tilde{D} \supset H^\infty$, $\tilde{D}$ weak-$*$ closed subalgebra of $L^\infty\} \subset L^\infty$.

Case III) is technical, and it reduces to Case I below.

Case II). If D H^∞ then there exists a nonconstant inner function q such that $[H^\infty, \bar{q}] \subset D$. By Lemma 2.5, $B \subset [H^\infty, \bar{q}] \subset D$. Thus $B = H^\infty$.

Case I). We need some auxiliary results.

LEMMA 2.5. *If* $H^\infty \subset B$ *then there exists a unimodular function* ϕ *such that* $\phi, \bar{\phi} \in B \setminus H^\infty$.

LEMMA 2.6. *If* $\phi, \bar{\phi} \in B$ *and* $|\phi| = 1$ a.e. (m), *then* $\|\phi + H^\infty\| < 1$.

LEMMA 2.7. *If* H^∞ *is maximal (as a weak-* closed subalgebra of* L^∞*) and* $\phi \in L^\infty$ *is such that* H_ϕ *attains its norm, then there exists* $\psi_0 \in \phi + H^\infty$ *such that* $|\psi_0| = \|\phi + H^\infty\|$ a.e. (m).

LEMMA.2.8. *If* H^∞ *is maximal and* $\psi \in B$ *is unimodular, then* $\|H_\psi\| = 1$ *only if* $\overline{\psi} \in B$.

With the aid of the above lemmas, we can now obtain Case I as follows:

If $H^\infty \subsetneq B$, pick a function ϕ as in Lemma 2.5. Since H_ϕ is compact, we can use Lemma 2.7 to get $\psi_0 \in \phi + H^\infty$ such that $|\psi_0| = \|\phi + H^\infty\|$ a.e. (m). Of course $\|\phi + H^\infty\| > 0$, so that we can form $\psi = \psi_0 / \|\phi + H^\infty\|$. $\psi \in B$, $|\psi| = 1$ a.e. and $\|\psi + H^\infty\| = 1$. Since by Corollary 2.3, $\|H_\psi\| = \|\psi + H^\infty\| = 1$, we can use Lemma 2.8 to conclude that $\overline{\psi} \in B$. Lemma 2.6 now says that $\|\psi + H^\infty\| < 1$, a contradiction.

§3. INVARIANT SUBSPACES FOR SUPERALGEBRAS OF THE BIDISC ALGEBRA

In Section 1 we have studied Hankel operators associated with a subsemigroup Γ_+ of a discrete group Γ. In particular we showed that if $\Gamma_+ = \{(k,l) \in \mathbf{Z}^2 : l > 0 \text{ or } (l = 0 \text{ and } k \geq 0)\}$ then $H^\infty(G) \subsetneq B^{(1)} = B^{(2)}$. Another such example can be obtained by considering

$$\Gamma_+^{(n)} = \{(k,l) \in \mathbf{Z}^2 : l > 0 \text{ and } (0 \leq l \leq n-1 \Longrightarrow k \geq 0)\},$$

for $n = 0, 1, \ldots$ (Γ_+ above is just $\Gamma_+^{(1)}$; $\mathbf{Z} \times \mathbf{Z}_+$ is $\Gamma_+^{(0)}$). Notice that unless $n = 0$ or 1, $\mathbf{Z}^2 \neq \Gamma_+^{(n)} \cup (-\Gamma_+^{(n)})$ so that we cannot directly conclude that $B^{(1)} = B^{(2)}$ (for $E = \Gamma_+^{(n)}$). By using Theorem 1.14 we can conclude, however, that $H^\infty(G) \subsetneq B^{(1)}$ and $H^\infty(G) \subsetneq B^{(2)}$. Now, if $\gamma_0 = (0,-1)$ ($n \geq 2$, of course) we see that $e_{\gamma_0} \in B^{(2)}$ but $e_{\gamma_0} \notin B^{(1)}$, so that $B^{(1)} \neq B^{(2)}$. (The reader is asked to compare this example with Example 1.7.)

In connection with the above example, we would like to study the invariant subspace theory associated with algebras of continuous functions on $\mathbf{T}^2$ whose Fourier transforms are supported in $\Gamma_+^{(n)}$. For $n \geq 1$, let A_n be the uniform closure in $C(\mathbf{T}^2)$ of the algebra of polynomials in $z^k w^l$, where $l \geq 0$ and if $0 \leq l \leq n-1$, then $k \geq 0$. We study the subspaces M of $L^2(\mathbf{T}^2)$ which are invariant under multiplication by the functions in A_n.

THEOREM 3.1. *Let* $M \subset L^2(\mathbf{T}^2)$ *be an invariant subspace for* A_n. *Then* M *is simply invariant (i.e.,* $[A_{n,0}M]_2 \neq M$, *where* $A_{n,0} = \{f \in A_n : \int f d\sigma = 0\}$ *(*σ *is the Lebesgue measure on* $\mathbf{T}^2$*)) if and only if* $zM \subsetneq M$.

Using results of the third author [6] we obtain the following corollary.

COROLLARY 3.2. *Let M be an invariant subspace for A_n. Then M is not simply invariant if and only if $M = \chi_{E_1} qH^2 \oplus \chi_{E_2} L^2$, where χ_{E_1}, χ_{E_2} denote the characteristic functions of two measurable sets E_1 and E_2, χ_{E_1} is independent of w, $\chi_{E_1} + \chi_{E_2} \leq 1$, and $|q| = 1$ a.e. (σ). Here $\mathbf{H}^2$ stands for the subspace of L^2 consisting of those functions with Fourier series supported in $\{(k,m) : m \geq 0\}$.*

To describe all simply invariant subspaces of A_n, we need to introduce a definition.

DEFINITION 3.3. Let M be an invariant subspace. Then, by definition, $M_{-\infty} :=$
$:= [\bigcup_{k\geq 0} \bar{z}^k M]_2$ and $M_\infty := [\bigcap_{k\geq 0} z^k M]_2$.

Clearly $M_\infty \subset M \subset M_{-\infty}$ and both $M_{-\infty}$ and M_∞ are nonsimply invariant.

PROPOSITION 3.4. *Let M be simply invariant. Then $M_{-\infty} = q_1 \mathbf{H}^2$ and $M_\infty =$ $= q_2 \mathbf{H}^2$, where q_1 and q_2 are unimodular.*

In other words, when M is simply invariant, the "L^2-pieces" in the decomposition for M_∞ and $M_{-\infty}$ given by Corollary 3.2 are absent.

If N is a simply invariant subspace for A_n such that $N_\infty = w^l \mathbf{H}^2$ and $N_{-\infty} = \mathbf{H}^2$ $(1 \leq l \leq n)$, then N is of the following form:

$$N = [z; f_1, \ldots, f_j]_2 \oplus w^l \mathbf{H}^2,$$

where $f_1, \ldots, f_j$ $(j \leq l)$ are nonzero functions in $\sum_{k=0}^{n-1} \oplus L^2(\mathbf{T}) w^k$ such that $f_i = \sum_{k=0}^{n-1} f_{ik} w^k$, $1 \leq i \leq j$, and $\sum_{k=0}^{n-1} f_{ik} \overline{f_{mk}} = \delta_{im}$, $1 \leq i, m \leq j$, and where $[z; f_1, \ldots, f_j]_2$ stands for the subspace of L^2 generated by $f_1, \ldots, f_j$, invariant under multiplication by z (cf. [4, VI.3, p.60]). If F is a unimodular function and $M = FN$, where N is as above, then M is again simply invariant. Our main result of this section establishes that the converse is true.

THEOREM 3.5. *Let M be simply invariant for A_n. Then $M = FN$, where F is unimodular and N is a simply invariant subspace with $N_\infty = w^l \mathbf{H}^2$ and $N_{-\infty} = \mathbf{H}^2$, for some l between 1 and n. Moreover, $M \cap F w^{l-1} \mathbf{H}^2 = F w^{l-1} q H_1^2$, where q is unimodular in $L^\infty(\mathbf{T})$ (independent of w), and $H_1^2 = [A_1]_2$.*

The proofs of 3.1, 3.2, 3.4 and 3.5 appear in [2]. We mention here that our results can be extended to the case when Γ_+ is obtained from $\Gamma_+^{(n)}$ by perturbing the edges by a finite staircase; for instance, let $p_1, \ldots, p_k, q_1, \ldots, q_{k+1}$ be positive integers and let Γ_+ be the $\mathbf{Z}_+^2$-module whose boundary consists of the integer points of the segments

$$([0,\infty) \times \{0\}) \cup (\{0\} \times [0, q_1]) \cup ([-p_1, 0] \times \{q_1\}) \cup \ldots$$
$$\ldots \cup (\{-p_k\} \times [q_1 + \ldots + q_k, q_1 + \ldots + q_{k+1}]) \cup ((-\infty, -p_1 - \ldots - p_k] \times \{q_1 + \ldots + q_{k+1}\}).$$

Then the invariant subspaces corresponding to this subset of $\mathbf{Z}^2$ can be described in terms of the invariant subspaces for $A_{q_1 + \ldots + q_{k+1}}$ together with a finite number of additional constraints. An interesting situation would arise if one considers sets with infinite staircases as parts of their boundaries.

REFERENCES

1. **Curto, R. ; Muhly, P. ; Nakazi, T. ; Xia, J. :** Uniform algebras and Hankel operators, *Archiv. der Mathematik*, to appear.
2. **Curto, R. ; Muhly, P. ; Nakazi, T. ; Yamamoto, T. :** On superalgebras of the polydisc algebra, preprint.
3. **Hartman, P. :** On completely continuous Hankel matrices, *Proc. Amer. Math. Soc.* **9**(1958), 862-866.
4. **Helson, H. :** *Lectures on invariant subspaces*, Academic Press, New York, 1964.
5. **Hewitt, E. ; Ross, K. :** *Abstract harmonic analysis.II*, Springer-Verlag, 1970.
6. **Nakazi, T. :** Invariant subsapces of weak-* Dirichlet algebras, *Pacific J. Math.* **69**(1977), 151-167.
7. **Salas, H. :** *C^*-algebras of isometries with commuting range projections*, Ph.D. Thesis, University of Iowa, July 1983.
8. **Srinivasan, T. ; Wang, J.-K. :** Weak-* Dirichlet algebras, in *Function algebras*, Proceedings of an International Symposium held at Tulane Univ., F.T.Birtel, ed., Scott-Foresman, 1966.

R. E. Curto and **P. S. Muhly**
Department of Mathematics
University of Iowa
Iowa City, IA 52240
U.S.A.

T. Nakazi
Department of Mathematics
Hokkaido University
Saporo
Japan

Operator Theory:
Advances and Applications, Vol.17

CONTRACTIONS GÉNÉRIQUES

J. Dazord

D. Sarason a montré que pour qu'un opérateur de Toeplitz analytique $T_\phi = \phi(S)$ ait les mêmes sous-espaces fermés invariants que la translation unilatérale S, il faut et il suffit que la fonction $\phi \in H^\infty$ soit un générateur de H^∞, c'est-à-dire que H^∞ soit l'algèbre, fermée pour la topologie faible de dual, engendrée par les fonctions ϕ et $f \equiv 1$ ([4], Proposition 1, p.515). D'autre part, si ϕ est un générateur de H^∞ et si T est une contraction complètement non unitaire (c.n.u.) sur un espace de Hilbert $\boldsymbol{H}$, les opérateurs T et $\phi(T)$ ont les mêmes sous-espaces fermés invariants.

DÉFINITION. On dit qu'une contraction c.n.u. T sur un espace de Hilbert $\boldsymbol{H}$ est *générique* si pour toute fonction ϕ de H^∞, la condition $\mathbf{Lat}\,\phi(T) = \mathbf{Lat}\,T$ implique que ϕ est un générateur de H^∞.

($\mathbf{Lat}\,X$ désigne le treillis des sous-espaces fermés invariants par un opérateur X).

Un générateur ϕde H^∞ est une fonction univalente, autrement dit la fonction $\phi : \mathbf{D} \to \mathbf{C}$, où $\mathbf{D}$ désigne le disque unité ouvert, est injective ([4], Proposition 3, p.516).

THÉORÈME 1. *Le spectre d'une contraction générique contient le cercle unité.*

PREUVE. Désignons par $\sigma(T)$ le spectre d'une contraction c.n.u. T. Supposons qu'on ait $1 \notin \sigma(T)$. Il existe alors un réel $a \in [-1,1[$ tel que le spectre $\sigma(T)$ soit contenu dans le demi-plan $\{z \in \mathbf{C} \; ; \mathbf{Re}\, z \leq a\}$. Posons $b = (1 + a)/2$. Le polynôme $p(z) = (z - b)^2$ est univalent dans le demi-plan ouvert $\{z \in \mathbf{C} \; ; \mathbf{Re}\, z < b\}$. On en déduit l'égalité $\mathbf{Lat}\,p(T) =$ $= \mathbf{Lat}\,T$ alors que le polynôme p n'est pas univalent dans $\mathbf{D}$. ∎

Etant donné un opérateur T, désignons par $\boldsymbol{A}(T)$ l'algèbre ultrafaiblement fermée engendrée par T et l'opérateur I. Notons $\Phi^T : H^\infty \to \boldsymbol{A}(T)$ le calcul fonctionnel de Sz.-Nagy et Foiaş ([16, Chapitre 3).

THÉORÈME 2. *Soit* T *une contraction générique. Le calcul fonctionnel* Φ^T *est injectif et sa restriction* $\Phi^T|A(\mathbf{D})$ *à l'algèbre du disque est isométrique.*

PREUVE. Supposons que le calcul fonctionnel Φ^T ne soit pas injectif; soit $\theta \in H^\infty$ une fonction telle que $\Phi^T(\theta) = 0$, i.e. $\theta(T) = 0$. La fonction θ n'étant pas constante, considérons deux points $a,b \in \mathbf{D}$ tels qu'on ait $\theta(a) \neq \theta(b)$. Soit u la fonction identique dans $\mathbf{D}$ ($u(z) \equiv z$). Posons $\phi = u + \lambda\theta$ avec $\lambda = -(b-a)(\theta(b) - \theta(a))^{-1}$. On a $\phi(T) = T$ et $\phi(a) = \phi(b)$, ce qui établit la première assertion.

Considérons maintenant une fonction $\phi \in A(\mathbf{D})$. En utilisant l'égalité $\phi(\sigma(T)) = \sigma(\phi(T))$, l'inégalité de von Neumann et le théorème 1, on obtient les inégalités:

$$\|\phi(T)\| \leq \|\phi\|_\infty = \sup\{|\phi(z)|;\ |z| = 1\} \leq \sup\{|\phi(z)|;\ z \in \sigma(T)\} \leq$$

$$\leq \sup\{|\lambda|;\ \lambda \in \sigma(\phi(T))\} = r(\phi(T)) \leq \|\phi(T)\|$$

où $\|\phi\|_\infty$ désigne la norme de ϕ dans $A(\mathbf{D})$ et $r(X)$ le rayon spectral d'un opérateur X. ■

Précisons qu'à l'aide d'une preuve similaire à celle qui vient d'être donnée, on peut établir que le calcul fonctionnel Φ^T est isométrique si le spectre de T est dominant ([3], p.129).

Introduisons quelques notations; pour plus de détails on se reportera à [2], [3] et [1]. Notons $\mathcal{L}(\mathcal{H})$ l'espace des opérateurs sur $\mathcal{H}$ et τc l'espace des opérateurs à trace; la trace d'un opérateur $K \in \tau c$ est notée $\operatorname{tr} K$. Considérons la forme bilinéaire sur $\mathcal{L}(\mathcal{H}) \times \tau c$ définie par $\langle X, K\rangle = \operatorname{tr}(XK)$, $X \in \mathcal{L}(\mathcal{H})$, $K \in \tau c$. Etant donné un opérateur $L \in \tau c$, notons $[L]_{Q(T)}$ sa classe dans $\tau c / {}^\perp\mathcal{A}(T) = Q(T)$, où ${}^\perp\mathcal{A}(T)$ désigne l'orthogonal de $\mathcal{A}(T)$ dans τc pour la forme bilinéaire considérée. La dualité entre $\mathcal{A}(T)$ et $Q(T)$ est définie par la forme bilinéaire $\langle X, [L]\rangle = \operatorname{tr}(XL)$, $X \in \mathcal{A}(T)$, $L \in \tau c$. Enfin, étant donnés deux vecteurs h et k dans $\mathcal{H}$, $h \otimes k$ désigne l'opérateur de rang 1 défini par $(h\otimes k)(\ell) = \langle \ell, k\rangle h$, $\ell \in \mathcal{H}$. Nous utiliserons le lemme suivant:

LEMME. *Soit* T *une contraction c.n.u. et soit* ϕ *une fonction dans* H^∞ *telle qu'on ait* $\operatorname{Lat}\phi(T) = \operatorname{Lat} T$. *Alors pour tout couple de vecteurs* (h,k) *vérifiant* $[h\otimes k]_{Q(\phi(T))} = 0$, *on a* $\langle Th, k\rangle = 0$.

PREUVE. Soit $h \in \mathcal{H}$. Comme les opérateurs T et $\phi(T)$ ont les mêmes sous-espaces fermés invariants, les sous-espaces fermés engendrés par les suites $\{h, Th, \ldots, T^n h, \ldots\}$ et $\{h, \phi(T)h, \ldots, \phi^n(T)h, \ldots\}$ sont égaux. En particulier il existe une suite (p_n) de polynômes telle que la suite $((p_n \circ \phi)Th)$ converge au sens de la norme de $\mathcal{H}$ vers le vecteur Th. Or si h et k sont deux vecteurs vérifiant $[h\otimes k]_{Q(\phi(T))} = 0$, on a pour tout polynôme p:

$$\langle (p \circ \phi)Th , k \rangle = \langle (p \circ \phi)T , [h \otimes k]_{Q(\phi(T))} \rangle = 0.$$

On en deduit l'egalite cherchee $\langle Th , k \rangle = 0$. ■

Rappelons qu'un sous-ensemble E du disque unité fermé $\bar{\mathbf{D}}$ est dit dominant si pour toute fonction ϕ de H^∞ la norme de ϕ dans H^∞ est donnée par $\|\phi\|_\infty = \sup\{ |\phi(z)| ; z \in E \cap \mathbf{D}\}$. Une contraction c.n.u. T est dite de classe **(BCP)** si son spectre essentiel est dominant. L'étude de ces contractions a été abordée dans [3] et poursuivie dans [1].

THÉORÈME 3. *Une contraction de classe* **(BCP)** *est générique.*

PREUVE. Notons $W(X)$ l'algèbre fermée pour la topologie faible des opérateurs engendrée par un opérateur X et par l'opérateur identique I. Soit T une contraction de classe **(BCP)** et soit ϕ une fonction de H^∞ vérifiant l'égalité $\mathbf{Lat}\, \phi(T) = \mathbf{Lat}\, T$.

Etablissons tout d'abord l'égalité $W(T) = W(\phi(T))$. Il suffit, compte tenu du lemme 1 de [1], de prouver que pour tout entier $n \geq 1$ et pour toute paire de suites finies $\{h_1, \ldots, h_n\}$ et $\{k_1, \ldots, k_n\}$ de vecteurs de $\mathcal{H}$, la relation $\sum_{i=1}^{n} [h_i \otimes k_i]_{Q(\phi(T))} = 0$ implique $\sum_{i=1}^{n} \langle Th_i , k_i \rangle = 0$. Posons $L = \sum_{i=1}^{n} h_i \otimes k_i$. Puisque T est de classe **(BCP)**, il existe deux vecteurs h et k dans $\mathcal{H}$ tels qu'on ait $[L]_{Q(T)} = [h \otimes k]_{Q(T)}$ ([3], lemme 4.9, p.134). On en déduit $[L]_{Q(\phi(T))} = [h \otimes k]_{Q(\phi(T))} = 0$ donc, en utilisant le lemme $\langle Th , k \rangle = 0$. Un argument élémentaire de linéarité permet d'établir la relation cherchée: $\sum_{i=1}^{n} \langle Th_i , k_i \rangle = 0$.

On sait que pour un opérateur T de classe **(BCP)** les algèbres $W(T)$ et $\mathcal{A}(T)$ sont égales et que la topologie ultrafaible coïncide avec la topologie faible d'opérateurs sur l'algèbre $\mathcal{A}(T) = W(T)$ ([1], Corollary 1, p.372). On en déduit l'égalité: $W(\phi(T)) = \mathcal{A}(\phi(T))$, d'où finalement $\mathcal{A}(\phi(T)) = \mathcal{A}(T)$. Ainsi il existe une suite généralisée (p_α) de polynômes telle que la suite généralisée d'opérateurs $((p_\alpha \circ \phi)T)$ converge ultrafaiblement vers T. Or le spectre $\sigma(T)$ est dominant et il en résulte que le calcul fonctionnel $\Phi^T : H^\infty \to \mathcal{A}(T)$ est un homéomorphisme lorsque H^∞ et $\mathcal{A}(T)$ sont munis respectivement des topologies faible de dual et ulftrafaible ([3], Theorem 3.2, p.128). On en conclut que la suite $(p_\alpha \circ \phi)$ converge faiblement dans H^∞ vers la fonction identique u, donc que ϕ est un générateur de H^∞. ■

Enoncons un théorème de stabilité de la classe des contractions génériques; la preuve en est immédiate.

THÉORÈME 4. *Si la restriction d'une contraction c.n.u.* T *à un sous-espace invariant par* T *est générique, alors* T *est générique.*

Le théorème 3 fournit un exemple de contraction générique dont le spectre est dominant. Cependant, le spectre d'une contraction générique n'est pas nécessairement dominant. Ainsi la translation bilatérale T de poids $w_n = 1$, $n \geq 0$; $w_{-n} = 1 - (n+1)^{-2}$, $n > 0$ est générique en vertu du théorème 4. Or l'opérateur T est inversible et son spectre est la couronne $\{z \in \mathbf{C} \,;\, r(T^{-1})^{-1} \leq |z| \leq r(T)\}$ ([5], Theorem 5, p.67). Un calcul élémentaire montre qu'on a $r(T) = r(T^{-1}) = 1$ et donc que l'opérateur T admet pour spectre le cercle unité.

Mentionnons pour terminer que le lemme que nous utilisons pour établir le théorème 3 figure avec la même démonstration dans un article à paraître de Bercovici, Chevreau, Foiaş et Pearcy.

BIBLIOGRAPHIE

1. **Bercovici, H. ; Foiaş, C. ; Langsam, J. ; Pearcy, C. :** (BCP)-operators are reflexive, *Michigan Math. J.* **29**(1982), 371-379.
2. **Brown, S. :** Some invariant subspaces for subnormal operators, *Integral Equations Operator Theory* **1**(1978), 310-333.
3. **Brown, S. ; Chevreau, B. ; Pearcy, C. :** Contractions with rich spectrum have invariant subspaces, *J. Operator Theory* **1**(1979), 123-136.
4. **Sarason, D. :** Invariant subspaces and unstarred operator algebras, *Pacific J. Math.* **17**(1966), 511-517.
5. **Shields, A. :** *Weighted shift operators and analytic function theory*, Math. Surveys AMS **13**(1974), 49-128.
6. **Sz.-Nagy, B. ; Foiaş, C. :** *Harmonic analysis of operators on Hilbert space*, North-Holland, Amsterdam, 1970, translation and revision of the french edition 1967.

Jean Dazord
U.E.R. de Mathématiques
Université Claude-Bernard Lyon I
43, Boulevard du 11 Novembre 1918
69622, Villeurbanne Cedex
France.

Operator Theory:
Advances and Applications, Vol.17

HILBERT MODULES OVER FUNCTION ALGEBRAS *)

Ronald G. Douglas

0. INTRODUCTION

Much of the early motivation for the study of operator theory came from integral equations although early in this century both operator theory and functional analysis took on a life of their own. Self-adjoint operators, both bounded and unbounded, occupied center stage for several decades either singly or in algebras. During the last two or three decades various approaches to the non-selfadjoint theory have been introduced with considerable success at least in the case of a single operator. The generalization to several operators, whether commuting or non-commuting, has largely eluded us. In this note we want to outline a different point of view which may assist in guiding developments in this area.

In the spectral theory for a single self-adjoint operator, a functional calculus is introduced based on the algebra of continuous functions on **R** or a closed subset of **R**. In the extension to normal operators, one considers continuous functions on a closed subset of **C**. Finally, for a commutative algebra of normal operators, the functional calculus is based on the algebra of continuous functions on a locally compact Hausdorff space. At a first level of inquiry, there is little difference between a closed subset of **C** and an arbitrary locally compact Hausdorff space although more refined questions certainly detect a difference. The basic functional calculus for a non-selfadjoint operator is an algebra of holomorphic functions on some domain containing its spectrum. As soon as one considers more than one operator, one is dealing with holomorphic functions in more than one variable, which is a very different situation.

Everyone who has set out to learn about functions of several complex variables is struck by the fact that beyond the basics, very few results extend from the one variable case. In order to proceed one must start all over and take a different point of view. It seems likely that the same is true in the study of non-selfadjoint operator theory. Although in the theory of operator algebras, there are new phenomena which sometimes capture analytic features of the context, the truly non-selfadjoint situation

*) Work partially supported by a grant form the National Science Foundation.

portends to be even more like the theory of several complex variables and even that of classical algebraic geometry. In order to facilitate this we recast some results from operator theory into the language of modules over function algebras. Although we have some expectation of what might be proved in this context, we shall be content with a brief description here hoping that this will cause others to think about this.

The introduction of the language of modules into operator theory is not new dating at least from the development of Hochschild cohomology for Banach algebras (cf. [7]). We do not believe it has appeared in this context except for the earlier work of Foiaş and the author [10]. Before continuing we want to acknowledge the contributions of Ciprian Foiaş, Vern Paulsen and others in helping to formulate and clarify these ideas. Also we want to acknowledge the kind hospitality of the Institute des Hautes Études Scientifiques where the first draft of this paper was written.

1. HILBERT MODULES

We shall give the definition of Hilbert module for general Banach algebras although we shall consider only function algebras.

DEFINITION: The Hilbert space M will be said to be a *Hilbert module* for the unital Banach algebra A if there exists a map

$$A \times M \to M$$

which satisfies:

(1) $1h = h$ for $h \in M$;

(2) $a(bh) = (ab)h$ for $a,b \in A$ and $h \in M$;

(3) $a(\alpha_1 h_1 + \alpha_2 h_2) = \alpha_1 ah_1 + \alpha_2 ah_2$ for $a \in A$, $\alpha_1,\alpha_2 \in \mathbf{C}$ and $h_1,h_2 \in M$; and

(4) $(a_1 + a_2)h = a_1 h + a_2 h$ for $a_1,a_2 \in A$ and $h \in A$.

It is said to be *bounded* if for some $K > 0$

(5) $\|ah\| \leq K\|a\|\,\|h\|$ for $a \in A$ and $h \in M$,

and *contractive* if K can be taken equal to 1.

The existence of a finite K can be shown to be equivalent to the joint continuity of the module multiplication using the principle of uniform boundedness.

If M is a bounded Hilbert module over A and for a in A, one sets

$$\sigma(a)h = ah \quad \text{for } h \text{ in } M,$$

then $\sigma(a)$ is a bounded linear operator on M. Hence, σ is a bounded representation of A into $L(M)$. Conversely, if $\sigma : A \to L(M)$ is a bounded representation of A, then M can be made into a bounded Hilbert module over A by defining

$$ah = \sigma(a)h \quad \text{for a in } A \text{ and h in } M.$$

Thus, bounded Hilbert modules over A correspond to bounded representations of A as operators on a Hilbert space. Such objects have been studied since at least the book of Hoffman [15]. One should consult [20] for a more recent treatment. Despite the fact that this topic has been studied in the language of algebra representations, we feel there is merit in the module point of view which is what is used in algebra and homological algebra.

Although one can consider modules over any Banach algebra, we shall restrict attention in this note to modules over function algebras. Indeed we have in mind restricting the class even further in subsequent work. Further, we shall restrict attention to the use of *separable* algebras and *separable* Hilbert modules.

Let X be a compact metrizable space and C(X) be the algebra of complex-valued continuous functions on X. A subalgebra A of C(X) is said to be a *function algebra* if it contains the constant functions and is uniformly closed. There are two canonical choices for the space X, if we assume that the functions of A separate the points of X. The largest space M_A is the maximal ideal space of A or the Gelfand space, while the Šilov boundary ∂A is the smallest such space. In particular, one has

$$A \subset C(\partial A), \quad A \subset C(M_A)$$

and

$$\partial A \subset X \subset M_A$$

for every space X such that $A \subset C(X)$ and A separates the points of X (cf. [12] for more details).

Let us begin by considering the contractive Hilbert modules over C(X) for an arbitrary compact metrizable space X. If M is a contractive Hilbert modules over C(X), then there exists a Borel measure μ on X and a direct integral such that

$$M \simeq \int_X \oplus M_x \, d\mu$$

where $\simeq$ denotes *unitary equivalence.* The action of a function in C(X) on a cross-section of the direct integral is just pointwise multiplication. Two direct integrals

$$\int_X \oplus M_x \, d\mu \quad \text{and} \quad \int_X \oplus N_x \, d\nu$$

represent unitarily equivalent modules over C(X) if and only if μ and ν are mutually absolutely continuous and $\dim M_x = \dim N_x$ a.e.. This is just a standard reinterpretation of classical multiplicity theory.

Moreover, if M is a bounded Hilbert module over C(X), then a result due to Dixmier yields the existence of a contractive Hilbert module N over C(X) and a bounded invertible module map $S: M \to N$. Thus every bounded Hilbert module over C(X) is *similar* to a contractive Hilbert module over C(X). Further, since two contractive Hilbert modules over C(X) are known to be similar if and only if they are unitarily equivalent, it follows that two bounded Hilbert modules over C(X) are similar if and only if their associated multiplicity functions are equal (cf. [9] for more details).

The preceding brief summary shows that all the basic facts about contractive and bounded Hilbert modules over C(X) follow directly from known results. Unfortunately for no other function algebra are the results so complete.

Let us consider what is probably the next simplest function algebra, namely the disk algebra. This algebra can be described in various ways; perhaps most simply $A(\mathbf{D})$ is the closure of the analytic polynomials p(z) in $C(\partial \mathbf{D})$. In this case the maximal ideal space is the closed unit disk $\mathbf{D}^-$, while the Šilov boundary is the unit circle $\partial \mathbf{D}$, the ordinary topological boundary of $\mathbf{D}$. This algebra sits at the confluence of several branches of functional analysis and has been much studied. We want to consider the contractive and bounded Hilbert modules over $A(\mathbf{D})$. We begin by reformulating the theorem of von Neumann on spectral sets.

THEOREM. *The contractive Hilbert modules over* $A(\mathbf{D})$ *correspond in a one-to--one fashion to the contraction operators on Hilbert space.*

If M is a contractive Hilbert module over $A(\mathbf{D})$, then by defining

$$Th = zh \quad \text{for h in } M$$

one obtains a contraction operator since $\|z\| = 1$. Conversely, if T is a contraction operator on the Hilbert space H, then we define the module action of a polynomial p(z) as follows:

$$p(z)\cdot h = p(T)h \quad \text{for h in } H.$$

The inequality of von Neumann states that

$$\|p(T)h\| \leq \|p\|_\infty \|h\|$$

and hence the action can be extended to all of $A(\mathbf{D})$ making H into a contractive Hilbert module over $A(\mathbf{D})$. That two modules are unitarily equivalent or similar if and only if the corresponding contractions are, is obvious.

Reducing the study of contractive Hilbert modules over $A(\mathbf{D})$ to that of contraction operators does not solve all problems, however, since our understanding of the latter is only partial, at best. It does provide, however, one connection between the study of modules and classical operator theory.

If we try to extend the preceding result to bounded Hilbert modules over $A(\mathbf{D})$, we run into one of the unsolved problems of operator theory.

PROPOSITION. *The bounded Hilbert modules over* $A(\mathbf{D})$ *correspond in a one-to--one fashion to the polynomially bounded operators on Hilbert space.*

Indeed, the definition of polynomially bounded is that there exist a constant K such that $\|p(T)\| \leq K\|p(z)\|$ for each analytic polynomial p(z). Hence the proposition is obvious.

The problem of Sz.-Nagy [21] as reformulated by Halmos [13] becomes in the language of modules.

PROBLEM. Is every bounded Hilbert module over $A(\mathbf{D})$ similar to a contractive one?

This question can be raised for every function algebra A. The answer is affirmative for C(X) but unknown for any other example. It seems that it should depend on the "complexity" of the algebra and, in particular, I suspect that the answer is positive for the disk algebra and other planar algebras. Further, I suspect that the answer is negative for the polydisk algebra $A(\mathbf{D}^n)$.

PROBLEM. Find a function algebra A and a bounded Hilbert module over A which is not similar to a contractive one.

2. ŠILOV MODULES

Although the connections between operator theory and Hilbert modules over $A(\mathbf{D})$ are interesting and provide motivation, we want to try to develop the theory independently. Hence, how would one go about studying Hilbert modules over $A(\mathbf{D})$. We shall introduce a class of "nice modules" for any function algebra which will be especially useful for the case of $A(\mathbf{D})$.

DEFINITION. Let A be a function algebra and N be a contractive Hilbert module over $C(\partial A)$. Then N is also a module for A and a submodule M of N *for* A will be said to be a *Šilov module for* A. A module similar to a Šilov module will be said to be a *quasi-Šilov module for* A.

An example of a Šilov module for $A(\mathbf{D})$ is the Hardy module $H^2(\mathbf{D})$ which consists of the norm closure in $L^2(\mathbf{D})$ of the analytic polynomials. That all Šilov modules over $A(\mathbf{D})$ are obtained from the Hardy module and $C(\partial\mathbf{D})$ modules is the content of the decomposition theorem of Wold and von Neumann as reformulated in the module setting.

THEOREM. *If M is a Šilov module over* $A(\mathbf{D})$, *then* $M = N \oplus H^2_E(\mathbf{D})$, *where N is a contractive Hilbert module over* $C(\partial\mathbf{D})$ *and* $H^2_E(\mathbf{D})$ *denotes the Hardy module with values in the Hilbert space E.*

In the correspondence between contractive modules over $A(\mathbf{D})$ and contraction operators, Šilov modules correspond to isometries.

It would seem an interesting problem to classify the Šilov modules or to provide a moduli space for them, at least for certain natural function algebras. If Ω is a finitely connected domain in $\mathbf{C}$ and $R(\Omega)$ is the algebra obtained from the norm closure of the rational functions with poles outside Ω^-, then results of Abrahamse and the author [1] yield the following:

THEOREM. *If M is a Šilov module for* $R(\Omega)$, *then*

$$M \simeq N \oplus H^2_E(\Omega),$$

where N is a contractive Hilbert module for $C(\partial\Omega)$ *and* $H^2_E(\Omega)$ *is the Hardy module defined by the holomorphic cross-sections of the flat unitary holomorphic vector bundle E over* Ω *with boundary values in* $L^2_{E|\partial\Omega}(\partial\Omega)$.

The results of Helson and Lowdenslager [14] can be interpreted to classify the Šilov modules over various function algebras of almost periodic functions on $\mathbf{R}$. In this case the moduli space is a cohomology group. That there exists non-unitarily equivalent Šilov modules of the same "multiplicity" for these algebras is equivalent to the failure of inner functions to represent all invariant subspaces.

Before leaving this topic we should emphasize that classifying Šilov modules is not the same as classifying the corresponding invariant subspaces (cf. [8]). In the latter, one seeks to specify each subspace of functions, while in the former one is interested in the module action. The first problem should be easier and is more relevant for operator theoretic applications. Still one should not expect good answers for the moduli space for Šilov modules in many cases. For example, the recent results of Agrawal, Clark and the author [3] show that the Šilov modules over the polydisk algebra $A(\mathbf{D}^n)$ given by submodules of $H^2(\mathbf{D}^n)$ of finite codimension are all non-unitarily equivalent. Hence the moduli space for just these Šilov modules consists of the space of ideals in $\mathbf{C}[z_1, z_2, \ldots, z_n]$ of finite codimension [4] or equivalently of those ideals having a 0-dimensional variety of zeros.

3. ŠILOV RESOLUTIONS

Let us return now to the disk algebra. The Šilov modules over A(**D**) do not exhaust all the contractive Hilbert modules over A(**D**). This is obvious since the correspoding contraction operators in this case are just the isometries. How can one obtain other modules? An algebraist might consider modules which have Šilov resolutions. (The question of which category and, in particular, which maps to allow in the following definition does not have a clear answer and hence the choices made should be regarded as tentative.)

DEFINITION. The contractive Hilbert module M over the function algebra A is said to have a *Šilov resolution* if there exists an exact sequence

$$0 \leftarrow M \leftarrow S_0 \leftarrow S_1 \leftarrow 0$$

where S_0 and S_1 are Šilov modules over A and the maps are partially isometric module maps.

If one has $S_1 = (0)$, then M is unitarily equivalent to S_0 and hence is itself a Šilov module. Since a submodule of a Šilov module is itself a Šilov module, it follows that if a Šilov resolution exists of length greater than one, then one exists of length less than or equal to one. Hence, we have chosen to consider Šilov resolutions of length one. We will have more to say about this later.

If T is a contraction on the Hilbert space M, then the dilation theorem of Sz.-Nagy [22] exhibits a Hilbert space G and an isometry V on $M \oplus G$ of the form

$$V = \begin{bmatrix} T & 0 \\ X & U \end{bmatrix}.$$

If we set $S_0 = M \oplus G$ and $S_1 = G$ with the module multiplication defined by

$$\Psi \cdot h = \Psi(V)h \quad \text{for h in } S_0 \text{ and}$$

$$\Psi \cdot k = \Psi(U)k \quad \text{for k in } S_1,$$

then S_0 and S_1 are Šilov modules over A(**D**) and we have

$$0 \leftarrow M \leftarrow S_0 \leftarrow S_1 \leftarrow 0$$

where $S_0 \rightarrow M$ is projection and $S_1 \rightarrow S_0$ is inclusion. Thus we can reformulate the dilation theorem of Sz.-Nagy to state:

THEOREM. *Every contractive Hilbert module over* A(**D**) *has a unique Šilov resolution.*

The uniqueness follows from that of the result of Sz.-Nagy and means that given two Šilov resolutions

$$0 \leftarrow M \leftarrow S_0 \leftarrow S_1 \leftarrow 0$$

$$0 \leftarrow M \leftarrow \tilde{S}_0 \leftarrow \tilde{S}_1 \leftarrow 0$$

where S_0 and $\tilde{S}_0$ are minimal (that is, no submodule of S_0 or $\tilde{S}_0$ maps onto M), then there exist unitary module maps

$$W_i : S_i \to \tilde{S}_i \qquad i = 0,1$$

such that the following diagram commutes:

$$\begin{array}{ccccc} & \swarrow & S_0 & \leftarrow & S_1 \\ M & & \downarrow W_0 & & \downarrow W_1 \\ & \nwarrow & \tilde{S}_0 & \leftarrow & \tilde{S}_1 \end{array} .$$

In a completely analogous manner, the theorem of Arveson [6] can be reformulated to yield the same result for Dirichlet algebras. Recall that the function algebra A is said to be a *Dirichlet algebra* if the linear manifold $A + \bar{A}$ is norm dense in $C(\partial A)$.

THEOREM. *Every contractive Hilbert module over a Dirichlet algebra has a unique Šilov resolution.*

The class of function algebras for which the preceding result holds (not requiring uniqueness) is not known. Using a result of Ando [5] one knows that Šilov resolutions exist for contractive Hilbert modules over $A(\mathbf{D}^2)$ although uniqueness fails in this case. Moreover, using the example of Parrott [17] and the result of Varopoulos [24], one knows that the Šilov resolutions fail to exist for all contractive Hilbert modules for either $A(\mathbf{D}^n)$ or $A(\mathbf{B}^n)$ for $n \geq 3$, where $\mathbf{B}^n$ denotes the unit ball in $\mathbf{C}^n$. A recent result of Agler [2] shows that Šilov resolutions always exist for R(**A**), where **A** is an annulus in **C**.

There is a larger class of modules which might seem at first to be better adapted for resolutions than the class of Šilov modules which we have defined. That is, rather than considering submodules of modules over $C(\partial A)$ one could consider instead submodules of modules over $C(M_A)$. We will show that nothing is gained by this after we obtain a different characterization of the modules which possess a Šilov resolution.

If A is a function algebra contained in C(X), then a norm can be defined on the algebra $M_n(A)$ of $n \times n$ matrices with entries from A by setting

$$\|(\phi_{ij})\| = \sup_{x \in X} \|(\phi_{ij}(x))\|$$

where the norm on the right is the usual operator norm on $M_n(\mathbf{C})$. If (ξ^i) and (η^j) are unit vectors in $\mathbf{C}^n$, then

$$\sup_{x \in X} \left| \sum_{i,j=1}^{n} \phi_{ij}(x)\xi^i \overline{\eta^j} \right| = \sup_{x \in \partial A} \left| \sum_{i,j=1}^{n} \phi_{ij}(x)\xi^i \overline{\eta^j} \right|$$

since the function in absolute values lies in A. Therefore, the norm defined on $M_n(A)$ does not depend on the space X and is intrinsic.

If M is a bounded Hilbert module for A with bound K_1, then $M \otimes \mathbf{C}^n$ is a bounded Hilbert module for $M_n(A)$ with finite bound K_n and $K_1 \le K_n$ with strict inequality possible. (The norm on $M \otimes \mathbf{C}^n$ is the usual tensor product norm for Hilbert spaces with

$$\left\| \sum_{i=1}^{n} \phi_i \otimes e_i \right\| = \left\{ \sum_{i=1}^{n} \|\phi_i\|^2 \right\}^{\frac{1}{2}}$$

where $\{e_i\}_{i=1}^n$ is an orthonormal basis for $\mathbf{C}^n$.)

The Hilbert module M for A is said to be *completely contractive* if $K_n = 1$ for all positive integers n. The result of Arveson [6] characterizing completely contractive maps can be reformulated to yield:

THEOREM. *A Hilbert module over* A *has a Šilov resolution if and only if it is completely contractive.*

One direction of the theorem is transparent. Let

$$0 \leftarrow M_0 \leftarrow M_1$$

be an exact sequence of Hilbert modules over A, where the module maps are partial isometries. Then the sequence obtained by tensoring with $\mathbf{C}^n$

$$0 \leftarrow M_0 \otimes \mathbf{C}^n \leftarrow M_1 \otimes \mathbf{C}^n$$

is also exact with partially isometric module maps. If the bounds for $M_0 \otimes \mathbf{C}^n$ and $M_1 \otimes \mathbf{C}^n$ are $K_{0,n}$ and $K_{1,n}$, respectively, then $K_{0,n} \le K_{1,n}$. Therefore, if M_1 is completely contractive, then so must be M_0. Moreover, if A is a subalgebra of B, N is a completely contractive module for B and R a submodule of N for A, then R is completely contractive. Since all contractive modules over C(X) are completely contractive, we see that every module having a Šilov resolution is completely contractive. Moreover, the argument also proves the result promised earlier.

THEOREM. *Let M be a contractive Hilbert module over the function algebra* A, *R be a submodule for* A *of the contractive module N for* $C(M_A)$ *and assume that*

$$0 \leftarrow M \leftarrow R$$

is exact, where the maps are partially isometric module maps. Then M has a Šilov resolution.

4. QUASI-ŠILOV RESOLUTIONS

In the category of bounded Hilbert modules one should consider at least the following kind of resolution.

In a manner completely analogous to what was done in the contractive case, a bounded Hilbert module M over A is said to be *completely bounded* if

$$\sup K_n = K_\infty < \infty.$$

Using a result of Paulsen [18] one can prove:

THEOREM. *A Hilbert module M for A has a quasi-Šilov resolution if and only if it is completely bounded.*

Again, if a Hilbert module M_0 is dominated by a completely bounded Hilbert module M_1 for A, then M_0 is completely bounded. That is, if one has the sequence

$$0 \leftarrow M_0 \leftarrow M_1$$

which is (strict) exact (the module map X form M_1 to M_0 is onto) and M_1 is completely bounded, then M_0 is completely bounded. In this case one has

$$K_{0,\infty} \leq \|X\| K_{1,\infty}/m$$

where m is the lower bound for X.

5. HYPO-PROJECTIVE MODULES

In algebra one usually seeks resolutions in which the modules are either projective or injective. Let us consider the former and recall that the module P is said to be *projective* if for every diagram

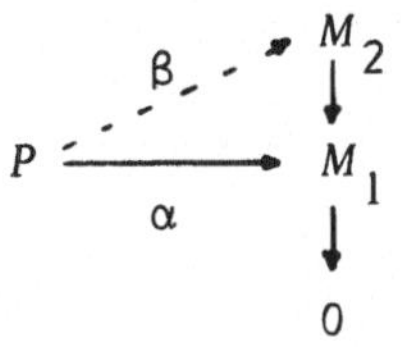

in the category, the map α lifts to the dotted map β. What does this mean in our setting? If the given maps are allowed to be arbitrary contractive module maps and the lifting is also required to be contractive, then only the zero module is projective as the following diagram shows

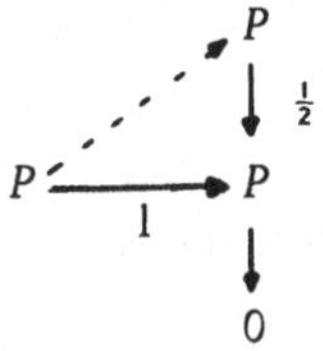

since the lifted map must be twice the identity which is not contractive unless $P = (0)$. A reasonable restriction is to require that the module mapping $M_2 \to M_1$ be partially isometric. In that case modules possessing this lifting property were said to be *hypo--projective* by Foiaş and the author [10] where it was proved that

THEOREM. *A contractive Hilbert module over* A(**D**) *is hypo-projective if and only if it is a Šilov module.*

Although this can be shown to follow from the lifting theorem of Sz.-Nagy and Foiaş [23], an independent proof of this is given in [10] which is essentially constructive. Conversely, one can obtain the lifting theorem as a consequence of the preceding theorem.

THEOREM. *If M and $\tilde{M}$ are contractive Hilbert modules over the function algebra* A, $X : M \to \tilde{M}$ *is a contractive module map and*

$$0 \leftarrow M \leftarrow P_0 \leftarrow P_1 \leftarrow 0$$

$$0 \leftarrow \tilde{M} \leftarrow \tilde{P}_0 \leftarrow \tilde{P}_1 \leftarrow 0$$

are hypo-projective resolutions, then there exist contractive module maps

$$Y_i : P_i \to \tilde{P}_i \quad \textit{from } i = 0,1$$

such that

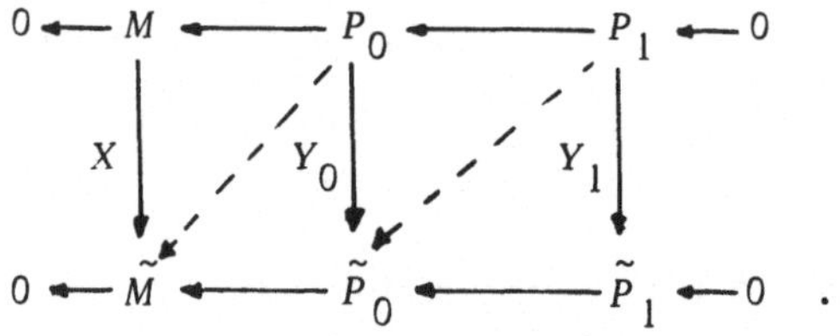

The proof consists in putting in the dotted arrows and then using the definition of the hypo-projectivity of P_0.

Since the lifting theorem is known to fail in many situations and, in particular, for the bidisk algebra, one obtains the following result.

COROLLARY. *Not every Šilov module over* $A(\mathbf{D}^2)$ *is hypo-projective.*

Partial results characterizing the hypo-projective modules of $A(\mathbf{D}^2)$ are given in [10]. All hypo-projective Hilbert modules for $A(\mathbf{D}^2)$ are Šilov modules. It is not known whether this is true for other function algebras.

If resolutions are to be useful in understanding Hilbert modules over a large class of function algebras, then Šilov modules are not the right class. On the one hand, it is too large since not all Šilov modules are hypo-projective, while on the other hand there are not enough Šilov modules to resolve all contractive modules. Moreover, because the class of Šilov modules is closed under the operation of taking submodule, no resolution of true length greater than one can exist. Clearly, the "homological dimension" of the algebras $A(\mathbf{D}^n)$ is greater than one for $n > 1$. We will not speculate further on how to solve these problems at this time but we do not believe that the situation is hopeless.

There is another way to proceed, however, and that is to consider bounded Hilbert modules and quasi-Šilov resolutions. Although the answers may be more difficult to obtain, the results may be more satisfactory. Even in this case, however, quasi-Šilov modules can not be the right class.

DEFINITION. The bounded Hilbert module M over the function algebra A is said to have a *quasi-Šilov resolution* if there exists an exact sequence

$$0 \leftarrow M \leftarrow Q_0 \leftarrow Q_1 \leftarrow 0$$

where Q_0 and Q_1 are quasi-Šilov modules over A and the maps are bounded module maps with closed range (that is, exactness is strictly interpreted).

The statement that all bounded Hilbert modules over A(**D**) have quasi-Šilov resolutions is easily seen to be equivalent to the statement that all such modules are similar to contractive ones. Hence, we will have nothing more to say about this for a general function algebra. We can, however, reformulate a recent result of Paulsen and the author [11] to yield the following for hypo-Dirichlet algebras. Recall that the function algebra A is said to be a *hypo-Dirichlet algebra* if the closure of the linear manifold $A + \bar{A}$ has finite codimension in $C(\partial A)$. Classical examples of such algebras are obtained by considering $R(\Omega)$ for Ω a finitely connected bounded domain in **C**.

THEOREM. *Every* ***contractive*** *Hilbert module for a hypo-Dirichlet algebra has a quasi-Šilov resolution.*

The result one would like to prove is that this is true for bounded Hilbert modules but as we stated above, this is unknown even for the disk algebra. The result of Agler [2] for the annulus is, of course, stronger but Agler's techniques do not seem to extend even to a two-holed domain.

6. HILBERT MODULES AND SPECTRAL SETS

We conclude with some observations on a related phenomena typified by von Neumann's theorem that the closed unit disk is a spectral set for contraction operators. If one proceeds naively, one would expect the annulus

$$\mathbf{A}_T = \{z \in \mathbf{C} : \|T^{-1}\|^{-1} \leq |z| \leq \|T\|\}$$

to be a spectral set for any invertible operator T. That it is not can be seen by considering various bilateral weighted shifts [6]. That it fails even for two by two matrices was shown by Misra [16]. In the language of modules, this means that although T makes H into a module over $\mathrm{Rat}(\mathbf{A}_T)$ by defining

$$r \cdot h = r(T)h \quad \text{for } h \text{ in } H$$

and each rational function r(z) with poles off $\mathbf{A}_T$, this can not always be extended to a contractive Hilbert modules for $R(\mathbf{A}_T)$. It is somewhat surprising, therefore, that it does extend to yield a *bounded* Hilbert module as was shown by Paulsen and the author in [11]. Moreover, the constant K in this case is always less than or equal to 3 and analogous results hold for any finitely connected domain Ω in **C**. Further, such a module is *similar* to a contractive Hilbert module for $R(\Omega)$ and does have a quasi-Šilov resolution.

It is unknown whether similar results hold for the polydisk algebra $A(\mathbf{D}^n)$ for $n \geq 3$. It has been shown [24] for each $n \geq 3$ that there exists an n-tuple $(T_1, T_2, \ldots, T_n)$ of commuting contraction operators on H such that the module multiplication defined by

$$ph = p(T_1, T_2, \ldots, T_n)h$$

for h in H and p(z) in $\mathbf{C}[z_1, \ldots, z_n]$ does not extend to a contractive Hilbert module for $A(\mathbf{D}^n)$. Moreover, the construction yields polynomials $p_n(z)$ such that

$$\|p_n(T_1, \ldots, T_n)\| / \|p_n(z_1, z_2 \ldots, z_n)\| \to \infty \quad \text{as } n \to \infty.$$

However, it is unknown whether for a fixed n-tuple $(T_1, T_2 \ldots, T_n)$ the module multiplication for $\mathbf{C}[z_1, \ldots, z_n]$ can be extended to yield a *bounded* Hilbert module for $A(\mathbf{D}^n)$. And if so, is it similar to a contractive one?

As we stated in the introduction, this paper consists largely in the reinterpretation of known results in the language of modules. In doing this we have raised many questions and perhaps have provided a framework in which to profitably place the few pieces of the puzzle we possess.

REFERENCES

1. **Abrahamse, M.B. ; Douglas, R.G. :** A class of subnormal operators related to multiply-connected domains, *Adv. Math.* **19**(1976), 106-148.

2. **Agler, J. :** Rational dilation on an annulus, *Indiana Univ. Math. J.*, to appear.

3. **Agrawal, O.P. ; Clark, D.N. ; Douglas, R.G. :** Invariant subspaces in the polydisk, *Pacific J. Math.*, to appear.

4. **Ahren, P.R. ; Clark, D.N. :** Invariant subspaces and analytic continuation in several variables, *J. Math. Mech.* **19**(1970), 963-969.

5. **Ando, T. :** On a pair of commutative contractions, *Acta Sci. Math. (Szeged)* **24**(1963), 88-90.

6. **Arveson, W.B. :** Subalgebras of C^*-algebras, *Acta Math.* **123**(1969), 141-224.

7. **Bonsall, F.F. ; Duncan, J. :** *Complete normed algebras*, Springer-Verlag, New York, Heidelberg, Berlin, 1973.

8. **Cowen, M.J. ; Douglas, R.G. :** On moduli for invariant subspaces, in *Topics in modern operator theory*, Birkhäuser Verlag, Basel, 1982, pp.65-73.

9. **Dixmier, J. :** *Les C*-algèbres et leurs representations*, Gauthier-Villars, Paris, 1964.

10. **Douglas, R.G. ; Foiaş, C. :** A homological view of dilation theory, *INCREST Preprint* **15**(1976).

11. **Douglas, R.G. ; Paulsen, V.I. :** Completely bounded maps and hypo-Dirichlet algebras, *Acta Sci. Math. (Szeged)*, to appear.

12. **Gamelin, T.W.** : *Uniform algebras,* Prentice-Hall, Englewood Cliffs, New Jersey, 1969.

13. **Halmos, P.R.** : Ten problems in Hilbert space, *Bull. Amer. Math. Soc.* **70**(1970), 887-933.

14. **Helson, H. ; Lowdenslager, D.** : Invariant subspaces, in *Proc. Int. Symp. Linear Spaces*, Jerusalem, 1960, MacMillan (Pergamon) 1961, pp.251-262.

15. **Hoffman, K.** : *Banach spaces of analytic functions,* Prentice-Hall, Englewood Cliffs, New Jersey, 1962.

16. **Misra, G.** : Curvature inequalities and extremal properties of bundle shifts, *J. Operator Theory* **11**(1984), 305-317.

17. **Parrott, S.** : Unitary dilations for commuting contractions, *Pacific J. Math.* **34**(1973), 481-490.

18. **Paulsen, V.I.** : Every completely polynomially bounded operator is similar to a contraction, *J. Functional Analysis* **55**(1984), 1-17.

19. **Shields, A.L.** : Weighted shift operators and analytic function theory, in *Math. Surveys Amer. Math. Soc.* **13**(1974), 49-128.

20. **Suciu, I.** : *Function algebras,* Editura Academiei, Bucharest, 1973.

21. **Sz.-Nagy, B.** : On uniformly bounded linear transformations in Hilbert space, *Acta Sci. Math. (Szeged)* **11**(1947), 152-157.

22. **Sz.-Nagy, B.** : Sur les contractions de l'espace de Hilbert, *Acta Sci. Math. (Szeged)* **15**(1953), 87-92.

23. **Sz.-Nagy, B. ; Foiaş, C.** : *Harmonic analysis of operators on Hilbert space,* American Elsevier, 1970.

24. **Varopoulos, N Th.** : On an inequality of von Neumann and an application of the metric theory of tensor products to operator theory, *J. Functional Analysis* **16**(1974), 83-100.

R.G.Douglas

Department of Mathematics
State University of New York
Stony Brook, NY 11794
U.S.A.

Operator Theory:
Advances and Applications, Vol.17

DIFFERENTIAL EQUATION OF AN INNER FUNCTION. II

Henry Helson

1.

H is separable Hilbert space, $M = M(t)$ a function on **R** whose values are self-adjoint operators in **H**. We assume that M is norm-continuous and bounded. The differential equation

(1) $$F' = iMF$$

is soluble with arbitrary value $F(0)$ in **H**, and F has constant norm. Let U be a fundamental solution: U is an operator function satisfying the equation, normalized to have unitary values with $U(0) = I$ (the identity operator in **H**).

B_o is the operator in L^2 (the space of vector functions on **R** with square-summable norms) defined by

(2) $$B_oF = -iF',$$

on the domain consisting of all absolutely continuous F in L^2 such that F' is in L^2. Then B_o is self-adjoint, and $B = B_o - M$ is self-adjoint on the same domain.

B_o and B are related; an easy calculation shows that

(3) $$UB_oU^* = B_o - M = B.$$

Thus B, like B_o, has continuous spectrum.

The positive subspace of L^2 for B_o is the Hardy space H^2 of vector functions with analytic extensions to the upper half-plane. From (3), the corresponding space for B is $U \cdot H^2$ (the set of all UF where F is in H^2). This is an *invariant subspace* of L^2 in the sense of [2]: if e_u is the function on **R** with values **exp** iut, then for each F in the subspace and for each positive u, e_uF lies in the subspace. U is determined by this subspace up to multiplication on the right by a constant unitary function.

U is an *inner function* if $U \cdot H^2$ is contained in H^2. That is, U has an extension to the upper half-plane that is analytic and bounded, and (under our hypotheses) continuous on the closed half-plane.

The *invariant subspace problem* (whether every continuous linear operator in

Hilbert space of dimension greater that 1 has a proper closed invariant subspace) has a positive answer if and only if this statement is true [2]: *for every inner function* U, *with trivial exceptions, there is an invariant subspace* $V \cdot H^2$ *lying properly between* $U \cdot H^2$ *and* H^2. The exceptional case is where $U \cdot H^2$ has codimension one in H^2.

Let U and V be any unitary functions. $U \cdot H^2$ is contained in $V \cdot H^2$ if and only if $W = V^* U$ is inner. If U is inner and differentiable, and V inner, then necessarily V and W are differentiable. In order to settle the invariant subspace problem it suffices to factor inner functions that are actually analytic on a neighborhood of the real axis, in which case the factors V, W have the same property.

Thus the invariant subspace problem leads to the following question: *for which self-adjoint operator functions* M, *under restrictions such as those mentioned above, is it the case that every solution* F *of the differential equation* (1) *has a bounded analytic extension to the upper half-plane*? Aside from the invariant subspace problem, this is an attractive question in the field of ordinary differential equations. A first paper on the subject appeared a number of years ago [3]. In this paper those results are summarized, and new ones are presented.

2.

THEOREM 1. *Let* U *and* V *be unitary functions on the line, with bounded coefficient functions* M *and* N, *respectively.* $U \cdot H^2$ *is contained in* $V \cdot H^2$ *if and only if*

(4) $$f(B_o - M) \leq f(B_o - N)$$

for each bounded, increasing real function f *on* **R**.

Take $f(t) = 0$ for $t < 0$, $= 1$ for $t > 0$. Then $f(B_o - M)$ is the orthogonal projection of L^2 on $U \cdot H^2$, and similarly for N and V. Thus (4), for this function alone, is equivalent to saying that $U \cdot H^2$ is contained in $V \cdot H^2$.

Let P and Q be the spectral measures of $B_o - M$, $B_o - N$, respectively. Denote the operation of multiplication by e_u in L^2 by S_u. Then we have easily

(5) $$S_u P_v S_{-u} = P_{u+v}, \quad S_u Q_v S_{-u} = Q_{u+v}.$$

From this fact it follows that if (4) holds when f is the characteristic function of $(0, \infty)$, then it holds for the characteristic function of (u, ∞) for any real u. Furthermore adding a constant to f does not affect the relation. Any increasing bounded function f can be approximated by positive linear combinations of constants and such characteristic functions, so (4) is true for all increasing bounded functions. This proves the theorem.

COROLLARY. *If* $U \cdot H^2$ *is contained in* $V \cdot H^2$ *then* $M(t) \geq N(t)$ *for each* t. *In particular, if* U *is inner,* $M(t) \geq 0$ *for all* t.

Approximate the function $f(u) = u$ by bounded increasing functions; we find that $(B_0 - M) \leq (B_0 - N)$ as operators in L^2. It follows that $M \geq N$ as operators in L^2, and therefore also in the pointwise sense.

The necessary condition given by the corollary is far from being sufficient.

The results to this point were proved in [3].

THEOREM 2. *In order to have* $U \cdot H^2 \subset V \cdot H^2$ *it is necessary and sufficient that* $f(B_0 - M) \leq f(B_0 - N)$ *for the function* $f(t) = 0$ $(t \leq 0)$, $= t$ $(t > 0)$.

Approximation by bounded increasing functions shows that the inequality of operators is necessary for the inclusion.

Suppose the inequality holds. Denote the complement of H^2 in L^2 by K^2 (the space of square-summable vector functions with analytic extensions to the lower half--plane). Let F be any element of $V \cdot K^2$. Thus F belongs to the subspace of L^2 on which $B_0 - N$ is non-positive, where $f(B_0 - N)$ is zero. Hence

(6) $$(f(B_0 - M)F\,,\, F) \leq (f(B_0 - N)F\,,\, F) = 0.$$

Since $f(B_0 - M)$ is a positive operator, this implies that F belongs to its negative subspace, which is $U \cdot K^2$. Hence $V \cdot K^2 \subset U \cdot K^2$, which is equivlent to the inclusion of the theorem.

THEOREM 3. *The differentiable unitary function* U *is inner if and only if*

(7) $$|B_0 - M| \leq |B_0| + M\,.$$

For any self-adjoint operator T, let T^+ be the positive part of T. The last theorem says that U is inner if and only if

(8) $$(B_0 - M)^+ \leq B_0^+\,.$$

Multiply by 2 and subtract $B_0 - M$ from both sides; we find (7).

It is tempting to square the inequality (7). On simplifying we find

(9) $$B_0^+ M + M B_0^+ \geq 0\,.$$

Loewner's theorem on monotone matrix functions says that (9) implies (7), so that any self-adjoint function M satisfying (9) is associated with an inner function. Unfortunately only constant positive M do, so that no interesting inner functions are obtained in this way.

3.

Let M be a bounded norm-continous self-adjoint function on the line, and U the unitary function satisfying $U' = iMU$ with $U(0) = I$. We show now that if U is inner, then M has an analytic extension to a strip containing the real axis; or equivalently, that U is analytic and invertible in such a strip.

THEOREM 4. *If* $\|M(x)\| \leq 1$ *for real* x, *then* M *has an analytic extension to the strip* $|\mathbf{Im}\, z| \leq 1$, *and is bounded in each interior strip; and* U *is analytic and invertible in that strip.*

As we showed in [3], $U'(z)$ is bounded in the upper half-plane by the bound of M on the line. Hence for real x and positive t

$$\|U(x+it) - U(x)\| = \left\|\int_0^t iU'(x+iu)\,du\right\| \leq t. \tag{10}$$

Since $U(x)$ is unitary, $U(x+it)$ is invertible for $0 \leq t < 1$. It follows that $iM = U'U^{-1}$ is analytic for $0 < \mathbf{Re}\, z < 1$, and continuous on the closed half-plane. By analytic continuation it is analytic in the similar strip below the axis. (10) gives a bound for U^{-1} depending only on t, so that M is also bounded for $0 \leq \mathbf{Re}\, z < t < 1$, and by symmetry in the lower strip as well.

To complete the proof we show that U is invertible also on the line $\mathbf{Im}\, z = 1$. Suppose that U is not invertible at i. In (10) we take $x = 0$, $t = 1$. The left side is at least 1 (otherwise $U(i)$ would be invertible); it follows that $\|U'(iu)\| = 1$ for $0 \leq u \leq 1$. Now $U'(i)$ is represented by its Poisson integral:

$$U'(i) = \pi^{-1} \int_{-\infty}^{\infty} U'(x)(1+x^2)^{-1}\,dx\,. \tag{11}$$

The norm of each side is 1. Hence given $\varepsilon > 0$, we can find elements ϕ, ψ of norm 1 in **H** such that

$$\pi^{-1}\left| \int_{-\infty}^{\infty} (U'(x)\phi, \psi)(1+x^2)^{-1}\,dx\right| \geq 1 - \varepsilon\,. \tag{12}$$

This is possible only if, given ε and $a < b$, there are ϕ and ψ such that

$$(13) \qquad |((U(b) - U(a))\phi , \psi)| = |\int_a^b (U'(x)\phi , \psi)\,dx| \geq b - a - \varepsilon .$$

But U(b), U(a) are unitary, so this is impossible if $b - a > 2$. This shows that U(i) was invertible, and completes the proof.

THEOREM 5. *Assume as above that* M *is bounded and continuous on the real axis and* U *is inner. There is an operator function* A *analytic and bounded on the closed upper half-plane such that* $M(x) = A(x)A^*(x)$ *for real* x.

We suppose that M(x) has trivial null space for each real x. The result of the next section shows that this restriction is immaterial.

Let N(x) be the positive square root of M(x) for each x. Denote by **N** the smallest closed subspace of L^2 containing NF for all F in K^2, and by $\mathbf{N}_t$ the set of all functions e_tF for F in **N**. Then $(\mathbf{N}_t)$ is a nested family of closed subspaces of L^2, increasing with t.

LEMMA. *The intersection of all* $\mathbf{N}_t$ *is* (0), *and their union is dense in* L^2.

Let F and G be any elements of H^2 and of K^2, respectively. Then

$$(14) \qquad i\int_{-\infty}^{\infty} (NUF , NG)\,dx = i\int_{-\infty}^{\infty} (MUF , G)\,dx = \int_{-\infty}^{\infty} (U'F , G)\,dx .$$

Now U'F is in H^2, because U' is analytic and bounded in the upper half-plane. Since H^2 and K^2 are complements of each other, the last integral vanishes. That is, $NU{\cdot}H^2$ is orthogonal to **N**.

Suppose H is in the intersection of all $\mathbf{N}_t$. Then e_tH is in **N** for every t and we have

$$(15) \qquad 0 = \int_{-\infty}^{\infty} e^{-itx}(NUF , H)\,dx = \int_{-\infty}^{\infty} e^{-itx}(F , U^*NH)\,dx$$

for all F in H^2 and all real t. Hence the inner product (F , U^*NH) vanishes a.e. . This is true in particular if $F = f\phi$, where f is in the scalar space H^2 and ϕ is any vector in **H**. We deduce easily that U^*NH is the null function. But U is unitary and N has trivial null space, so H is null. That is, the intersection of the $\mathbf{N}_t$ is (0).

The closure of the union of the $\mathbf{N}_t$ is invariant under multiplication by all exponentials. Such subspaces consist of all functions H in L^2 such that H(x) lies in a certain closed subspace J(x) of **H** for a.e. x [2]. The range of N(x) is dense at each point, so that $\mathcal{J}(x)$ must be all of **H** for each x, and the second assertion of the lemma is proved.

The lemma furnishes the hypotheses for the basic structure theorem for invariant subspaces [2]: there is a unitary function V on the line such that $\mathbf{N} = V \cdot K^2$. Thus $N \cdot K^2$ is dense in $V \cdot K^2$, or equivalently $V^* N \cdot K^2$ is dense in K^2.

This means the bounded operator function V^*N is conjugate-analytic. Call it A^*; then A is bounded and analytic. And we have, on **R**,

$$AA^* = NVV^*N = N^2 = M. \tag{16}$$

Furthermore $A^* \cdot K^2$ is contained and dense in K^2; it follows that $A \cdot H^2$ is contained in H^2 and that

$$F \text{ in } L^2 \text{ and } AF \text{ in } H^2 \text{ imply } F \text{ in } H^2. \tag{17}$$

From the first statement about A we conclude that A has a bounded analytic extension to the upper half-plane, and the theorem is proved. We shall discuss the property (17) presently.

The factoring asserted in the theorem is on the real line, but it extends into the complex plane as far as M is analytic. Observe that $AA^*U = -iU'$ is bounded and analytic. By (17), A^*U is itself analytic. Hence A^* has an analytic extension B throughout the region in the upper half-plane where U is invertible. This domain includes a strip above the real axis, by the last theorem. Similarly, A is analytic across the real axis. The equality $M(z) = A(z)B(z)$, valid on the real axis, persists throughout the region where M is analytic.

An analytic operator function was called *outer* [2] if $A \cdot H^2$ is dense in H^2. We have proved that $A^* \cdot K^2$ is dense in K^2; thus A^* is outer in the conjugate-analytic sense. It does not follow that A is outer; the dual assertion is that A satisfies the condition (17). For scalar functions, these properties are equivalent. In the case of vector functions they are not the same, but each has a claim to be the definition of outer function.

It is not obvious that M need have a factoring B^*B with B analytic.

4.

We continue with a simple proof, due to D.Sarason, of a theorem of S.Campbell [1]. We are grateful to Sarason for permission to reproduce the proof.

THEOREM 6. (Campbell). *With* M *bounded and* U *inner, as hitherto, the null spaces of* $U'(x)$ *and of* $M(x)$ *are independent of* x.

Say $U(0) = I$. Suppose that $U'(0)\phi = 0$ for some ϕ of norm 1 in **H**. We shall show that $U'(x)\phi$ and $M(x)\phi$ vanish identically. The full statement follows easily from this fact.

The Taylor coefficients of $U(z)$ about 0 are operators U_k. Thus we have the expansion

$$(U(z)\phi, \phi) = 1 + z(U_1\phi, \phi) + z^2(U_2\phi, \phi) + \dots \quad (18)$$

The second term vanishes by hypothesis. Let n be the first integer, if any, such that $U_n\phi$ is not orthogonal to ϕ. Then

$$(U(z)\phi, \phi) = 1 + z^n(U_n\phi, \phi) + O(z^{n+1}). \quad (19)$$

This quantity has modulus at most 1 for z in the upper half-plane, because U is inner. But $n > 1$, so this is only possible if $(U_n\phi, \phi) = 0$, contrary to assumption. Therefore all the terms after the first vanish, and $U(z)\phi$ must equal ϕ for all z in the upper half-plane.

Hence $U'(z)\phi$ vanishes in the half-plane, and $0 = U'(x)\phi = iM(x)U(x)\phi = iM(x)\phi$ for all real x. This concludes the proof.

If the null space of M is a non-trivial subspace of **H**, then $U(x)$ is constant on that subspace, and U can be regarded as an inner function on the complementary subspace. This makes it possible to reduce most problems (such as that of the last section) to the case where the null space of $M(x)$ is trivial for all x.

5.

Let **M** be the invariant subspace $U{\cdot}H^2$, where U is a unitary function with coefficient function M, and $B = B_o - M$, as above.

THEOREM 7. *M is contained in* H^2 *if and only if* $\exp(-tB_o) \leq \exp(-tB)$ *for all positive* t.

If F is any element of **M** it belongs to the positive subspace of B, and therefore $\|[\exp -tB]F\|$ is bounded for positive t. The condition of the theorem implies that $\|[\exp -tB_o]F\|$ is bounded, which is possible only if F is in the positive subspace of B_o, that is in H^2. Thus H^2 contains **M**.

If H^2 contains **M** the order relation follows from Theorem 1, and the theorem is proved.

For F in H^2, or with spectral energy sufficiently weak on the negative axis, $[\mathbf{exp}\,{-tB_0}]F(x) = F(x + it)$ is the analytic extension of the boundary function F into the upper half-plane. Differentiation with respect to t gives

$$(20) \qquad (\partial/\partial t)(F(x + it)) = -B_0 e^{-tB_0} F = i(\partial/\partial x)(F(x + it)), \quad \text{or} \quad F_x + iF_t = 0,$$

the Cauchy-Riemann equation for a complex function. In exactly the same way, the function $G(x + it) = [\mathbf{exp}\,{-tB}]F(x)$ satisfies the differential equation

$$(21) \qquad G_x + iG_t = iM(x)G ,$$

a perturbed Cauchy-Riemann equation. (The differential operator on the left is the famous $\bar\partial$.) The inequality of Theorem 7 means that if a boundary function F admits an extension to the upper half-plane satisfying (21) and bounded in the norm of L^2 on horizontal lines, then the analytic extension of F is similarly bounded. This seems to be an interesting description of the coefficient functions M that are associated to inner functions.

The operator inequality of the last theorem is equivalent to

$$(22) \qquad \| e^{-tB_0} e^{tB} \| \leq 1, \quad t > 0.$$

Let $H(t) = (\mathbf{exp}\,{-tB_0})(\mathbf{exp}\; tB)$. Differentiation shows that H satisfies the differential equation

$$(23) \qquad H'(t) = -B_0 H(t) + H(t)B, \quad H(0) = I .$$

The boundedness of the solution of this equation for positive t is equivalent to (22), and to the condition of Theorem 7. We know that boundedness at $t = 0$ merely means that $M(x) \geq 0$; we would like to interpret boundedness for large t. If we set $K(t) = H(1/t)$, (23) becomes

$$(24) \qquad t^2 K'(t) = B_0 K(t) - K(t)B .$$

For each t, H(t) is a multiplication operator in L^2 (so the same is true of K(t)). Indeed from (3) we have

$$(25) \qquad H(t) = e^{-tB_0} U e^{tB_0} U^* = U(x + it)U^*(x).$$

Moreover if U is analytic and invertible at ∞, which is the case when U arises from a strict contraction, then H(t) has the limit $U(\infty)U^*(x)$, a unitary operator which is the initial value K(0) for (24).

The multiplication operator $K(t)$ is $K(t,x)$, operating in **H** for each (t,x) with $t > 0$. Simplification of (24) leads to

(26) $$it^2K_t - K_x = iKM ,$$

analogous to (21). For which M has (26) bounded solutions for $t > 0$?

6.

Suppose that U is a smooth inner function that is not factorable. What consequences can be deduced about M ?

The product MU is bounded and analytic, because U' is analytic. That is, MUF is in H^2 for every F in H^2; equivalently, $M(U \cdot H^2) \subset H^2$. Thus the set of all F in H^2 such that MF is in H^2 is a closed subspace of H^2 that contains $U \cdot H^2$. If U is not factorable, this subspace coincides with H^2 or with $U \cdot H^2$. The first possibility would imply that M is an analytic operator function; this is the case only if M is constant. We reject this case (which includes the exceptional inner functions mentioned above), and conclude: *if* U *is not factorable and* M *non-constant, then* $U \cdot H^2$ *consists exactly of the functions* F *in* H^2 *such that* MF *is analytic.* Roughly, $M(z)$ has singularities in the upper half-plane of the same strength as $U(z)^{-1}$. Perhaps therefore U' is outer in some sense, but no theorem of this sort has been proved.

REFERENCES

1. **Campbell, S.** : Inner functions analytic at a point, *Illinois J. Math.* **16** (1972), 651-652.
2. **Helson, H.** : *Lectures on invariant subspaces*, Academic Press, 1964.
3. **Helson, H.** : The differential equation of a inner function, *Studia Math.* **35** (1970), 311-321.

H. Helson
Department of Mathematics
University of California
Berkeley, California 94720
U. S. A.

Operator Theory:
Advances and Applications, Vol.17

ON A CLASS OF UNITARY OPERATORS IN KREIN SPACE

Peter Jonas

INTRODUCTION

Let H be a Krein space and U a unitary operator in H. Assume that for a certain open subset Γ of the unit circle $\mathbf{T}$ no point of Γ is accumulation point of the nonunitary spectrum $\sigma(U) \setminus \mathbf{T}$ of U and that every point of Γ can be connected with 0 and ∞ by curves in the resolvent set $\rho(U)$. Such an operator is called definitizable over Γ, if, roughly speaking, it has a spectral function over Γ (in the sense of the theory of definitizable operators). This class of operators, of course, contains the definitizable unitary operators in H. The main concern of this paper is to study some aspects of the behaviour of the spectral functions under compact perturbations of the corresponding operators within this class.

Unitary operators in Krein space which are definitizable over an open subset of $\mathbf{T}$ naturally arise from perturbations of some unitary operators with a simple structure. In this connection we first mention the following result of H.Langer ([13]) for bounded selfadjoint operators A in a Krein space (there is an analogous result for unitary operators which is proved in the same way): Assume that there exists a fundamentally reducible bounded selfadjoint operator A_0 such $A - A_0$ belongs to the Macaev ideal $\mathbf{S}_\omega$. Then A has a spectral function over every open interval containing no point of $\sigma(A_0 | H_+) \cap \sigma(A_0 | H_-)$, where $H = H_+ \dot{+} H_-$ is a fundamental decomposition of H with $A_0 H_\pm \subset H_\pm$. In the special case of a definitizable "unperturbed" operator A_0 this result is slightly generalized in Proposition 2.5 (for unitary operators). We prove that a unitary operator in H obtained from a definitizable unitary operator U_0 with $c_\infty(U_0) = c_{r,\infty}(U_0)$ (see Section 2.2) by a perturbation of class $\mathbf{S}_\omega$ is definitizable over $\mathbf{T} \setminus c_\infty(U_0)$.

Section 1 of this paper is devoted to some preliminary material on function spaces and a special class of bounded operators. In Section 2 we give the definition of unitary operators definitizable over an open subset of $\mathbf{T}$ and construct their spectral function (via a functional calculus). These considerations contain an approach to the functional calculus and the spectral function for definitizable unitary operators different from the original one ([12]), see also [15] (selfadjoint operators)). In Section 3 we prove that for two unitary operators U_1 and U_2 definitizable over Γ with compact difference $U_1 - U_2$ the signs of the corresponding spectral functions coincide on neighbourhoods of almost all points of Γ. We admit that U_1 and U_2 act in different Krein spaces with the same underlying linear space such that their Gram operators have compact difference. Further, some stability properties of the spectral functions are proved. These results (which are new also for definitizable operators) supplement some results of [8] and [7] and have points of contact with [16]. The considerations of Section 3 have some consequences for the investigation of trace formulas for unitary operators, which will be dealt with in a subsequent paper.

We mention that by means of the Cayley transform all definitions and assertions can be carried over to selfadjoint operators in Krein space. Then in the perturbation considerations one has to assume that the difference of the resolvents of the selfadjoint operators is compact.

1. PRELIMINARIES

1.1. The spaces A_K^p and their duals. Let $\mathbf{C}$ denote the complex plane and $\bar{\mathbf{C}}$ the extended complex plane. Set $\mathbf{D} := \{z \in \mathbf{C} : |z| < 1\}$, $\bar{\mathbf{D}} := \{z \in \mathbf{C} : |z| > 1\} \cup \{\infty\}$. In this section K denotes a compact subset of $\mathbf{C}$ with the following property.

(*K*) $\mathbf{D} \setminus K$ and $\bar{\mathbf{D}} \setminus K$ are simply connected domains of $\bar{\mathbf{C}}$, and $0 \notin K$.

Denote by A_K^p, $p = 0, 1, \ldots, \infty$, the linear space of complex functions defined on $K \cup \mathbf{T}$ which are C^p functions on $\mathbf{T}$ and locally holomorphic on K. The spaces denoted by the same symbols in [4] are special cases of those considered here. We equip A_K^p, $p = 0, 1, \ldots$, with a locally convex topology in the following way.

Assume first that $K \neq \emptyset$. Let (O_n), $n = 1, 2, \ldots$, be a descending sequence of bounded open subsets of $\mathbf{C}$ such that for every $n = 1, 2, \ldots$ the following holds:

(i) $K \subset \bar{O}_{n+1} \subset O_n$.

(ii) O_n lies in a $1/n$-neighbourhood of K.

(iii) O_n is a union of a finite number of smooth domains and $\bar{O}_n$ fulfils the condition (*K*).

Denote by $D_{K,n}^p$, $n = 1, 2, \ldots$, the linear subset of A_K^p of functions f which can be extended analytically to O_n such that the functions $f^{[\nu]}$, $0 \leq \nu \leq p$, defined by

$$f^{[0]}(z) := f(z), \quad f^{[\nu]}(z) := iz(df^{[\nu-1]}/dz)(z), \qquad \nu \geq 1,\ z \in O_n,$$

have continuous boundary values on ∂O_n. If $z \in \mathbf{T} \cap O_n$ we have

$$f^{[\nu]}(z) = (d^\nu/d\theta^\nu) f(e^{i\theta})\,\big|_{e^{i\theta} = z}, \qquad 0 \leq \nu \leq p.$$

For $z \in \mathbf{T} \setminus K$ we define $f^{[\nu]}(z)$ by the latter relation. The linear space $D^p_{K,n}$ with the norm

$$\|f\|^p_n := \sup_{x \in O_n \cup \mathbf{T},\ \nu \leq p} |f^{[\nu]}(x)|$$

is a Banach space. We furnish $A^p_K = \bigcup_{n=1}^{\infty} D^p_{K,n}$ with the topology of the inductive limit of the spaces $D^p_{K,n}$, $n = 1, 2, \ldots$.

We set $A^p_{\emptyset} := C^p(\mathbf{T})$, $p = 0, 1, \ldots$, and equip this space with the norm $\sup_{x \in \mathbf{T},\ \nu \leq p} |f^{[\nu]}(x)|$.

The space of locally holomorphic functions on a compact subset F of $\bar{\mathbf{C}}$ provided with the usual topology (see [10]) is denoted by H(F). The same symbol H(G) is used for the space of locally holomorphic functions on an open subset G of $\bar{\mathbf{C}}$, and we write $f \in H_o(G)$ if $f \in H(G)$ and $f(\infty) = 0$ in the case when $\infty \in G$.

It is easy to see that $H(K\ \ \mathbf{T})$ is dense in A^p_K, $p = 0, 1, \ldots$. Evidently, the imbedding of $H(K \cup \mathbf{T})$ in A^p_K is continuous. Therefore, $(A^p_K)'$ can be regarded as subspace of $H'(K \cup \mathbf{T})$. Denote the duality between A^p_K and $(A^p_K)'$ ($H(K \cup \mathbf{T})$ and $H'(K \cup \mathbf{T})$) by u.f, $u \in (A^p_K)'$, $f \in A^p_K$ (resp. $u \in H'(K \cup \mathbf{T})$, $f \in H(K \cup \mathbf{T})$).

To every $u \in H'(K \cup \mathbf{T})$ there corresponds a complex function $\tilde{u}$ defined on $\bar{\mathbf{C}} \setminus (K \cup \mathbf{T})$ by $\tilde{u}(\lambda) := u.f_\lambda$, where $f_\lambda(z) := (2\pi)^{-1} z(z - \lambda)^{-1}$, $\lambda \in \bar{\mathbf{C}} \setminus (K \cup \mathbf{T})$. By a result of G.Kothe ([9]) the mapping $u \to \tilde{u}$ defined on $H'(K\ \ \mathbf{T})$ is a bijection onto $H_o(\bar{\mathbf{C}} \setminus (K \cup \mathbf{T}))$ and for every $f \in H(K \cup \mathbf{T})$ we have

$$u.f = - \int_C f(\lambda)\tilde{u}(\lambda)(d\lambda/i\lambda), \tag{1.1}$$

where C is the oriented boundary of a smooth ring domain containing $K \cup \mathbf{T}$ such that f is defined on its closure.

PROPOSITION 1.1. *Assume that* $\mathbf{T} \setminus K \neq \emptyset$. *Then the following holds:*

(i) *If* $u \in (A^m_K)'$, $m \geq 0$ *integer, then for every closed arc* $\gamma \subset \mathbf{T} \setminus K$ *there exists a constant* N *such that*

$$|\tilde{u}(re^{i\theta})| \leq N|1 - r|^{-(m+1)}$$

for all $e^{i\theta} \in \gamma$ *and all* $r \neq 1$ *from some neighbourhood of* 1.

(ii) *If* $u \in H'(K \cup \mathbf{T})$, $m \in [1, \infty)$, *and for every closed arc* $\gamma \subset \mathbf{T} \setminus K$ *there exists a constant* N *such that*

$$|\tilde{u}(re^{i\theta})| \leq N|1 - r|^{-m}$$

for all $e^{i\theta} \in \gamma$ *and all* $r \neq 1$ *from some neighbourhood of* 1, *then* $u \in (A_K^{[m]+1})'$, *where* $[\,\cdot\,]$ *denotes the integral part.*

PROOF. The proof is similar to that of Hilfssatz 1.4 in [4]. One easily verifies the assertion (i) in the special case when $K = \emptyset$. Then (i) follows by a partition of unity.

Let $u \in H'(K \cup \mathbf{T})$ and, hence, $u_0 := u - (2\pi)^{-1}u.\mathbf{1}$ fulfil the assumptions of (ii). We have $\tilde{u}_0(0) = 0$. For all positive integers k we define (see [4])

$$(u_0^{[o]})_i^\sim(\mu) := \tilde{u}_0(\mu), \quad (u_0^{[-k]})_i^\sim(\mu) := \int_0^\mu (u_0^{[-k+1]})_i^\sim(\zeta)(d\zeta/i\zeta), \qquad \mu \in \mathbf{D} \setminus K,$$

and

$$(u_0^{[o]})_e^\sim(\mu) := \tilde{u}_0(\mu), \quad (u_0^{[-k]})_e^\sim(\mu) := \int_\infty^\mu (u_0^{[-k+1]})_e^\sim(\zeta)(d\zeta/i\zeta), \qquad \mu \in \bar{\mathbf{D}} \setminus K.$$

For $f \in H(K \cup \mathbf{T})$ and a curve C as in (1.1) we have

$$u_0.f = -\int_C f(\lambda)\tilde{u}_0(\lambda)(d\lambda/i\lambda) = -(-1)^q \int_{C_i} f^{[q]}(\zeta)(u_0^{[-q]})_i^\sim(\zeta)(d\zeta/i\zeta) - (-1)^q \int_{C_e} f^{[q]}(\zeta)(u_0^{[-q]})_e^\sim(\zeta)(d\zeta/i\zeta),$$

where $C_i = C \cap \mathbf{D}, C_e = C \cap \bar{\mathbf{D}}$. On account of the assumption of (ii) the functions $(u_0^{[-q]})_i^\sim$ and $(u_0^{[-q]})_e^\sim$ with $q = [m]+1$ possess continuous boundary values on Γ. Hence for sufficiently large n we can replace the curves C_i and C_e by the boundaries C_i' and C_e' of $\mathbf{D} \setminus \bar{O}_n$ and $\bar{\mathbf{D}} \setminus \bar{O}_n$, respectively, with converse orientations:

$$(1.2) \qquad u_0.f = -(-1)^q \int_{C_i'} f^{[q]}(\zeta)(u_0^{[-q]})_i^\sim(\zeta)(d\zeta/i\zeta) - (-1)^q \int_{C_e'} f^{[q]}(\zeta)(u_0^{[-q]})_e^\sim(\zeta)(d\zeta/i\zeta).$$

Therefore, u_0 and, hence, u are continuous with respect to A_K^q. We remark that the relation (1.2) holds also for $f \in A_K^q$. The proposition is proved.

The following class of functions will be needed below. Assume that $\mathbf{T} \setminus K \neq \emptyset$ and let γ be a finite union of open subarcs of $\mathbf{T}$ such that $K \cap \mathbf{T} \subset \gamma$. Then $X_K^\infty(\gamma)$ denotes the set of functions $f \in A_K^\infty$ with the following properties:

(a) $f(K \cup \mathbf{T}) = \mathbf{T}$ and $f|\mathbf{T}$ preserves orientation on every arc where it is not constant.

(b) On every component of $K \cup \gamma$ the function f is equal to a constant.

1.2. A class of operators. Let H be a complex Hilbert space. Throughout this section Γ is an open subset of $\mathbf{T}$.

Assume first that $\Gamma \neq \mathbf{T}$. A bounded operator T in H is said to belong to the class $S^m(\Gamma)$, $m \in [1,\infty)$, if the following holds.

(i) There exists a compact set $K \subset \mathbf{C}$ with $\mathbf{T} \setminus K = \Gamma$ which satisfies condition (K) such that the spectrum $\sigma(T)$ of T is contained in $\mathbf{T} \cup K$.

(ii) For every closed arc $\gamma \subset \Gamma$ there exists a constant N such that

$$\|(T - re^{i\theta}I)^{-1}\| \leq N|1 - r|^{-m}$$

for all $e^{i\theta} \in \gamma$ and all $r \neq 1$ from some neighbourhood of 1.

The collection of compact sets satisfying (i) for a given bounded operator T is denoted by $K(\Gamma;T)$. By $K_s(\Gamma;T)$ we denote the collection of all sets of $K(\Gamma;T)$ which lie symmetrically with respect to $\mathbf{T}$.

PROPOSITION 1.2. *The Riesz-Dunford functional calculus for an arbitrary* $T \in S^m(\Gamma)$ *can be extended by continuity to each of the algebras* A_K^p, $p \geq [m] + 1$, $K \in K(\Gamma;T)$.

Moreover, for arbitrary $x,y \in H$, $f \in A_K^p$ *we have*

$$(1.3)\quad \begin{aligned}(f(T)x,y) &= -(2\pi)^{-1}\lim_{r\uparrow 1}\{\int_{C_i'} f(\zeta)(T(T - r\zeta I)^{-1}x,y)(d\zeta/i\zeta) + \int_{C_e'} f(\zeta)(T(T - r^{-1}\zeta I)^{-1}x,y)(d\zeta/i\zeta)\} = \\ &= -(2\pi i)^{-1}\lim_{r\uparrow 1}\{\int_{C_i'} f(\zeta)((T - r\zeta I)^{-1}x,y)\,d\zeta + \int_{C_e'} f(\zeta)((T - r^{-1}\zeta I)^{-1}x,y)\,d\zeta\},\end{aligned}$$

where C_i' *and* C_e' *are the boundaries of* $\mathbf{D} \setminus \bar{O}_n$ *and* $\bar{\mathbf{D}} \setminus \bar{O}_n$, *respectively, with converse orientations, for sufficiently large* n.

PROOF. Define $u \in H'(K \cup \mathbf{T})$ by

$$u : H(K \cup \mathbf{T}) \ni h \mapsto (h(T)x,y)$$

for some $x,y \in H$. We have

$$\tilde{u}(\zeta) = (2\pi)^{-1}(T(T - \zeta I)^{-1}x,y), \qquad \zeta \in \bar{\mathbf{C}} \setminus (K \cup \mathbf{T}).$$

Then on account of Proposition 1.1 we have $u \in (A_K^p)'$. From the principle of uniform boundedness we obtain the first assertion of the proposition.

Set $u_o := u - (2\pi)^{-1}u.\mathbf{1}$. Then

$$\tilde{u}_o(\zeta) = \tilde{u}(\zeta) + \begin{cases} -(2\pi)^{-1}(x,y) & \text{if } \zeta \in \mathbf{D} \setminus K \\ 0 & \text{if } \zeta \in \bar{\mathbf{D}} \setminus K. \end{cases}$$

Using the notations of the proof of Proposition 1.1 from (1.2) we obtain

$$(f(T)x,y) = (2\pi)^{-1}(x,y)\int_0^{2\pi} f(e^{i\theta})d\theta - (-1)^q \lim_{r\uparrow 1}\{\int_{C'_i} f^{[q]}(\zeta)(u_o^{[-q]})_i^{\sim}(r\zeta)(d\zeta/i\zeta) +$$

$$(1.4) \qquad + \int_{C'_e} f^{[q]}(\zeta)(u_o^{[-q]})_e^{\sim}(r^{-1}\zeta)(d\zeta/i\zeta)\} = (2\pi)^{-1}(x,y)\int_0^{2\pi} f(e^{i\theta})\,d\theta -$$

$$- \lim_{r\uparrow 1}\{\int_{C'_i} f(\zeta)\tilde{u}_o(r\zeta)(d\zeta/i\zeta) + \int_{C'_e} f(\zeta)\tilde{u}_o(r^{-1}\zeta)(d\zeta/i\zeta)\}.$$

This implies the relations (1.3).

We will say that T belongs to $S^m(\mathbf{T})$, $m \in [1,\infty)$, if the following holds:

(i') The sets $\sigma(T)\cap \mathbf{D}$ and $\sigma(T)\cap \bar{\mathbf{D}}$ are closed.

(ii') $0 \in \rho(T)$. The points 0 and ∞ lie in the same component of $\rho(T)\cup\mathbf{T}\cup\{\infty\}$.

(iii') $\|(T - \lambda I)^{-1}\| = O(|1 - |\lambda||^{-m})$, $\quad |\lambda| \to 1$.

If the spectrum of $T \in S^m(\mathbf{T})$ is contained in $\mathbf{T}$ we will write $T \in \mathring{S}^m(\mathbf{T})$. It is not difficult to verify that for open subsets Γ_1 and Γ_2 of $\mathbf{T}$ we have

$$S^m(\Gamma_1)\cap S^m(\Gamma_2) = S^m(\Gamma_1 \cup \Gamma_2).$$

Assume that $T \in S^m(\Gamma)$, $\Gamma \neq \emptyset, \mathbf{T}$. We are going to construct spectral subspaces of T. We fix some $K \in K(\Gamma;T)$. The following notations will be useful.

The set ring of all finite unions of (open, closed, half open) arcs of $\mathbf{T}$ whose endpoints are contained in Γ is denoted by $B_\Gamma(\mathbf{T})$. For arbitrary $\Delta \in B_\Gamma(\mathbf{T})$, let $Z(\Delta;K)$ denote the union of Δ and all components k of K such that $k\cap\Delta \neq \emptyset$. Then

$$B_\Gamma(K\cup\mathbf{T}) := \{Z(\Delta;K) : \Delta \in B_\Gamma(\mathbf{T})\}$$

is a set ring of subsets of $K\cup\mathbf{T}$. By $B_\Gamma(T)$ we denote the Boolean ring of Borel subsets M of $\bar{\mathbf{C}}$ for which there exist a set $Z \in B_\Gamma(K\cup\mathbf{T})$ and subsets σ_1, σ_2 of $\sigma(T)$ with $(\sigma_1\cup\sigma_2)\cap\mathbf{T} = \emptyset$, $\sigma_1 \subset Z$, $\sigma_2\cap Z = \emptyset$, which are closed and open in $\sigma(T)$ such that

$$(1.5) \qquad M\cap\sigma(T) = ((Z\cap\sigma(T))\setminus\sigma_1)\cup\sigma_2.$$

$B_\Gamma(T)$ depends only on Γ and T and not on the choice of K.

Now for a closed $f \in B_\Gamma(K\cup\mathbf{T})$ we define

$$H_T(F) := \{x \in H : g(T)x = 0 \text{ for every } g \in A_K^\infty \text{ with supp } g\cap F = \emptyset\}.$$

Evidently, $H_T(F)$ is closed and invariant for every bounded operator which commutes with T. By the functional calculus we have

$$\sigma(T \mid H_T(F)) \subset F \cap \sigma(T).$$

Moreover, $H_T(F)$ is a spectral maximal space for T (see [2]). This can be proved without difficulty from the relation (1.3).

For arbitrary $Z = Z(\Delta;K)$, $\Delta \in B_\Gamma(T)$, we define

$$H_T(Z) := \overline{\bigcup_i H_T(Z(\Delta_i;K))}$$

where (Δ_i) denotes a non-decreasing sequence of closed sets belonging to $B_\Gamma(\mathbf{T})$ such that $\bigcup_i \Delta_i = \Delta$. $H_T(Z)$ is a closed T-invariant subspace.

Now for every $M \in B_\Gamma(T)$ as in (1.5) we define

$$H_T(M) := (I - E(\sigma_1))H_T(Z) + E(\sigma_2)H. \tag{1.6}$$

Here $E(\sigma)$ denotes the Riesz-Dunford projection correponding to T and a set $\sigma \subset \sigma(T)$ which is closed and open in $\sigma(T)$. It is easy to see that $H_T(M)$ does not depend on the choice of Z (i.e., on the choice of K). $H_T(M)$ is closed and invariant for every bounded operator which commutes with T, and we have

$$\sigma(T \mid H_T(M)) \subset \bar{M} \cap \sigma(T).$$

Making use of the spectral mapping theorem we obtain the following lemma which will be needed below.

LEMMA 1.3. *Given a bounded operator* T *with* $0 \notin \sigma(T)$ *such that the accumulation points of* $\sigma(T) \setminus \mathbf{T}$ are contained in $\mathbf{T} \cup F_o$ *where* F_o *is a finite (possibly empty) subset of* $\mathbf{C} \setminus \mathbf{T}$. *Let* f *be a holomorphic function on a domain containing* $\sigma(T) \cup \mathbf{T}$ *such that* $f \mid \mathbf{T}$ *is a bijection of* $\mathbf{T}$ *onto* $\mathbf{T}$. *Let* Γ *be an open arc of* $\mathbf{T}$, $\Gamma \neq \mathbf{T}$, *such that* f' *does not vanish on* Γ.

Then $f(T) \in S^m(f(\Gamma))$ *implies* $T \in S^m(\Gamma)$. *If* $f(T) \in S^m(f(\Gamma))$ *and, for a closed subarc* γ *of* Γ, *the set* $s := f^{-1}(f(\gamma)) \setminus \Gamma$ *is closed and open in* $\sigma(T)$, *then*

$$H_{f(T)}(f(\gamma)) = H_T(\gamma) + E(s;T)H. \tag{1.7}$$

2. A CLASS OF UNITARY OPERATORS IN KREIN SPACE

2.1. Resolvents of definitizable unitary operators. Let there be given a Krein space $(H,[\cdot,\cdot])$. As usual all topological notions are understood with respect to some

Hilbert norm $||\cdot||$ on H such that $[\cdot,\cdot]$ is $||\cdot||$-continuous. For a Krein subspace L of H, by $\kappa_+(L)$ ($\kappa_-(L)$) we denote the least upper bound ($\leq \infty$) of the dimensions of positive (resp.negative) definite subspaces of L. We set

$$\kappa(L) := \min(\kappa_+(L), \kappa_-(L)).$$

For $T \in L(H)$ the Krein space adjoint is denoted by T^+. A unitary operator U in the Krein space H (i.e., $UU^+ = U^+U = I$) is called *definitizable* if there exists a locally holomorphic function g on $\sigma(U)$ such that $[g(U)x,x] \geq 0$, $x \in H$. Such functions g are called *definitizing functions* for U. The spectrum of a unitary operator U lies symmetrically with respect to **T** ([1]). If, additionally, U is definitizable, $\sigma_0(U) := \sigma(U) \setminus \mathbf{T}$ consists of no more than a finite number of points which are **poles** of the resolvent ([12]; see also for selfadjoint operators: [6], [15]).

In this section we consider the resolvent $R(z;U) = (U - zI)^{-1}$, $z \in \rho(U)$, of a definitizable unitary operator U near **T**. Let g be a definitizing function for U. We assume that g is real on the intersection of **T** and the domain D_g of g. This is no restriction.

Consider the quotient space $H/N(g(U))$ and write $x' := x + N(g(U))$ for every $x \in H$. Denote by H' the completion of $H/N(g(U))$ with respect to the quadratic norm $x' \mapsto [g(U)x,x]^{\frac{1}{2}} =: (x',x')'^{\frac{1}{2}}$, $x \in H$. The operator $U' : x' \mapsto (Ux)'$ is unitary in the Hilbert space H'. For every $z \in \rho(U) \cap D_g$ with $g(z) \neq 0$ and every $x \in H$ we have ([12]; see also [15])

$$\begin{aligned}[R(z;U)x,x] &= g(z)^{-1}[g(U)R(z;U)x,x] + g(z)^{-1}[Q(z;U)x,x] = \\ &= g(z)^{-1}(R(z;U')x',x')' + g(z)^{-1}[Q(z;U)x,x],\end{aligned} \tag{2.1}$$

where $Q(\cdot;U)$ is a locally holomorphic function on $\sigma(U)$ with values in $L(H)$. This implies that U belongs to $S^m(\mathbf{T})$ for some $m \in [1,\infty)$.

Let γ be an open arc of **T** contained in D_g such that $g(e^{i\theta}) > 0$ for $e^{i\theta} \in \gamma$. Set $h(z) := g(z)^{-1}$ for all $z \in D_g$ with $g(z) \neq 0$. According to (2.1) for arbitrary $x \in H$, $e^{i\theta} \in \gamma$, $r \in (0,1)$ such that $re^{i\theta} \in \rho(U) \cap D_g$ and $g(re^{i\theta}) \neq 0$ hold, we have

$$\begin{aligned}&[U\{R(re^{i\theta};U) - R(r^{-1}e^{i\theta};U)\}x,x] = h(e^{i\theta})(U'\{R(re^{i\theta};U') - R(r^{-1}e^{i\theta};U')\}x',x')' + \\ &+ (h(re^{i\theta}) - h(e^{i\theta}))(U'R(re^{i\theta};U')x',x')' - (h(r^{-1}e^{i\theta}) - h(e^{i\theta}))(U'R(r^{-1}e^{i\theta};U')x',x')' + \\ &+ h(re^{i\theta})[UQ(re^{i\theta};U)x,x] - h(r^{-1}e^{i\theta})[UQ(r^{-1}e^{i\theta};U)x,x].\end{aligned} \tag{2.2}$$

By the functional calculus for unitary operators in Hilbert space it is easy to see that the second and the third term on the right in (2.2) are uniformly bounded for $e^{i\theta} \in \gamma_0$, $r \in (1 - \delta, 1)$, where γ_0 is an arbitrary closed subarc of γ and $\delta \in (0,1)$ is sufficiently small. Moreover, for almost all $e^{i\theta} \in \gamma$ the sum of the last four terms of (2.2) converges to 0 for $r \uparrow 1$. On the other hand we have

$$h(e^{i\theta})(U'\{R(re^{i\theta};U') - R(r^{-1}e^{i\theta};U')\}x',x')' \geq 0$$

for $e^{i\theta} \in \gamma$, $r \in (0,1)$ such that $re^{i\theta} \in \rho(U)$. Hence for every $x \in H$ the following holds.

(a_+) $\quad \underline{\lim}_{r\uparrow 1}[U\{R(re^{i\theta};U) - R(r^{-1}e^{i\theta};U)\}x,x] \geq 0$ for almost every $e^{i\theta} \in \gamma$.

(b_+) For every closed subarc γ_0 of γ and every $\delta \in (0,1)$ such that $e^{i\theta} \in \gamma_0$ and $r \in [1 - \delta, 1)$ imply $re^{i\theta} \in \rho(U)$ there exists a positive real number M such that

$$[U\{R(re^{i\theta};U) - R(r^{-1}e^{i\theta};U)\}x,x] \geq -M$$

for $e^{i\theta} \in \gamma_0$ and $r \in (1 - \delta, 1)$.

If g is negative on γ we obtain similar conditions (a_-) (with ≥ 0 replaced by ≤ 0) and (b_-) (with $\geq -M$ replaced by $\leq M$).

Let U be an arbitrary unitary operator in the Krein space H belonging to $S^m(\Gamma)$, $\Gamma \subset \mathbf{T}$ open. If for some open subarc γ of Γ and every $x \in H$ the conditions (a_+) and (b_+) ((a_-) and (b_-)) hold we shall say that γ is of *positive* (resp. *negative*) *type* with respect to U. If an arc γ is of positive or of negative type we shall say that it is of *definite type.*

PROPOSITION 2.1. *Let* U *be a unitary operator in the Krein space H belonging to* $S^m(\Gamma)$, $\Gamma \subset \mathbf{T}$ *open, and let* $K \in K(\Gamma;U)$. *Then for an arbitrary open arc* γ, $\overline{\gamma} \subset \Gamma$, *the following conditions are equivalent.*

(i) γ *is of positive (negative) type with respect to* U.

(ii) *For every* $x \in H$ *and every nonnegative* $f \in A_K^\infty$ *with* supp $f \subset \gamma$ *we have* $[f(U)x,x] \geq 0$ *(resp.* $[f(U)x,x] \leq 0$).

(iii) $(H_U(\gamma_0),[\cdot,\cdot])$ *(resp.* $(H_U(\gamma_0), -[\cdot,\cdot])$) *is a Hilbert space for every closed arc* γ_0 *with* $\gamma_0 \subset \gamma$.

PROOF. It is sufficient to consider the case when γ is of positive type.

(i)$\Longrightarrow$(ii). Assume that (i) and the assumptions of (ii) hold. Let (ϕ_1,ϕ_2) be a real interval, $|\phi_1 - \phi_2| < 2\pi$, which is mapped onto γ by $\phi \mapsto e^{i\phi}$. Then, in view of (1.3), we have

$$[f(U)x,x] = (2\pi)^{-1} \lim_{r\uparrow 1} \int_{\phi_1}^{\phi_2} f(e^{i\theta})[U\{R(re^{i\theta};U) - R(r^{-1}e^{i\theta};U)\}x,x]d\theta.$$

According to (a_+) and (b_+) this expression is nonnegative.

(ii)$\Longrightarrow$(iii). Condition (ii) implies that the functional calculus $A_K^\infty \ni f \to f(U)$ with $\operatorname{supp} f \subset \gamma$ can be extended to continuous functions by continuity and to the characteristic functions of all relatively compact Borel subsets of γ in the usual way. For a closed subarc γ_0 of γ the projection $E(\gamma_0)$ corresponding to the characteristic function of γ_0 is nonnegative, i.e., $(E(\gamma_0)H,[\cdot,\cdot])$ is a Hilbert space. It is easy to see that $E(\gamma_0)H$ and $H_U(\gamma_0)$ coincide, which implies (iii).

(iii)$\Longrightarrow$(i). Let γ_0 be an arbitrary closed subarc of γ. From (iii) we conclude that H is the J-orthogonal direct sum of $H_U(\gamma_0)$ and $H_U(\gamma_0)^{[\perp]}$. U is reduced by this decomposition. Then from well-known properties of unitary operators in Hilbert space we obtain (i) with γ replaced by the interior of γ_0. Since γ_0 was arbitrary we get (i).

We have seen above that, for a definitizable unitary operator U, an arc $\gamma \subset \mathbf{T}$ is of definite type with respect to U if some definitizing function g defined on γ does not vanish there. Thus from the general functional calculus (Proposition 1.2) and the considerations of this section (observe part (ii)$\Longrightarrow$(iii) of the proof of Proposition 2.1) we obtain a functional calculus defined for continuous functions which on some neighbourhoods of the zeros of g have continuous derivatives up to a certain order. Moreover, in this way we obtain the spectral function of U ([12]). We shall consider a functional calculus and the spectral function for a more general class of unitary operators in the following section.

2.2. Operators definitizable over an open subset of the unit circle. Let U be a unitary operator in the Krein space H and let Γ be an open subset of $\mathbf{T}$.

DEFINITION. U is called *definitizable over* Γ if the following conditions are satisfied.

(i) For every open arc γ with $\overline{\gamma} \subset \Gamma$ and $\overline{\gamma} \neq \mathbf{T}$ there is an $m \in [1,\infty)$ with $U \in S^m(\gamma)$.

(ii) For every $\alpha \in \Gamma$ there exist reduced one-sided neighbourhoods (α',α) and (α,α'')*) of α in Γ of definite type.

In the following proposition we give a useful characterization of this class of operators.

---

*) Arcs of the unit circle are denoted similarly to real intervals. For example, (α_1,α_2) denotes the open arc run over by a point moving form α_1 to α_2 in counterclockwise direction.

PROPOSITION 2.2. *Assume that* $\Gamma \neq \mathbf{T}$ *and that the condition* (i) *of the above definition is satisfied. Then* (ii) *is equivalent to the following.*

(ii*) *For every* $K \in K_s(\Gamma;U)$, *every open* $\Gamma_0 \in B_\Gamma(\mathbf{T})$ *such that* $K \cap \mathbf{T} \subset \Gamma_0$ *and every* $\chi \in X_K^\infty(\Gamma_0)$ *the operator* $\chi(U)$ *is unitary and definitizable and we have* $\sigma(\chi(U)) \subset \mathbf{T}$.

PROOF. (ii)$\Longrightarrow$(ii*). For an arbitrary trigonometric polynomial h, we have $\tilde{h}(U) = (h(U))^+$, where $\tilde{h}(z) := \overline{h(\bar{z}^{-1})}$, $z \neq 0, \infty$. Since χ can be approximated in the topology of A_K^q (for every positive integer q) by trigonometric polynomials, we find that

$$(1/\chi)(U) = \tilde{\chi}(U) = (\chi(U))^+.$$

Hence $\chi(U)$ is unitary. By the functional calculus of U we derive that $\sigma(\chi(U)) \subset \mathbf{T}$. On account of (i) and the assumptions on χ the mapping

$$C^p(\mathbf{T}) \ni g \mapsto g(\chi(U)) \in L(H)$$

is continuous for some sufficiently large integer p. Therefore, according to Proposition 1.1,

$$\chi(U) \in \overset{\circ}{S}{}^{p+1}(\mathbf{T}). \tag{2.3}$$

Let γ be a component of $\rho(U) \cap \Gamma$ such that both endpoints of γ belong to Γ, one of the endpoints belongs to an open arc of positive type and the other to an open arc of negative type. In every such arc γ we choose a point α_γ. The union of the set of points obtained in this way and the set of all points of Γ which do not belong to an open arc of definite type is denoted by c'. This set is at most countable and its accumulation points are not in Γ. Therefore the set $c'' := \chi(\Gamma_0) \cup \chi(c')$ is finite. By definition of c', for every component γ'' of $\mathbf{T} \setminus c''$ the inverse image $\chi^{-1}(\gamma'')$ is an open arc of definite type. If $\chi^{-1}(\gamma'')$ is of positive (negative) type then for every $x \in H$ the linear functional

$$C^\infty(\mathbf{T}) \ni f \mapsto [f(\chi(U))x,x]$$

restricted to γ'' (i.e., $\operatorname{supp} f \subset \gamma''$) is positive (resp. negative). On account of a result from [5] (Satz 5) this fact and (2.3) imply that $\chi(U)$ is definitizable.

(ii*)$\Longrightarrow$(ii). For every $\alpha \in \Gamma$ we can find a function χ as in (ii*) such that χ restricted to some open arc containing α is bijective. Making use of Proposition 2.1 we obtain the assertion (ii).

The class of unitary operators definitizable over Γ is denoted by $UD(\Gamma)$. By [5, Satz 5], the operators $U \in UD(\mathbf{T})$ with the additional property that $\sigma(U) \setminus \mathbf{T}$ consists of no more than a finite number of poles of the resolvent are precisely the definitizable operators.

The maximal open subset of $\mathbf{T}$ over which U is definitizable is denoted by $\Gamma(U)$.

A point $\alpha \in \Gamma(U)$ is called a *critical point* of U if it is not contained in an open arc of definite type. The set of critical points of U is denoted by c(U). Evidently, c(U) is at most countable and the accumulation points of c(U) are not contained in $\Gamma(U)$.

To define spectral projections we first assume that $\Gamma(U) \neq \mathbf{T}$. Let $K \in K(\Gamma(U);U)$. Applying Proposition 2.1 and using a partition of unity we conclude that the functional calculus for U can be extended by continuity to $A^{o}_{K \cup c(U)}$ even to the bounded Borel measurable functions on $\mathbf{T} \cup K$ which are locally holomorphic on $K \cup c(U)$. If χ_Z is the characteristic function of a set $Z \in B_{\Gamma(U) \setminus c(U)}(K \cup c(U))$ we define

$$E(Z) := \chi_Z(U).$$

It is easy to see that the range of the projection E(Z) coincides with $H_U(Z)$ (see Section 1.2). Define the set ring

$$B(U) := B_{\Gamma(U) \setminus c(U)}(U)$$

(see Section 1.2). In general, $B(U)$ is no σ-ring. If $b \in B(U)$, $Z \in B_{\Gamma(U) \setminus c(U)}(K \cup c(U))$ and $\sigma_1, \sigma_2 \subset \sigma(U)$ are closed and open in $\sigma(U)$ with $(\sigma_1 \cup \sigma_2) \cap \mathbf{T} = \emptyset$, $\sigma_1 \subset Z$, $\sigma_2 \cap Z = \emptyset$ such that

$$b \cap \sigma(U) = ((Z \cap \sigma(U)) \setminus \sigma_1) \cup \sigma_2,$$

then in accordance with (1.6) we define

$$E(b) := (I - E(\sigma_1))E(Z) + E(\sigma_2).$$

If U belongs to $UD(\mathbf{T})$ then $B(U)$ denotes the Boolean ring of all Borel subsets b' of $\bar{\mathbf{C}}$ such that

$$b' \cap \sigma(U) = (\gamma' \cap \sigma(U)) \cup \sigma$$

for some $\gamma' \in B_{\mathbf{T} \setminus c(U)}(\mathbf{T})$ and some subset σ of σ(U) with $\sigma \cap \mathbf{T} = \emptyset$ which is closed and open in σ(U). We define

$$E(b') := E(\gamma') + E(\sigma),$$

where $E(\gamma') = \chi_{\gamma'}(U)$.

In the following theorem we formulate some properties of the spectral function E of U, which are easy consequences of the functional calculus.

THEOREM 2.3. *Let* U *be a unitary operator in the Krein space* H *which is definitizable over some open subset of* **T**.

Then the Riesz-Dunford functional calculus can be extended by continuity to $A^{o}_{K \cup c(U)}$ *for every* $K \in K(\Gamma(U);U)$ *if* $\Gamma(U) \neq \mathbf{T}$ *or to* $A^{o}_{c(U)}$ *if* $\Gamma(U) = \mathbf{T}$. *The mapping* E *is* *a strongly (i.e. with respect to the strong operator topology) σ-additive homomorphism of $B(\mathrm{U})$ into a Boolean ring of projections in H such that for $\mathrm{b} \in B(\mathrm{U})$ the following holds.*

(a) *If* $\mathrm{b} \cap \sigma(\mathrm{U})$ *is closed and open in* $\sigma(\mathrm{U})$ *then* $\mathrm{E(b)}$ *coincides with the Riesz-Dunford projection corresponding to* $\mathrm{b} \cap \sigma(\mathrm{U})$. *In particular,* $\mathrm{E}(\sigma(\mathrm{U})) = \mathrm{I}$.

(b) *If* b *is symmetric with respect to* $\mathbf{T}$, *then* $\mathrm{E(b)}$ *is selfadjoint and, hence,* $\mathrm{E(b)}H$ *is a Krein space.*

(c) *If* $\mathrm{TU} = \mathrm{UT}$ *holds for a bounded operator* T, *then* $\mathrm{TE(b)} = \mathrm{E(b)T}$.

(d) $\sigma(\mathrm{U} \,|\, \mathrm{E(b)}H) \subset \bar{\mathrm{b}}$.

(e) $\alpha \in \Gamma(\mathrm{U})$ *belongs to* $\mathrm{c(U)}$ *if and only if for all open arcs* $\gamma \in B(\mathrm{U})$ *containing* α *the subspace* $\mathrm{E}(\gamma)H$ *is indefinite.*

(f) *For every arc* $\gamma \in B(\mathrm{U})$, $\bar{\gamma} \subset \Gamma(\mathrm{U})$ *the operator* $\mathrm{U} \,|\, \mathrm{E}(\gamma)H$ *is definitizable.*

Every strongly σ-additive homomorphism of $B(\mathrm{U})$ into a Boolean ring of projections in H with the properties (a), (c) *and* (d) *coincides with* E.

The critical points of a unitary operator U definitizable over some open subset of $\mathbf{T}$ are classified in the same way as those of a definitizable operator: A critical point α of U is called *regular* if there exists an open arc $\gamma_0 \ni \alpha, \gamma_0 \in B(\mathrm{U})$, with $\bar{\gamma}_0 \cap (\mathrm{c(U)} \setminus \{\alpha\}) = \emptyset$ such that the projections $\mathrm{E}(\gamma)$, $\gamma = \bar{\gamma} \subset \gamma_0 \setminus \{\alpha\}$, are uniformly bounded. Let $\mathrm{c_r(U)}$ denote the set of regular critical points of U. The elements of $\mathrm{c_s(U)} := \mathrm{c(U)} \setminus \mathrm{c_r(U)}$ are called *singular* critical points. Denote by $\mathrm{c_\pi(U)}$ the set of those critical points α such that there exists an open arc $\gamma \ni \alpha, \gamma \in B(\mathrm{U})$, with $\kappa(\mathrm{E}(\gamma)H) <$ $< \infty$. Further, we set

$$\mathrm{c_\infty(U)} := \mathrm{c(U)} \setminus \mathrm{c_\pi(U)}, \quad \mathrm{c_{r,\infty}(U)} := \mathrm{c_r(U)} \cap \mathrm{c_\infty(U)}.$$

The following lemma which will be needed in Section 3 is a consequence of the functional calculus.

LEMMA 2.4. *Assume that* U *belongs to* $UD(\Gamma)$, $\Gamma \neq \mathbf{T}$. *Let* χ *be a function as in Proposition* 2.2, (ii*). *Then we have*

$$\chi(\mathrm{c(U)}) \subset \mathrm{c}(\chi(\mathrm{U}))$$

and

$$\mathrm{E}(\gamma;\chi(\mathrm{U})) = \mathrm{E}(\chi^{-1}(\gamma);\mathrm{U})$$

for every arc γ *whose endpoints do not belong to* $\mathrm{c}(\chi(\mathrm{U}))$.

2.3. Perturbations of class S_ω of definitizable unitary operators. The following theorem shows that unitary operators definitizable over open subsets of **T** arise, for example, from definitizable operators if perturbations of class $\mathbf{S}_\omega$ are considered. Here $\mathbf{S}_\omega$ denotes the Macaev ideal ($T \in \mathbf{S}_\omega :\Leftrightarrow T$ is compact and $\sum_{j=1}^{\infty} (2j-1)^{-1} s_j < \infty$, where s_j are the s-numbers of T). $\mathbf{S}_\omega$ contains the ideals $\mathbf{S}_p$, $p \geq 1$. The essential part of the proof of this theorem is due to H.Langer ([13]).

We admit that the unperturbed and the perturbed operator are unitary with respect to different Krein spaces $H_1 = (H,[\cdot,\cdot]_1)$ and $H_2 = (H,[\cdot,\cdot]_2)$ with the same underlying linear space H. Assume that there is a Hilbert scalar product $(\cdot,\cdot)$ on H with respect to which $[\cdot,\cdot]_1$ and $[\cdot,\cdot]_2$ are continuous. We define bounded and boundedly invertible selfadjoint operators G_1 and G_2 in $(H,(\cdot,\cdot))$ by

$$(G_j x,y) = [x,y]_j, \qquad x,y \in H,\ j = 1, 2.$$

For a unitary operator U in a Krein space, the set of isolated eigenvalues of finite algebraic multiplicity belonging to $\sigma_0(U) := \sigma(U)\setminus \mathbf{T}$ is denoted by $\sigma_{0,\pi}(U)$. We set $\sigma_{0,\infty}(U) := \sigma_0(U)\setminus\sigma_{0,\pi}(U)$.

THEOREM 2.5. *Let* U *be a definitizable unitary operator in* H_1 *with* $c_\infty(U) = c_{r,\infty}(U)$. *Let* U' *be a unitary operator in* H_2 *such that*

$$U' - U =: V \in \mathbf{S}_\omega. \tag{2.4}$$

Assume that

$$G_1 - G_2 \in \mathbf{S}_\omega. \tag{2.5}$$

Then the following holds.

(i) $\sigma_0(U')$ *is a countable set whose accumulation points are contained in* $\sigma_{0,\infty}(U) \cup c_\infty(U)$, *and we have*

$$\sigma_0(U')\setminus \sigma_{0,\infty}(U) \subset \sigma_{0,\pi}(U').$$

(ii) U' *is definitizable over* $\mathbf{T}\setminus c_\infty(U)$, *and we have*

$$c(U')\setminus c_\infty(U) \subset c_\pi(U').$$

PROOF. (I) By (2.5) we have $\kappa(H_1) < \infty$ if and only if $\kappa(H_2) < \infty$. Since in a Pontrjagin space every unitary operator is definitizable we can assume that $\kappa(H_1) = \kappa(H_2) = \infty$. One verifies as in [8], Proposition 2, that there exists a unitary operator U_0 in H_1 such that $U - U_0$ is of finite rank and

$$\sigma_0(U_0) = \sigma_{0,\infty}(U_0) = \sigma_{0,\infty}(U), \quad c(U_0) = c_\infty(U_0) = c_\infty(U).$$

Therefore it is no restriction to assume that $\sigma_o(U) = \sigma_{o,\infty}(U)$ and $c(U) = c_\infty(U)$.

(2) Assume first, additionally, that $\sigma(U) \subset \mathbf{T}$ and that all points of

$$c(U) \cap \sigma_p(U) = c_{r,\infty}(U) \cap \sigma_p(U) =: \{\alpha_j : j = 1, \dots, n\}$$

are semisimple eigenvalues. Let P_j^+ and P_j^-, $j = 1, \dots, n$, be the selfadjoint projections in H_1 corresponding to some fundamental decomposition of $E(\{\alpha_j\})H$, where E denotes the spectral function of U.

Let γ_j^+, $j = 1, \dots, n^+$; γ_j^-, $j = 1, \dots, n^-$, be pairwise disjoint open arcs of $\mathbf{T}$ such that the following hold:

(a) $E(\gamma_j^+)$, $j = 1, \dots, n^+$, is nonnegative, $E(\gamma_j^-)$, $j = 1, \dots, n^-$, is nonpositive.

(b) $\sigma(U) \setminus c(U) \subset \bigcup_{j=1}^{n^+} \gamma_j^+ \cup \bigcup_{j=1}^{n^-} \gamma_j^-$.

With respect to the fundamental decomposition

$$H_1 = \Big(\sum_{j=1}^{n} P_j^+ + \sum_{j=1}^{n^+} E(\gamma_j^+)\Big)H + \Big(\sum_{j=1}^{n} P_j^- + \sum_{j=1}^{n^-} E(\gamma_j^-)\Big)H$$

the operator U can be written in the form

$$U = \begin{bmatrix} U_{11} & 0 \\ 0 & U_{22} \end{bmatrix}.$$

Then, on account of (2.4) and (2.5), one can prove as in [14], Theorem 6, that there is a fundamental decomposition of H_2 with the following properties. If the matrix form of U' with respect to this decomposition is

$$U' = \begin{bmatrix} U'_{11} & U'_{12} \\ U'_{21} & U'_{22} \end{bmatrix}$$

we have $U'_{12}, U'_{21} \in S_\omega$ and the relations $\tilde{\sigma}(U'_{11}) = \tilde{\sigma}(U_{11})$, $\tilde{\sigma}(U'_{22}) = \tilde{\sigma}(U_{22})$. Here $\tilde{\sigma}(T)$ denotes the set of points $\lambda \in \sigma(T)$ which are not isolated eigenvalues of finite algebraic multiplicity of an operator T.

Now the proof of (i) and (ii) in this case is essentially the same as the proof of the main result of [13].

(3) Assume now that $\sigma(U) \subset \mathbf{T}$ and that U has a non-semisimple eigenvalue μ_o in $c(U) = c_{r,\infty}(U)$, and $(c(U) \cap \sigma_p(U)) \setminus \{\mu_o\}$ consists of semisimple eigenvalues. Moreover, we assume that $\mu_o = 1$. This is no restriction. Set

(2.6) $\psi(z) := -(z-i)/(z+i), \quad \phi(z) := i(1-z)/(1+z), \quad z \in \bar{\mathbf{C}},$

and

$$p^{(n)}(z) := z^{4n+1}, \quad n = 0, 1, \ldots, z \in \bar{\mathbf{C}}.$$

We consider the following holomorphic mappings of $\bar{\mathbf{C}}$ onto itself:

$$\pi^{(n)} := \psi \circ p^{(n)} \circ \phi, \quad n = 0, 1, \ldots .$$

Evidently, $\pi^{(n)}(U)$ and $\pi^{(n)}(U')$ are unitary and we have $\pi^{(n)}(U) - \pi^{(n)}(U') \in \boldsymbol{S}_\omega$. There exists an integer $n_0 \geq 0$ such that $c(\pi^{(n_0)}(U)) \cap \sigma_p(\pi^{(n_0)}(U))$ consists of semisimple eigenvalues. Hence $\pi^{(n_0)}(U)$ and $\pi^{(n_0)}(U')$ fulfil the assumptions of part (2) of this proof and (i) and (ii) hold for $\pi^{(n_0)}(U)$ and $\pi^{(n_0)}(U')$. Now making use of the spectral mapping theorem and Lemma 1.3 one verifies without difficulty that (i) and (ii) hold with $c_\infty(U) \cup \{-1\}$ instead of $c_\infty(U)$.

Repeating this consideration with $\tau_a \circ \pi^{(n_0)}$ instead of $\pi^{(n_0)}$, where τ_a is a linear fractional transformation with $\tau_a(\mathbf{T}) = \mathbf{T}$, $\tau_a(1) = 1$ and $\tau_a(-1) = a \in \mathbf{T} \setminus \{-1, 1\}$ we obtain (i) and (ii).

In the case when $\sigma(U) \subset \mathbf{T}$ and we have more than one non-semisimple eigenvalue in $c(U)$ the assertions are proved by induction.

(4) Consider now the general case. Set $G^{(\delta)} := \{z : |z| \in (\delta, \delta^{-1})\}$ and choose $\delta \in (0,1)$ so that the relations $\sigma(U) \setminus \overline{G^{(\delta)}} = \sigma_0(U) = \sigma_{0,\infty}(U)$ and $\sigma(U') \cap \partial G^{(\delta)} = \emptyset$ hold. Define a locally holomorphic function f on $\mathbf{C} \setminus \partial G^{(\delta)}$ by $f|G^{(\delta)} = \mathrm{id}$ and $f|\mathbf{C} \setminus \overline{G^{(\delta)}} \equiv 1$. For the operators f(U) and f(U'), which also fulfil the assumptions, the theorem is already proved. Hence (i) and (ii) hold with $c_\infty(U) \cup \{1\}$ instead of $c_\infty(U)$. Repeating this consideration with a function f_b such that $f_b|G^{(\delta)} = \mathrm{id}$ and $f_b|\mathbf{C} \setminus \overline{G^{(\delta)}} \equiv b \in \mathbf{T} \setminus \{1\}$ we obtain (i) and (ii) in the general case.

REMARK. In general, under the assumptions of Theorem 2.5 one cannot expect that the perturbed operator U' is definitizable. Indeed, by Proposition 3 of [8], for every definitizable unitary operator U such that $\sigma_{0,\infty}(U) \cup c_\infty(U) \neq \emptyset$ and every $p > 1$, there exists a unitary operator U' (in the same Krein space) with $U - U' \in \boldsymbol{S}_p$ such that U' is not definitizable.

3. ON COMPACT PERTURBATIONS

Let $H_1 = (H, [\cdot,\cdot]_1)$, $H_2 = (H, [\cdot,\cdot]_2)$, $(\cdot,\cdot)$ and the operators G_1 and G_2 be given as in Section 2.2. We assume that $G_1 - G_2$ is compact. Let U_1 and U_2 be unitary operators in H_1 and H_2, respectively, which are definitizable over some open subsets of

T and whose difference is compact. For example, this situation occurs if U_1 and U_2 are different perturbations of class $\boldsymbol{S}_\omega$ of a fixed definitizable unitary operator U with $c_\infty(U) = c_{r,\infty}(U)$.

In the following theorem we consider relations between the signs of the spectral functions E_1 and E_2 of U_1 and U_2, respectively, on $\Gamma(U_1) \cap \Gamma(U_2)$.

THEOREM 3.1. *Assume that*

$$G_1 - G_2 \in \boldsymbol{S}_\infty . \tag{3.1}$$

Let U_1 *and* U_2 *be unitary operators in* H_1 *and* H_2*, respectively, which are definitizable over some open subsets of* **T**. *Assume that*

$$U_1 - U_2 \in \boldsymbol{S}_\infty .$$

Let Γ_o *be an open arc of* **T** *such that* $\Gamma_o \subset \Gamma(U_1) \cap \Gamma(U_2)$, $\Gamma_o \cap c(U_1) = \emptyset$ *and all projections* $E_1(\Gamma')$, $\Gamma' = \overline{\Gamma'} \subset \Gamma_o$, *are nonnegative (nonpositive).*

Then for every closed arc $\Gamma \subset \Gamma_o$ *whose endpoints are not in* $c(U_2)$ *the subspace* $E_2(\Gamma)H_2$ *is a Pontrjagin subspace of* H_2 *with finite-dimensional (possibly trivial) negative (resp. positive) part.*

PROOF. We assume that the projections $E_1(\Gamma')$, $\Gamma' = \overline{\Gamma'} \subset \Gamma_o$, are nonnegative. If they are nonpositive, a similar reasoning applies.

Let $\Gamma \in B(U_2)$ be a fixed closed arc contained in Γ_o. Without loss of generality we may assume, additionally, that the following holds: $\overline{\Gamma}_o \subset \Gamma(U_1) \cap \Gamma(U_2)$, $\overline{\Gamma}_o \cap c(U_1) = \emptyset$, $E_1(\Gamma_o)$ is nonnegative, $\Gamma_o \in B(U_2)$ and, denoting by γ and γ' the two open arcs such that $\Gamma_o \setminus \Gamma = \gamma \cup \gamma'$, $E_2(\overline{\gamma})H_2$ and $E_2(\overline{\gamma'})H_2$ are definite (possibly trivial) subspaces of H_2.

Let Γ_1 be an open arc of **T**, $\mathbf{T} \setminus \Gamma_o \subset \Gamma_1$, $\overline{\Gamma}_1 \cap \Gamma = \emptyset$. Since U_1 and U_2 differ by a compact operator we can choose some $K \in K(\Gamma_o;U_1) \cap K(\Gamma_o;U_2)$. Let χ be a function from $X_K^\infty(\Gamma_1)$, $\chi(\Gamma_1) = \{\alpha_o\}$, $\alpha_o \in \mathbf{T} \setminus \Gamma$, which coincides with the identity on a neighbourhood of Γ. According to Proposition 2.2, $\chi(U_1)$ and $\chi(U_2)$ are definitizable unitary operators in H_1 and H_2, respectively. $\chi(U_1)$ is similar to a unitary operator in a Hilbert space. Since χ can be approximated by functions locally holomorphic on $K \cup \mathbf{T}$ in a topology with respect to which the functional calculi for U_1 and U_2 are continuous, we have

$$\chi(U_1) - \chi(U_2) \in \boldsymbol{S}_\infty . \tag{3.2}$$

Let F_1 be a fundamental symmetry of the Krein space $((I - E_1(\Gamma_o))H, [\cdot,\cdot]_1)$. Then the operator

$$E_1(\Gamma_0) + F_1(I - E_1(\Gamma_o)) \tag{3.3}$$

is a fundamental symmetry of H_1. Let $H_1 = H_{1,+} + H_{1,-}$ be the corresponding fundamental decomposition of H_1. Evidently, $\chi(U_1)$ has diagonal form with respect to this decomposition:

$$\chi(U_1) = \begin{bmatrix} T_{++} & 0 \\ 0 & T_{--} \end{bmatrix}.$$

T_{++} and T_{--} are unitary operators in $H_{1,+}$ and $H_{1,-}$, respectively, and

$$T_{--} = \alpha_0 I_{H_{1,-}}. \tag{3.4}$$

On account of (3.1) and Theorem 6 from [14] there is a maximal nonpositive subspace M_- of H_2 which is invariant under $\chi(U_2)$ and satisfies the condition $\tilde{\sigma}(\chi(U_2)|M_-) = \tilde{\sigma}(T_{--})$ (see part (2) of the proof of Theorem 2.5). Hence, according to (3.4), the linear span $M_-(\Gamma)$ of the root spaces of $\chi(U_2)|M_-$ corresponding to eigenvalues in Γ is finite-dimensional (possibly trivial).

It is easy to verify that $E_2(\Gamma)M_- \subset M_-$. Therefore $M_-(\Gamma) = E_2(\Gamma)M_-$. Since $E_2(\Gamma)M_-$ is a maximal nonpositive subspace of $(E_2(\Gamma)H_2, [\cdot,\cdot]_2)$ and finite-dimensional, $(E_2(\Gamma)H_2, [\cdot,\cdot]_2)$ is a Pontrjagin space with a finite-dimensional (possibly trivial) negative part.

REMARK. It is easy to see that if, in particular, U_1 and U_2 are obtained by perturbations of class S_ω from a fixed definitizable unitary operator U with $c_\infty(U) = c_{r,\infty}(U)$, Theorem 3.1 can also be proved along the lines of the proof of Theorem 2.3, that is by making use of [13].

COROLLARY. *If* U_1 *and* U_2 *are given as in Theorem* 3.1, *then*

$$c_\infty(U_1) \cap \Gamma(U_1) \cap \Gamma(U_2) = c_\infty(U_2) \cap \Gamma(U_1) \cap \Gamma(U_2).$$

PROOF. Assume that $\lambda \in c_\infty(U_1) \setminus c_\infty(U_2)$. As in [8], Proposition 2, one verifies that there exists a unitary operator U_2' in H_2 definitizable on $\Gamma(U_2)$ such that $U_2 - U_2'$ is of finite rank and $\lambda \notin c(U_2')$. By Theorem 3.1 this contradicts $\lambda \in c_\infty(U_1)$.

Now let $H_j := (H, [\cdot,\cdot]_j)$, $j = 1, 2, \ldots, \infty$, be Krein spaces with the same underlying linear space H. Let $(\cdot,\cdot)$ be a Hilbert scalar product on H with respect to which all sesquilinear forms $[\cdot,\cdot]_j$ $j = 1, 2, \ldots, \infty$, are continuous. Assume that the operators G_j defined by

$$(G_j x,y) = [x,y]_j, \quad x,y \in H,\ j = 1, 2, \ldots, \infty,$$

satisfy the following conditions:

$$G_j - G_\infty \in \mathcal{S}_\infty,\ j = 1, 2, \ldots, \tag{3.5}$$

$$\lim_{j\to\infty} \|G_j - G_\infty\| = 0. \tag{3.6}$$

The following theorem has some points of contact with some results from [16].

THEOREM 3.2. *Consider a sequence of Krein spaces* H_j *such that* (3.5) *and* (3.6) *hold. Let* U_j, $j = 1, 2, \ldots, \infty$, *be unitary operators in* H_j, *respectively, which are definitizable over some open subsets of* $\mathbf{T}$ *and fulfil the following conditions:*

$$\begin{aligned} &U_j - U_\infty \in \mathcal{S}_\infty,\ j = 1, 2, \ldots, \\ &\lim_{j\to\infty} \|U_j - U_\infty\| = 0. \end{aligned} \tag{3.7}$$

Assume that there exist an open arc $\Gamma_0 \subset \mathbf{T}$, *and real numbers* $\eta \in (0,1)$, $k \geq 1$, *and* N *such that* $\overline{\Gamma}_0 \subset \Gamma(U_j)$, $j = 1, 2, \ldots, \infty$, *and the relations* $r \in (\eta,1) \cup (1,\eta^{-1})$ *and* $e^{i\theta} \in \Gamma_0$ *imply* $re^{i\theta} \in \rho(U_j)$, $j = 1, 2, \ldots, \infty$, *and*

$$\|(U_j - re^{i\theta}I)^{-1}\| \leq N|1 - r|^{-k}, \quad j = 1, 2, \ldots, \infty. \tag{3.8}$$

If Γ_i, $i = 1, 2, 3$, *are open arcs of* $\mathbf{T}$ *belonging to* $B(U_\infty)$ *such that* $\overline{\Gamma}_1 \subset \Gamma_2$, $\overline{\Gamma}_2 \subset \Gamma_3$, $\overline{\Gamma}_3 \subset \Gamma_0$, *then for sufficiently large* j *these arcs belong to* $B(U_j)$ *and the following relations hold:*

$$\begin{aligned} &\kappa_+(E_\infty(\Gamma_1)H) \leq \kappa_+(E_j(\Gamma_2)H) \leq \kappa_+(E_\infty(\Gamma_3)H), \\ &\kappa_-(E_\infty(\Gamma_1)H) \leq \kappa_-(E_j(\Gamma_2)H) \leq \kappa_-(E_\infty(\Gamma_3)H). \end{aligned} \tag{3.9}$$

Here E_j *denotes the spectral function of* U_j, $j = 1, 2, \ldots, \infty$.

PROOF. (a) Let Γ'_0 be an open arc of $\mathbf{T}$ with $\mathbf{T}\setminus\Gamma_0 \subset \Gamma'_0$, $\overline{\Gamma}_3 \cap \overline{\Gamma'_0} = \emptyset$. Evidently, $\bigcap\{K(\Gamma_0;U_j) : j = 1, 2, \ldots, \infty\}$ is non-void. Let K belong to this set and let χ be an arbitrary function from $X_K^\infty(\Gamma'_0)$. Then the operators $\chi(U_j)$, $j = 1, 2, \ldots, \infty$, are definitizable. As in the proof of Theorem 3.1 one verifies that

$$\chi(U_j) - \chi(U_\infty) \in \mathcal{S}_\infty, \quad j = 1, 2, \ldots . \tag{3.10}$$

(3.7) implies that we have

(3.11) $$\lim_{j\to\infty} \| R(z;U_j) - R(z;U_\infty) \| = 0$$

uniformly on compact subsets of $\mathbf{C} \setminus (T \cup K)$. We claim that

(3.12) $$\lim_{j\to\infty} \| \chi(U_j) - \chi(U_\infty) \| = 0.$$

Indeed, express the quantities $[(\chi(U_j) - \chi(U_\infty))x,x]$, $j = 1, 2, \ldots$, for arbitrary $x \in H$ with $\|x\| \leq 1$ by curve integrals as in (1.4). Then by (3.8) and (3.11) we find that

$$\lim_{j\to\infty} |[(\chi(U_j) - \chi(U_\infty))x,x]| = 0$$

uniformly for $\|x\| \leq 1$. Hence (3.12) holds.

(b) Let $\gamma_1 \in B(U_\infty)$ be an open arc such that $\overline{\gamma}_1 \subset \Gamma_0 \setminus \overline{\Gamma}'_0$. Assume that $E_\infty(\gamma_1)H_\infty$ is a Pontrjagin space. It is no restriction to assume that $\kappa := \kappa_-(E_\infty(\gamma_1)H_\infty) < \infty$. Then there is an open arc $\gamma_2 \in B(U_\infty)$, $\overline{\gamma}_2 \subset \gamma_1$, such that $\kappa_-(E_\infty(\gamma_1 \setminus \overline{\gamma}_2)H_\infty) = 0$. Let γ_3 be a further open arc with $\overline{\gamma}_2 \subset \gamma_3$ and $\overline{\gamma}_3 \subset \gamma_1$ and let χ be a function from $X^\infty_K(\Gamma'_o)$ with the following properties: χ is equal to a constant $\alpha \in \gamma_2$ on γ_2 and equal to a constant $\alpha_0 \notin \overline{\gamma}_1$ outside of γ_1. In small neighbourhoods of the endpoints of γ_3 the function χ coincides with the identity.

Consider a fundamental decomposition $H_\infty = H_{\infty,+} + H_{\infty,-}$ such that $E_\infty(\gamma_1 \setminus \overline{\gamma}_2)H_\infty$ and some maximal nonnegative Hilbert subspace of $E_\infty(\overline{\gamma}_2)H_\infty$ are contained in $H_{\infty,+}$. Then $\chi(U_\infty)$ is diagonal with respect to this decomposition and we have $\sigma(\chi(U_\infty) \,|\, H_{\infty,-}) \subset \{\alpha, \alpha_0\}$.

Now making use of the conditions (3.5) and (3.7) we follow the construction of a nonpositive invariant subspace $M_{j,-}$, $j = 1, 2, \ldots$, of $\chi(U_j)$ from [14]. It is easy to see, as a consequence of (3.6) and (3.12) that for sufficiently large j the following holds:

$$\sigma(\chi(U_j) \,|\, M_{j,-}) \subset \gamma_2 \cup (\mathbf{T} \setminus \overline{\gamma}_1).$$

The sum of the algebraic multiplicities of the eigenvalues of $\chi(U_j) \,|\, M_{j,-}$ in γ_2 is equal to κ.

Then, similarly to the proof of Theorem 3.1, one verifies that

$$\kappa_-(E(\gamma_3;\chi(U_j))H_j) = \kappa_-(E_j(\gamma_3)H_j) = \kappa$$

holds for sufficiently large j. For $\kappa = 0$ this proves that $\Gamma_1, \Gamma_2, \Gamma_3 \in B(U_j)$ for sufficiently large j.

(c) To prove the relations (3.9) it is sufficient to consider the first two inequalities. First we shall prove the relation

(3.13) $$\kappa_+(E_\infty(\Gamma_1)H_\infty) \leq \kappa_+(E_j(\Gamma_2)H_j)$$

for sufficiently large j.

Since by Theorem 3.1 (observe the proof of the Corollary after Theorem 3.1) $\kappa_+(E_\infty(\Gamma_1)H_\infty) = \infty$ and $\kappa_+(E_j(\Gamma_2)H_j) < \infty$ cannot hold simultaneously we can restrict ourselves to the case when $\kappa_+(E_\infty(\Gamma_1)H_\infty) < \infty$. Under this assumption (3.13) follows as in part (b) of this proof.

To prove

$$\kappa_+(E_j(\Gamma_2)H_j) \leq \kappa_+(E_\infty(\Gamma_3)H_\infty)$$

for sufficiently large j we can assume that $\kappa_+(E_\infty(\Gamma_3)H_\infty) < \infty$. Then this relation is proved as in (b).

REFERENCES

1. **Bognár, J.** : *Indefinite inner product spaces*, Berlin, 1974.
2. **Colojoară, I. ; Foiaş, C.** : *Theory of generalized spectral operators*, New York, 1968.
3. **Gohberg, I.C. ; Krein, M.G.** : *Introduction to the theory of linear nonselfadjoint operators* (Russian), Moscow, 1965.
4. **Jonas, P.** : Eine Bedingung für die Existenz einer Eigenspektralfunktion für gewisse Automorphismen lokalkonvexer Räume, *Math. Nachr.* **45**(1970), 145-160.
5. **Jonas, P.** : Zur Existenz von Eigenspektralfunktionen mit Singularitäten, *Math. Nachr.* **88**(1977), 345-361.
6. **Jonas, P.** : On the functional calculus and the spectral function for definitizable operators in Krein space, *Beiträge Anal.* **16**(1981), 121-135.
7. **Jonas, P** : Compact perturbations of definitizable operators.II, *J. Operator Theory* **8**(1982), 3-18.
8. **Jonas, P. ; Langer, H.** : Compact perturbations of definitizable operators, *J. Operator Theory* **2**(1979), 63-77.
9. **Köthe, G.** : Die Randverteilungen analytischer Funktionen, *Math. Z.* **57**(1952), 13-33.
10. **Köthe, G.** : *Topologische lineare Räume.I*, Berlin, 1960.
11. **Krein, M.G.** : *Introduction to the geometry of indefinite J-spaces and to the theory of operators of these spaces* (Russian), Second Summer School, Kiew, 1965.
12. **Langer, H.** : *Spektraltheorie linearer Operatoren in J-Räumen und einige Anwendungen auf die Schar* $L(\lambda) = \lambda^2 I + \lambda B + C$, Habilitationsschrift, Dresden, 1965.

13. **Langer, H.** : Spektralfunktionen einer Klasse J-selbstadjungierter Operatoren, *Math. Nach.* **33**(1967), 107-120.

14. **Langer, H.** : Factorization of operator pencils, *Acta Sci. Math. (Szeged)* **38**(1976), 83-96.

15. **Langer, H.** : Spectral functions of definitizable operators in Krein spaces, in *Functional Analysis*, Proceedings, Dubrovnik 1981; *Lecture Notes in Mathematics* **948**(1982), 1-46.

16. **Langer, H. ; Najman, B.** : Perturbation theory for definitizable operators in Krein spaces, *J. Operator Theory* **9**(1983), 297-317.

Peter Jonas

Karl-Weierstrass-Institut für Mathematik
Akademie der Wissenschaften der DDR
Mohrenstrasse 39, Berlin 1080
DDR.

Operator Theory:
Advances and Applications, Vol.17

NAIMARK DILATIONS, STATE-SPACE GENERATORS AND TRANSMISSION LINES *)

T.Kailath and A.M.Bruckstein

1. INTRODUCTION

With any positive Toeplitz form we can associate, via an inverse scattering algorithm, a discrete transmission-line model, parametrized by a so called "choice sequence" of local reflection coefficients. The infinite-dimensional state-space representation of this model then readily yields the structure of the Naimark dilation and of the state-space moment generator for the unit circle measure that corresponds to the Toeplitz form. This point of view motivates and considerably simplifies many operator-theoretic manipulations aimed at analyzing the structure of Naimark dilations and also connects this theory to some recent results on state-space generators for moment matrices associated with positive measures on the unit circle. The new insights readily yield several interesting matrix indentities and also suggest the extension of Naimark dilation results and state-space generator theory to the continuous case.

Suppose we are given a sequence, $\{I, R_1, R_2, R_3, \ldots\}$, of operators over a Hilbert space H so that the operators defined by the following Toeplitz matrices (over $H_N = H \oplus H \oplus H \oplus \ldots \oplus H$)

$$\mathbf{R}_N = \begin{bmatrix} I & R_1 & & R_N \\ R_1^* & I & \ddots & \\ & \ddots & \ddots & R_1 \\ R_N^* & & R_1^* & I \end{bmatrix} \tag{1.1}$$

are positive for all N. An objective of *Naimark dilation theory*, see e.g. [5], [6], [1], is to associate with the sequence $\{I, R_1, R_2, R_3, \ldots\}$ an operator Λ, over a high-dimensional

*) This work was supported in part by the Air Force Office of Scientific Research, Air Force Systems Command under contract AFOSR-83-0228 and by the U.S.Army Research Office, under contract DAAG 29-83-K-0028.

space $\mathbf{H} = H \oplus H \oplus \ldots$, such that

(1.2) $$R_i = [I \ 0 \ 0 \ldots \ ] \ \Lambda^i \begin{bmatrix} I \\ 0 \\ 0 \\ \cdot \\ \cdot \\ \cdot \end{bmatrix} .$$

This means that the restriction of the operator Λ^i to the first H-subspace yields the operator R_i.

It is further required that the construction of the operator Λ should proceed in a nested, recursive way. This means that to any finite (positive) sequence $\{I, R_1, R_2, R_3, \ldots, R_N\}$ we should be able to associate an operator Λ_N obeying (1.2) when restricted to $\mathbf{H}_N$. This requirement is important for solving the problem of continuation of a finite sequence of operators in a way that preserves positiveness, and plays an important role in characterizing the so-called maximum entropy extensions.

The theory of *state-space generators* for orthogonal polynomials, developed in [10] regards nested positive definite matrices of increasing size, $\mathbf{M}_N = (m_{ij})$ as the moment matrices associated to a measure on a curve in the complex plane, and constructs nested matrix-vector pairs $\{\mathbf{A}(N), \mathbf{B}(N)\}$ such that

(1.3) $$m_{ij} = \mathbf{B}^* (\mathbf{A}^*)^i (\mathbf{A})^j \mathbf{B} .$$

If the measure is on the unit circle, the moment matrices are Toeplitz forms, and then $\mathbf{B}^*$ is $[I \ 0 \ 0 \ldots]$, the matrices $\mathbf{A}(N)$ are "almost unitary", becoming exactly so as $N \to \infty$, and therefore $\mathbf{A}_N$ may be recognized as a nested sequence of Naimark dilations.

In this paper we shall show that a lossless discrete *transmission-line structure*, parametrized by a sequence of reflection coefficients having magnitude less than one (contraction operators), is a simple formal structural model of the Naimark dilation and the state-space generator constructs. Many results, quite difficult to derive operator-algebraically, are easily obtained by simply reading out various relationships on the transmission-line model. The association of a transmission-line structure with positive Toeplitz forms will be recognized as an *inverse scattering process* whereas the mapping from the reflection coefficients to the original positive sequence is easily seen to be a complementary process, the so-called *perfect reflection experiment*. The nestedness of the dilation and state-space generator constructs is implicit in the cascade structure of the transmission-line model.

This paper is organized as follows. In the next section we briefly discuss transmission-line models, perfect reflection experiments and the corresponding inverse scattering process. Section 3 presents the derivation of a series of dilation results via the transmission-line interpretation. Then in Section 4 we show that state-space generators are in fact the state-space representations of the cascade structures we are dealing with, in the perfect reflection experiment setting.

2. LOSSLESS TRANSMISSION-LINE MODELS

We shall first discuss some basic properties of discrete, lossless wave propagation structures. Such structures can arise as nonuniform transmission-lines with piecewise constant impedance (see e.g. [2]), as models for layered-earth acoustic media in geophysics (e.g. [12]), and also as implementations of fast algorithms in linear estimation theory ([9]). Such transmission-line structures extend over $[0,\infty)$, support "right-going" and "left-going" waves $W_R(x,t)$ and $W_L(x,t)$, and have the structure of cascades of elementary layers which contain a delay network and an *orthogonal* wave-interaction section, as depicted in Figure 1. (These are so-called "flow graphs" or "block-diagrams" in which the directed edges imply applying the operator indicated on it on the quantities appearing at the input end to produce the quantities at the output.) The *waves* that propagate are (discrete) time functions taking values in some space H. Therefore the elementary sections act as operators, transforming infinite (time indexed) sequences into other infinite sequences of elements of H. The basic operators of which the elementary sections are composed are very simple. The orthogonal wave-interaction, or *wave-scattering*, that occurs at each section of the transmission-line structure is parametrized by a contraction operator, K, (as shown in Figure 1) and has the (unitary) matrix representation

$$\Sigma(K) = \begin{bmatrix} K^C & -K^* \\ K & K^{*C} \end{bmatrix}, \quad K^C := (I - K^*K)^{\frac{1}{2}} \tag{2.1}$$

The operators K and $K^C := (I - K^*K)^{\frac{1}{2}}$, and the corresponding adjoints, act on the signals in a static way, i.e. on each of the elements of a time sequence separately. The *time delay operator* D acts as a shift of the time index. Therefore it commutes with the static gain operators and has the matrix representation

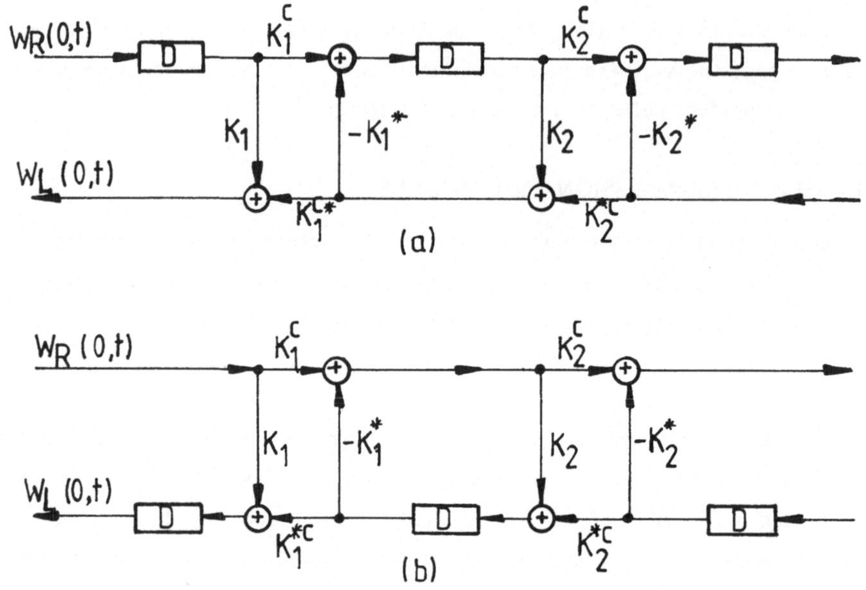

FIGURE 1: LOSSLESS TRANSMISSION LINE MODELS

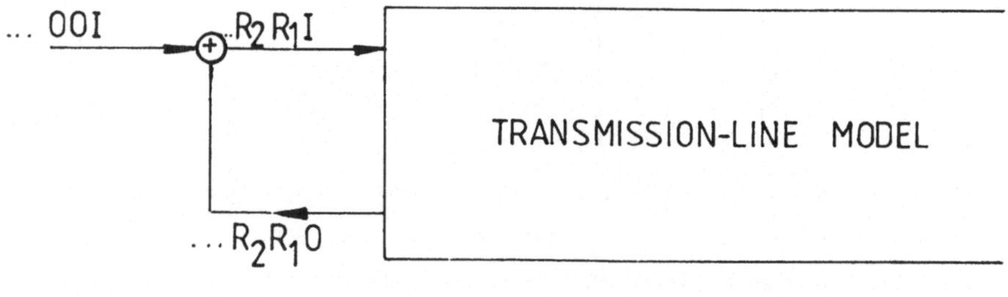

FIGURE 2: SCATTERING EXPERIMENT PROVIDING "PERFECT REFLECTION" DATA

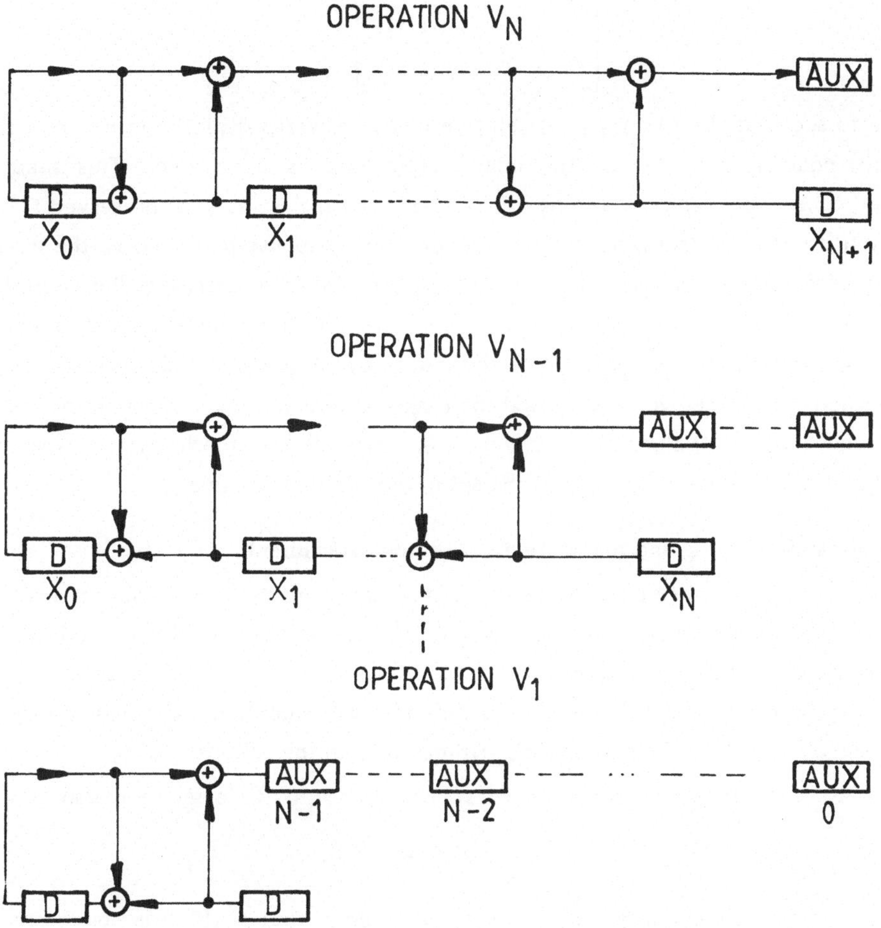

FIGURE 3 : DEFINITIONS OF V_i AND U_i EXPLAINED IN TERMS OF THE TRANSMISSION-LINE MODEL

$$D = \begin{bmatrix} 0 & 0 & 0 & 0 & . & . \\ I & 0 & 0 & 0 & . & . \\ 0 & I & 0 & 0 & . & . \\ 0 & 0 & I & 0 & . & . \\ . & . & . & . & . & . \\ . & . & . & . & . & . \end{bmatrix}. \qquad (2.2)$$

It is easy to see that the two types of relative delay networks depicted in Figure 1 are completely equivalent as far as the input-output map is concerned. (This map is unaffected when the relative timing of the propagating signals is preserved.) The input-output relation is characterized by a set of input-response pairs, where the inputs are right-propagating, probing sequences $W_R(0,t)$ and the corresponding outputs, $W_L(0,t)$, are causally evoked left propagating responses at the boundary $x = 0$. In order to obtain a complete characterization of the input-output operator that corresponds to a transmission-line model we need to find the response to a sequence of inputs from the space H such that their values at $x = 0$ and $t = 0$, $\{W_R(0,0)\}$, from a basis of this space. (We shall henceforth denote by $W_R(x,t)$ the *set of signals* thus obtained.)

Inverse scattering and the perfect reflection experiment

The inverse scattering problem is the following: given the input sequence and the corresponding response, determine the sequence of contractions K_N for $N = 1,2,3,\dots,$ that characterizes the transmission-line model. This problem is readily solved using causality arguments, via a so-called Schur, or downward continuation algorithm (see [13], [7], and also [4], [2]). Briefly this algorithm works in the following way:

1) By causality, we have that the first response, $W_L(0,0)$, is zero and also that

$$W_L(0,1) = K_1 W_R(0,0). \qquad (2.3\text{a})$$

Therefore, we can determine the operator K_1 via (recall that $W_R(0,0)$ is now a set of inputs that form a basis, hence it can be interpreted as an operator acting on the standard identity basis I)

$$K_1 = W_L(0,1) W_R^{-1}(0,0). \qquad (2.3\text{b})$$

2) Once K_1 is determined we can compute from the sequences $W_R(0,t)$ and $W_L(0,t)$ the waves at the level 1 inside the medium, since it is easy to see by some calculation from Figure 1 that

$$(2.4)\qquad \begin{bmatrix} W_R(1,t) \\ W_L(1,t) \end{bmatrix} = \begin{bmatrix} K^{-C} & -K^*K^{-*C} \\ K^{-*C}K & K^{-*C} \end{bmatrix} \begin{bmatrix} W_R(0,t) \\ W_L(0,t) \end{bmatrix} = \Theta(K) \begin{bmatrix} W_R(0,t) \\ W_L(0,t) \end{bmatrix}$$

where $\Theta(K_1)$ is the so-called transfer or *chain-scattering* operator, associated with $\Sigma(K)$.

3) Now note that the set of signals $\{W_R(1,t), W_L(1,t)\}$ formally characterizes the input-output map of a transmission-line model extending over $[2,\infty)$, and therefore the next reflection coefficient may now be identified by a similar procedure.

Proceeding this way, one can recursively compute the entire sequence of reflection coefficients.

Therefore, to an arbitrary sequence of scattering data, i.e., an input-response pair, the above algorithm associates a sequence of operators $\{K_N\}$. These operators will be contractive, if the data was generated by a lossless structure.

We shall be interested in this paper in a particular type of scattering data, the so-called perfect reflection data, defined by

$$(2.5)\qquad W_R(0,0) = I \text{ and } W_R(0,t) = W_L(0,t) \text{ for } t > 0.$$

See Figure 2 for a schematic description (which explains the name "perfect reflection") of the scattering experiment providing this type of data. Since $W_R(0,t) = W_L(0,t)$, for $t = 1, 2, 3, \ldots$ form a collection of transmission-line model responses to a basis of elements in H, (recall that $W_R(0,0) = I$), we whall identify these signal collections with operators and denote them by R_t. Then, a simple analysis of the underlying transmission-line structure shows that $\{I, R_1, R_2, R_3, \ldots\}$ will be a positive sequence of operators if the structure is lossless. The above discussed inverse scattering procedure then becomes the so-called Schur algorithm and it is not difficult to show that it provides the lower-upper (or causal-anticausal) factorization of the positive operator (1.1). Furthermore the so-called Krein system of equations for inverse scattering can be derived for the perfect reflection scattering experiment and its solution can be shown to provide an implicit upper-lower factorization of inverses of Toeplitz forms defined by $\{I,R_1,R_2,R_3,\ldots\}$. Such results are further discussed in [8].

In the sequel we shall be concerned with the problem of finding the mapping from the reflection coefficient sequence to the positive Toeplitz form. This is the problem solved by the "dilation" theories. In our context, this problem has a straightforward solution: simply feed the transmission-line structure with I as the first input, then feed the response sequence back as the future probing sequence inputs. This is indeed the meaning of performing a "perfect reflection experiment". We shall see that this interpretation, combined with a state-space description of transmission-line models, indeed provides a simple way to obtain dilation results on this problem.

3. STATE-SPACE DESCRIPTIONS AND DILATION RESULTS

In this section we shall consider transmission-line structures as general linear systems with a state-space description. The state $X(t) = [x_o(t)\ x_1(t)\ x_2(t)\ldots]$, of the system at time t is defined as the values of the inputs to the delay operators at that time. Once the state is defined, we can regard the transmission-line as a system described by a state transition operator **A**, an input-to-state operator **B** and a state-to-output or "read-out" operator **C**. The evolution of the state, from a quiescent, or all zero, initial condition when an input sequence $W_R(0,t)$ is applied, is given by

(3.1 a) $$X(t) = \mathbf{A}X(t-1) + \mathbf{B}W_R(0,t), \qquad X(0) = 0$$

and the output is

(3.1 b) $$W_L(0,t) = \mathbf{C}X(t).$$

Let us compute the explicit representations of the state-space operators for the two structures presented in Figure 1. This can be done fairly easily by tracing the signal flow in the block diagrams of Figure 1.

The first structure, having delay operators on the upper line (acting on the right propagating signals) has the following state-space description (for a line with N sections):

(3.2 a) $$\mathbf{A}_a = \begin{bmatrix} 0 & 0 & 0 & 0 & 0 \\ K_1^C & -K_1^*K_2 & -K_1^*K_2^{*C}K_3 & \cdot & -K_1^*K_2^{*C}K_3^{*C}\ldots K_N^{*C}K_{N+1} \\ 0 & K_2^C & -K_2^*K_3 & \cdot & -K_2^*K_3^{*C}\ldots K_N^{*C}K_{N+1} \\ 0 & 0 & K_3^* & \cdot & \cdot \\ \cdot & \cdot & \cdot & \cdot & \cdot \\ \cdot & \cdot & \cdot & \cdot & \cdot \\ \cdot & \cdot & \cdot & \cdot & -K_N^*K_{N+1} \end{bmatrix}$$

(3.2 b) $$\mathbf{B}_a = [I \quad 0 \quad 0 \quad 0 \quad . \quad . \quad . \quad 0]^*$$

(3.2 c) $$\mathbf{C}_a = [K_1,\ K_1^{*C}K_2,\ K_1^{*C}K_2^{*C}K_3,\ \ldots\ K_1^{*C}K_2^{*C}\ldots K_N^{*C}K_{N+1}].$$

The second structure, having delay operators on the bottom line acting on the left going wave, is also easily described as a state-space model with

(3.3 a)
$$\mathbf{A}_b = \begin{bmatrix} 0 & K_1^{*C} & 0 & \cdot & 0 \\ 0 & -K_2K_1^* & K_3^{*C} & \cdot & 0 \\ 0 & -K_3K_2^CK_1^* & -K_3K_2^* & \cdot & \cdot \\ \cdot & \cdot & \cdot & \cdot & \cdot \\ \cdot & \cdot & \cdot & \cdot & \cdot \\ \cdot & \cdot & \cdot & \cdot & 0 \\ 0 & -K_{N+1}K_N^C \cdots K_2^CK_1^* & -K_{N+1}K_N^C \cdots K_3^CK_2^* & \cdot & -K_{n+1}K_N^* \end{bmatrix}$$

(3.3 b)
$$\mathbf{B}_b = \begin{bmatrix} K_1 \\ K_2K_1^C \\ K_3K_2^CK_1^C \\ \cdot \\ \cdot \\ \cdot \\ K_{N+1}K_N^C \cdots K_1^C \end{bmatrix}$$

(3.3 c)
$$\mathbf{C}_b = [I \quad 0 \quad 0 \quad \ldots \quad 0].$$

Note the nice symmetry between the two state-space representations.

The perfect reflection experiment described in the previous section corresponds to setting the input sequence equal to the system output, which simply means "closing a feedback loop". The closed-loop system will have the state-transition matrices

(3.4 a)
$$\mathbf{F}_a = \mathbf{A}_a + \mathbf{B}_a\mathbf{C}_a =$$

$$= \begin{bmatrix} K_1 & K_1^{*C}K_2 & K_1^{*C}K_2^{*C}K_3 & \cdot & K_1^{*C}K_2^{*C} \cdots K_N^{*C}K_{N+1} \\ K_1^C & -K_1^*K_2 & -K_1^*K_2^{*C}K_3 & \cdot & -K_1^*K_2^{*C}K_3^{*C} \cdots K_N^{*C}K_{N+1} \\ 0 & K_2^C & -K_2^*K_3 & \cdot & -K_2^*K_3^{*C} \cdots K_N^{*C}K_{N+1} \\ 0 & 0 & K_3^* & \cdot & \cdot \\ \cdot & \cdot & \cdot & \cdot & \cdot \\ \cdot & \cdot & \cdot & \cdot & \cdot \\ \cdot & \cdot & \cdot & \cdot & -K_N^*K_{N+1} \end{bmatrix}$$

(3.4 b) $$\mathbf{F}_b = \mathbf{A}_b + \mathbf{B}_b\mathbf{C}_b =$$

$$= \begin{bmatrix} K_1 & K_1^{*C} & 0 & \cdot & 0 \\ K_2K_1^C & -K_2K_1^* & K_3^{*C} & \cdot & 0 \\ K_3K_2^CK_1^C & -K_3K_2^CK_1^* & -K_3K_2^* & \cdot & \cdot \\ \cdot & \cdot & \cdot & \cdot & \cdot \\ \cdot & \cdot & \cdot & \cdot & \cdot \\ \cdot & \cdot & \cdot & \cdot & 0 \\ K_{N+1}K_N^C\cdots K_1^C & -K_{N+1}K_N^C\cdots K_2^CK_1^* & -K_{N+1}K_N^C\cdots K_3^CK_2^* & \cdot & -K_{N+1}K_N^* \end{bmatrix}$$

It is clear from the nestedness of the structures depicted in Figure 1 that the system output up to time $t = N$ is entirely determined by the first N sections of the infinite cascade. Therefore, as N increases, the state-space representations of the finite cascades will match more and more time lags of the input-output map. (This can also be seen from the fact that the inverse scattering algorithm discussed in the previous section determines $\{K_1, K_2, \ldots, K_N\}$ from the input-output map up to time lag N.)

In the closed-loop system, corresponding to the perfect reflection experiment, the first input, $W_R(0,0) = I$, sets the state of the structure of Figure 1(a) to $X(0) = [I \ \ 0 \ \ 0 \ \ 0 \ \ \ldots]^*$, and there are no external inputs applied thereafter. Therefore the evolution of the state is determined by repeatedly applying $\mathbf{F}_a$ to the initial state X(0), i.e.

(3.5) $$X(i) = (\mathbf{F}_a)^i X(0).$$

However, the input sequence to the first delay element, $x_o(i)$ is recognized to be the sequence $\{I, R_1, R_2, R_3, \ldots\}$ and therefore we have immediately that

(3.6) $$R_i = [I \ \ 0 \ \ 0 \ \ 0\ldots]X(i) = [I \ \ 0 \ \ 0 \ \ 0\ldots](\mathbf{F}_a)^i \begin{bmatrix} I \\ 0 \\ 0 \\ 0 \\ \cdot \\ \cdot \\ \cdot \end{bmatrix}.$$

This is a dilation result of the type we were looking for. It shows that one can find a nested sequence of dilation operators, corresponding to cascades of transmission-line sections, each matching the positive sequence of operators further than the previous one. Conceptually, the infinite state-space transition operator is identifiable with Λ, and its nested structure is made clear by the cascade of elementary transmission-line section operators.

Note that the same result holds for the structure of Figure 1(b), provided we redefine the state variables as the outputs of the delay operators. Since the two structures have identical input-output maps we also have that

$$R_i = [I \quad 0 \quad 0 \quad 0 \ldots]X(i) = [I \quad 0 \quad 0 \quad 0 \ldots](\mathbf{F}_b)^i \begin{bmatrix} I \\ 0 \\ 0 \\ 0 \\ \cdot \\ \cdot \\ \cdot \end{bmatrix} . \tag{3.7}$$

It is clear, however that the time-histories of the entire state vectors differ significantly, since in the first structure the state vector fills up causally (i.e. $X(i) = [\times \ \times \ \times \ \ldots \ \times \ 0 \ 0 \ldots]$) whereas in the second one the state vectors are immediately completely filled with nonzero entries.

We shall next rederive in a straightforward way several identities involving various operators associated with positive definite sequences via their reflection coefficients (or choice operators).

Transmission-line derivation of some operator relations

Consider a cascade of transmission-line sections up to a maximal order of N + 1. Define the operator V_N represented by a $(N+1)\times(N+1)$ matrix as follows

$$V_N = \prod_{N}^{1} \{I_{j-1} \oplus \Sigma(K_j) \oplus I_{N-j}\} . \tag{3.8}$$

By inspection of the closed-loop (reflection experiment) structure of Figure 1(b), we realize that applying the operator V_N to an N + 1 - length state vector will produce the next state-vector, except for the last entry, where some auxiliary quantity appears. Applying to the resulting vector the operator $V_{N-1} \oplus I$ will likewise produce the next state-vector, except for the last two entries, and so forth. Therefore applying the operator

$$U_N = \prod_{N-1}^{0} \{V_{N-j} \oplus I_j\} \tag{3.9}$$

will produce the vector $[x_o^B(N)AUX(N-1)AUX(N-2)\ldots AUX(0)]$, where AUX(i) is produced after applying V_{N-i}. Now a *key observation* is the fact that this vector would be the state of the Figure 1(a) structure after N + 1 lags of a perfect reflection experiment. This is recognized easily by inspection of Figure 3. Therefore we have that the next output, which is R_{N+1}, has to be

$$(3.10) \qquad R_{N+1} = C_a U_N \begin{bmatrix} I \\ 0 \\ 0 \\ 0 \\ \cdot \\ \cdot \\ \cdot \end{bmatrix}$$

which yields the formula

$$(3.11) \quad R_{N+1} = [K_1, K_1^{*C} K_2, K_1^{*C} K_2^{*C} K_3, \ldots, K_1^{*C} K_2^{*C} \ldots K_N^{*C} K_{N+1}] \begin{bmatrix} U_{N-1} & 0 \\ 0 & I \end{bmatrix} \begin{bmatrix} K_1 \\ K_2 K_1^C \\ K_3 K_2^C K_1^C \\ \cdot \\ \cdot \\ \cdot \\ K_N^C \ldots K_1^C \end{bmatrix}$$

or, equivalently

$$(3.12) \qquad R_{N+1} = C_a(N) U_{N-1} B_b(N) + K_1^{*C} K_2^{*C} \ldots K_N^{*C} K_{N+1} K_N^C K_{N-1}^C \ldots K_1^C .$$

This formula is a key result in [5], and is derived there in a more algebraic way. Several other results of this paper, like the convolution relation between the sequence I, R_1, $R_2, \ldots$ and the open-loop impulse response sequence of the line, denoted there by $S_0, S_1, S_2 \ldots$, easily follow from the transmission-line interpretation.

We showed that applying the operator V_N to an arbitrary initial state of the structure of Figure 1(b) yields the next state of this structure except for the last entry. In order to also obtain the last state for the last entry, we have to multiply the resulting last entry by K_{N+1}. This proves that the closed-loop transition matrix $\mathbf{F}_b$ is given by

$$(3.13) \qquad \mathbf{F}_b = \begin{bmatrix} I_N & 0 \\ 0 & K_{N+1} \end{bmatrix} \prod_{N}^{1} \{I_{j-1} \oplus \Sigma(K_j) \oplus I_{N-j}\}$$

which is a nice decomposition of the matrix $\mathbf{F}_b$ into a product of (almost all) unitary matrices, corresponding to the cascaded sections in the transmission-line structure (see also [11]).

4. CONNECTION TO STATE-SPACE GENERATORS

In the paper [10] on state-space generators for orthogonal polynomials, it is shown how to determine a nested set of matrices $\{\mathbf{A}(N),\mathbf{B}(N)\}$ such that

$$R_{ij} = \mathbf{B}^*(N)(\mathbf{A}^*)^i(N)(\mathbf{A})^j(N)\mathbf{B}(N) \tag{4.1}$$

where R_{ij} are the moments of a positive measure on an arbitrary curve in the complex plane. In particular, when the curve is the unit circle, the moment matrices become Toeplitz forms. In this case one has $\mathbf{B}^*(N) = [I \quad 0 \quad 0 \quad 0 \ldots 0]$, and the state-space matrices are identical to $\mathbf{F}_a$, as given by (3.4a). These matrices, by the decomposition corresponding to (3.13), are almost orthogonal, (see also [10]), and obey

$$\mathbf{A}_a^*(N)\mathbf{A}_a(N) = \begin{bmatrix} I & 0 & 0 & 0 & . & 0 \\ 0 & I & 0 & 0 & . & 0 \\ 0 & 0 & I & 0 & . & 0 \\ . & . & . & I & . & . \\ . & . & . & . & . & 0 \\ 0 & 0 & 0 & 0 & . & I - K_{N+1}K_{N+1}^* \end{bmatrix} . \tag{4.2}$$

This readily shows, using (4.1), that

$$R_{ij} = \mathbf{B}^*(N)\mathbf{A}^{*i}(N)\mathbf{A}^j(N)\mathbf{B}(N) = \begin{cases} \mathbf{B}^*(N)\mathbf{A}^{j-i}(N)\mathbf{B}(N) & \text{for } j > i \\ \mathbf{B}^*(N)\mathbf{A}^{*j-i}(N)\mathbf{B}(N) & \text{for } j < i \end{cases} \tag{4.3}$$

since the vectors $\mathbf{B}(N)$ sift out the first entry of the matrix they right and left multiply. (4.3) is again recognized as a dilation result. In [10] it is also pointed out that the construction of nested state-space generators is intimately related to the triangular factorization of the positive definite Toeplitz form associated with the measure under consideration. Furthermore, the characteristic polynomials associated with the state-space transfer matrices are recognized to be *orthogonal polynomial* bases, with respect to the underlying measure. We refer the interested reader to [10] for further details.

The theory of inverse scattering has an immediate continuous parameter counterpart (see e.g. [3]). Therefore the connections between inverse scattering for transmission-line models, Naimark dilation and state-space generators for orthogonal polynomials suggest that similar results hold in the continuous parameter case. In this setting we will have that a continuous positive Toeplitz form, or displacement kernel, can be represented as the restriction of an extended operator, which describes the evolution of "signals" on a continuous transmission-line structure. This will also point out connections to the theory of Schrödinger operators and Krein's theory of vibrating strings and the continuous analogues of orthogonal polynomials.

REFERENCES

1. **Arsene, Gr. ; Constantinescu, T. :** The structure of Naimark dilations and Gaussian stationary processes, *Integral Equations Operator Theory* **8**(1985), 181--204.

2. **Bruckstein, A.M. ; Kailath, T. :** Inverse scattering for discrete transmission-line models, *SIAM Review*, 1985, to appear.

3. **Bruckstein, A.M. ; Levy, B.C. ; Kailath, T. :** Differential methods in inverse scattering, I.S.L. Report, Stanford University, 1983 (and *SIAM J. Appl. Math.*, 45(1985), 312-335).

4. **Bube, K.P. ; Burridge, R. :** The one-dimensional inverse problem of reflection seismology, *SIAM Review* **25**(1983), 497-559.

5. **Constantinescu, T. :** On the structure of positive Toeplitz forms, in *Dilation theory, Toeplitz operators, and related topics*, Birkhäuser Verlag (OT Series 11), 1983, pp.127-149.

6. **Constantinescu, T. :** On the structure of Naimark dilations, *J. Operator Theory* **12**(1984), 159-175.

7. **Dewilde, P. ; Vieira, A. ; Kailath, T. :** On a generalized Szegö-Levinson realization algorithm for optimal linear predictors based on a network synthesis approach, *IEEE Trans. Circuits Systems* **25**(1978), 663-675.

8. **Kailath, T. ; Bruckstein, A. ; Morgan,D. :** Fast matrix factorizations via discrete transmission lines, *Linear Alg. Appl.*, 1985, to appear.

9. **Kailath, T. ; Vieira, A. ; Morf, M. :** Inverses of Toeplitz operators, innovations and orthogonal polynomials, *SIAM Review* **20**(1978), 106-119.

10. **Kailath, T. ; Porat, B. :** State-space generators for orthogonal polynomials, in *Prediction Theory and Harmonic Analysis*, the Pesi Masani Volume, V.Mandrekar and H.Salehi (ed), 1983, pp.131-163.

11. **Kimura, H. :** Generalized Schwartz forms and lattice-ladder realizations of digital filters, Technical Report 84-03, Osaka Univ., Japan, 1984.

12. **Robinson, E.A. :** Spectral approach to geophysical inversion by Lorentz, Fourier and Radon transforms, *Proc.IEEE* **70**(1982), pp.1039-1054.

13. **Schur, I. :** Über Potenzreihen, die im innern das Einheitskreises beschrankt sind, *J. Reine Angew. Math.* **148**(1918), 122-145.

T.Kailath and **A.M.Bruckstein**
Department of Electrical Engineering
Stanford University, Stanford, CA 94305, U. S. A.

Operator Theory:
Advances and Applications, Vol.17

CONTRACTIONS BEING WEAKLY SIMILAR TO UNITARIES

László Kérchy

In this paper contractions being weakly similar in a certain sense to unitary operators are studied. Extending the investigations of [12], where only cyclic contractions were considered, it is examined when a contraction T, weakly similar to unitary possesses the bicommutant property **Alg** $T = \{T\}''$. Finally, a characterization of cyclic C_{11}-contractions is given.

0. PRELIMINARIES AND NOTATIONS

For a complex, separable Hilbert space H let L(H) denote the set of all bounded, linear operators acting on H. If $T \in L(H)$, then **Alg** T denotes the algebra, closed in the weak operator topology and generated by T and the identity operator I. Moreover, $\{T\}'$ and $\{T\}''$ stand for the commutant and bicommutant of T, respectively: $\{T\}' = \{A \in L(H) : AT = TA\}$, $\{T\}'' = \{B \in L(H) : BA = AB$ for every $A \in \{T\}'\}$. For a set of operators $S \subset L(H)$, **Lat** S denotes the set of invariant subspaces of S: **Lat** S = {M (closed) subspace of H : $AM \subset M$ for every $A \in S$}, while for a set L of subspaces of H, **Alg** L means the set of operators which leave invariant every element of L : **Alg** $L = \{A \in L(H) : AM \subset M$ for every $M \in L\}$. A subalgebra Q of L(H) is called reflexive, if **Alg Lat** Q = Q. An operator $T \in L(H)$ is reflexive, if so is the algebra **Alg** T (cf. [7]). We use the notations **Lat**'' T and **Hyplat** T for the subspace lattices **Lat**$\{T\}''$ and **Lat**$\{T\}'$, respectively ($T \in L(H)$).

In the present paper we are interested in contraction operators $T \in L(H)$, i.e. when for the norm of T : $\|T\| \leq 1$ holds. In connection with them we shall freely use the terminology of our main reference [15]. More precisely we shall deal with C_{11}-contractions, i.e. with contractions $T \in L(H)$ such that $\lim_{n\to\infty} \|T^n h\| \neq 0 \neq \lim_{n\to\infty} \|T^{*n} h\|$ for every non-zero vector $h \in H$. For a C_{11}-contraction T the subspace lattices $\mathbf{Lat}_1 T = \{M \in \mathbf{Lat}\, T : T|M \in C_{11}\}$ and $\mathbf{Hyplat}_1 T = \{M \in \mathbf{Hyplat}\, T : T|M \in C_{11}\}$ have a significant role. These are complete lattices with respect to set-inclusion, in which the greatest lower bound $\overset{(1)}{\bigwedge} L$ of a subset L is generally different form the intersection $\bigcap L$ (cf. [11]).

We remind that every completely non-unitary (c.n.u.) C_{11}-contraction T is quasi-similar to a (unique) absolutely continuous unitary operator U, which means that there are quasi-affinities in the sets I(T,U) and I(U,T). Here $I(T_1,T_2)$ stands for the system of intertwining operators: $I(T_1,T_2) = \{S : ST_1 = T_2S\}$, and quasi-affinity means an injective operator with dense range. We recall that being absolutely continuous U is unitarily equivalent to an operator of the form $\bigoplus_n M_{\alpha_n}$, where $\{\alpha_n\}_n$ is a decreasing sequence of Borel sets of the unit circle **C**, and M_{α_n} denotes the operator of multiplication by e^{it} in the Hilbert space $L^2(m,\alpha_n)$, m stands for the normalized Lebesgue measure on **C** (cf. [5]).

Let T be a c.n.u. C_{11}-contraction, Θ_T its characteristic function and $\Delta_T(e^{it}) = [I - \Theta_T(e^{it})^* \Theta_T(e^{it})]^{\frac{1}{2}}$. It is known (cf. [15], [19]) that there is a lattice-isomorphism $q_T : B_T(\mathbf{C}) \to \mathbf{Hyplat}_1 T$, where $B_T(\mathbf{C})$ is the set of equivalence classes of Borel subsets of **C** under the equivalence relation: $\alpha \simeq \beta$ if $\Delta_T(e^{it}) = 0$ a.e. on the symmetric difference $\alpha\Delta\beta$ (cf. also [11]). Moreover, q_T can be defined so that for any Borel set $\alpha\subset\mathbf{C}$ there corresponds a regular factorization $\Theta_T = \Theta_{2,\alpha}\Theta_{1,\alpha}$ to the invariant subspace $H_\alpha = q_T(\hat{\alpha})$ ($\hat{\alpha}$ is the equivalence class containing α) such that $\Theta_{1,\alpha}$ is outer, $\Theta_{1,\alpha}(e^{it})$ is isometric a.e. on $\mathbf{C}\setminus\alpha$ and $\Theta_{2,\alpha}(e^{it})$ is isometric a.e. on α. We note that a factorization of this type is essentially unique (cf. [15, Proposition V.4.3]). Since the characteristic function of $T|H_\alpha$ coincides with the purely contractive part of $\Theta_{1,\alpha}$ (cf. [15, Proposition VII.2.1]) we have

$$\| \Theta_{T|H_\alpha}(e^{it})^{-1} \| = \chi_\alpha(e^{it}) \| \Theta_T(e^{it})^{-1} \| + \chi_{\mathbf{C}\setminus\alpha}(e^{it}) \quad \text{a.e.}\,.$$

(Here χ_α stands for the characteristic function of α and if an operator A is non-invertible, then we define $\|A^{-1}\|$ to be infinity.)

1. CONTRACTIONS, WEAKLY SIMILAR TO UNITARIES

By a result of C.Apostol (cf. [1]) we know that for every C_{11}-contraction $T \in L(H)$ there exists a basic system $\{H_n\}_{n=1}^\infty$ consisting of invariant subspaces of T such that for every n, $T|H_n$ is similar to a unitary operator. We recall that a set $\{H_n\}_n$ of subspaces of H forms a basic system if $H_n + (\bigvee_{k\neq n} H_k) = H$ for every n, and $\bigcap_n(\bigvee_{k\geq n} H_k) = \{0\}$. Since the operator T is similar to the orthogonal sum $(T|H_n)\oplus$ $\oplus(T|\bigvee_{k\neq n} H_k)$, it follows that $H_n \in \mathbf{Lat}_1 T\subset\mathbf{Lat}''T$ (cf. [11]), for each $n \in \mathbf{N}$ (the set of natural numbers).

DEFINITION 1. We call a contraction T to be *weakly similar to unitary* if we can find a basic system $\{H_n\}_n$ such that $H_n \in$ **Hyplat** T and $T|H_n$ is similar to unitary for every n.

It is immediate that then T belongs to C_{11} and $H_n \in$ **Hyplat**$_1$ T for every n.

The following interesting example shows that there are C_{11}-contractions being not weakly similar to unitary.

PROPOSITION 2. *There exists a* C_{11}*-contraction* T *such that for every non-zero subspace* $M \in$ **Hyplat** T *the spectrum of the restriction of* T *to* M *covers the whole unit disc:* $\sigma(T|M) = \mathbf{D}^-$. *(Here* $\mathbf{D}$ *denotes the open unit disc:* $\mathbf{D} = \{\lambda \in \mathbf{C} : |\lambda| < 1\}$*).*

REMARK 3. This is also an example for a C_{11}-contraction having no proper spectral maximal spaces (cf. [6]). We note that weak contractions are decomposable (cf. [8] and [13]).

PROOF OF PROPOSITION 2. Let $\{\beta_n\}_{n=1}^{\infty}$ be a sequence of arcs on the unit circle $\mathbf{C}$ of measure $0 < m(\beta_n) < 1$ such that for any measurable set $\alpha \subset \mathbf{C}$ with positive measure $m(\alpha) > 0$ we have $m(\alpha \cap \beta_n) > 0$ for an appropriate $n \in \mathbf{N}$. Let us form a sequence $\{\alpha_n\}_{n=1}^{\infty}$ which contains each β_n infinitely many times.

Let θ_n be a (scalar-valued) outer function such that $|\theta_n(e^{it})| = (1/n)\chi_{\alpha_n}(e^{it}) + \chi_{\mathbf{C}\setminus\alpha_n}(e^{it})$ a.e. on $\mathbf{C}$, and consider the model-operator $S(\theta_n)$ (cf. [15, Chapter VI]). We define T to be the orthogonal sum of these operators:

$$T = \bigoplus_{n=1}^{\infty} S(\theta_n).$$

It is clear that T is a c.n.u. C_{11}-contraction whose characteristic function coincides with the contractive analytic function $\Theta(\lambda) = \mathbf{diag}(\theta_n(\lambda))$. Let us assume that M is a nonzero subspace from **Hyplat** T. Since the orthogonal projection P_n onto the domain of $S(\theta_n)$ commutes with T for every n, it follows that $M = \bigoplus_n M_n$, where $M_n = M \cap \mathbf{dom}\, S(\theta_n)$ $(n \in N)$. Taking into account that $\theta_n(e^{it})$ is an isometry on a set of positive measure, we infer by [21, Theorem 3.8] that $M_n \in \mathbf{Lat}_1 S(\theta_n) \subset \mathbf{Lat}_1 T$ for every n. Hence we obtain that $M = \bigvee_n M_n$ belongs to **Hyplat**$_1$ T (cf. [11]). Let $\alpha \subset \mathbf{C}$ be a Borel set such that $q_T(\hat{\alpha}) = M$, and let us consider the factorization $\Theta = \mathrm{diag}(\theta''_n)\,\mathrm{diag}(\theta'_n)$ where θ_n's are outer functions such that $|\theta'_n(e^{it})| = \chi_\alpha(e^{it})|\theta_n(e^{it})| + \chi_{\mathbf{C}\setminus\alpha}(e^{it})$ a.e. and $\theta''_n = \theta_n\theta'^{-1}_n$ $(n \in N)$. From the uniqueness of a factorization of this type it follows that the characteristic function of $T|M$ coincides with $\mathrm{diag}\,(\theta'_n)$.

Taking into account that M is non-zero subspace, we infer that $m(\alpha) > 0$, hence $m(\alpha \cap \beta_{n_o}) > 0$ for a suitable $n_o \in N$. Let $\{k_n\}_n$ be a subsequence of natural numbers such that $\alpha_{k_n} = \beta_{n_o}$ for every n. Then for any $\lambda = re^{i\phi} \in \mathbf{D}$ we have

$$\inf_n |\theta_n'(\lambda)| \leq \inf_n |\theta_{k_n}'(\lambda)| = \inf_n \exp[\int_0^{2\pi} P_r(t-\phi)\log|\theta_{k_n}'(e^{it})|\,dm(t)] =$$

$$= \inf_n \exp[\int_{\alpha \cap \beta_{n_o}} P_r(t-\phi)\,dm(t)\cdot\log k_n^{-1}] = 0,$$

where $P_r(t)$ denotes the Poisson-kernel. On account of [15, Theorem VI.4.1] this implies that $\lambda \in \sigma(T|M)$, and since $\lambda \in \mathbf{D}$ was arbitrary, we obtain that $\sigma(T|M) = \bar{\mathbf{D}}$.

Now we give a criterion for C_{11}-contractions to be weakly similar to a unitary in terms of their characteristic functions.

THEOREM 4. *A contraction* $T \in L(H)$ *is weakly similar to a unitary if and only if* T *belongs to* C_{11} *and its characteristic function* Θ_T *is (boundedly) invertible a.e. on the unit circle* $\mathbf{C}$.

REMARK 5. In virtue of this theorem a C_{11}-contraction T is weakly similar to a unitary both in the following two cases:

a) If every injection X in the commutant $\{T\}'$ has dense range; in particular if T is of finite multiplicity, e.g. if T has a cyclic vector (cf. [9]). (This statement has been proved before in [10, Proposition 5].)

b) If the characteristic function of T has a scalar multiple; in particular if T is similar to a unitary operator or T is a weak contraction.

To prove our theorem we need two lemmas. The first one is a reformulation of [15, Theorem IX.1.2].

LEMMA 6. *A contraction* T *is similar to a unitary operator if and only if* T *is of class* C_{11}, $\Theta_T(e^{it})$ *is invertible a.e. on* $\mathbf{C}$ *and* $\|\Theta_T^{-1}\|$ *belongs to* $L^\infty(\mathbf{C})$.

PROOF. If T is similar to a unitary then $T \in C_{11}$, $\Theta_T(\lambda)$ is invertible for every $\lambda \in \mathbf{D}$ and $\|\Theta_T(\lambda)^{-1}\|$ has a bound K independent of λ (cf. [15, Theorem IX.1.2]). Hence for every unit vector $f \in (\mathbf{ran}(I - T^*T))^-$ and $\lambda \in \mathbf{D}$ we have $\|\Theta_T(\lambda)f\| \geq K^{-1}$. Since $\Theta_T(e^{it})$ is the non-tangential limit of $\Theta_T(\lambda)$ a.e. in the strong operator topology, we infer that $\|\Theta_T(e^{it})f\| \geq K^{-1}$ a.e. . Taking into account that the same is true for $\Theta_{T^*}(\lambda) = \Theta_T(\bar{\lambda})^*$, we obtain that $\|\Theta_T(e^{it})^{-1}\| \leq K$ a.e..

Conversely, let us assume that $T \in C_{11}$ and $\|\Theta_T^{-1}\| \in L^\infty(\mathbf{C})$. Then on account of [15, Proposition V.7.1 and V.4.1] Θ_T has an outer function scalar multiple, which is invertible in the Hardy space H^∞. This yields that $\|\Theta_T(\lambda)^{-1}\|$ is bounded in $\mathbf{D}$ and so T is similar to a unitary.

Hyplat$_1$ T is generally not a sublattice of **Hyplat** T (cf. [11]). The following lemma provides an example when C_{11}-intersection and intersection coincide.

LEMMA 7. *Let* T *be a* C_{11}*-contraction and* $\{H_\alpha\}_{\alpha\in A} \subset$ **Hyplat**$_1$ T. *If* $\overset{(1)}{\bigcap_{\alpha\in A}} H_\alpha = \{0\}$, *then* $\bigcap_{\alpha\in A} H_\alpha = \{0\}$.

PROOF. Let U be a unitary operator quasi-similar to T, and $X \in I(T,U)$ a quasi-affinity. It is known (cf. e.g. [11]) that $K_\alpha = (XH_\alpha)^- \in$ **Hyplat** U = **Hyplat**$_1$ U for every $\alpha \in A$. For the subspace $L = \bigcap_{\alpha\in A} H_\alpha$ we clearly have $XL \subset (XH_\alpha)^- = K_\alpha$ $(\alpha \in A)$, hence $XL \subset (\bigcap_{\alpha\in A} K_\alpha)$. Taking into account that the mapping $M \to (XM)^-$ implements an isomorphism between the lattices **Hyplat**$_1$ T and **Hyplat** U (cf. [11]), the relation $\overset{(1)}{\bigcap_{\alpha\in A}} H_\alpha = \{0\}$ implies that $\bigcap_{\alpha\in A} K_\alpha = \overset{(1)}{\bigcap_{\alpha\in A}} K_\alpha = \{0\}$. Therefore $XL = \{0\}$, and by the injectivity of X we conclude that $L = \{0\}$.

PROOF OF THEOREM 4. On account of [10, Lemmas 1, 2] and Lemma 6 we may assume that T is c.n.u..

a) First we prove the sufficiency of our condition. So we assume that $T \in C_{11}$ and $\Theta_T(e^{it})$ is invertible a.e. on $\mathbf{C}$. Let $\{\rho_n\}_{n=1}^\infty$ be an increasing sequence of real numbers tending to infinity, where $\rho_1 = 1$. For every $n \in \mathbf{N}$ let α_n be the Borel set

$$\alpha_n = \{e^{it} \in \mathbf{C} : \rho_n \leq \|\Theta_T(e^{it})^{-1}\| < \rho_{n+1}\},$$

and $H_n = q_T(\hat{\alpha}_n)$ the corresponding subspace in **Hyplat**$_1$ T. Since

$$\|\Theta_{T|H_n}(e^{it})^{-1}\| = \chi_{\alpha_n}(e^{it})\,\|\Theta_T(e^{it})^{-1}\| + \chi_{\mathbf{C}\setminus\alpha_n}(e^{it}) \leq \rho_{n+1}$$

a.e., we infer by Lemma 6 that $T|H_n$ is similar to a unitary operator.

Let n be an arbitrary natural number. Taking into account that $q_T : B_T(C) \to$ **Hyplat**$_1$ T is a lattice isomorphism, the relations $(\alpha_n \cup (\bigcup_{k\neq n} \alpha_k))^\wedge = \mathbf{C}^\wedge$ and $\alpha_n \cap (\bigcup_{k\neq n} \alpha_k) = \emptyset$ imply that $H_n \vee H_n' = H$ and $H_n \overset{(1)}{\cap} H_n' = \{0\}$, where $H_n' := \bigvee_{k\neq n} H_k = q_T((\bigcup_{k\neq n} \alpha_k)^\wedge)$.

Similarly, in virtue of $\bigcap_n (\bigcup_{k\geq n} \alpha_k) = \emptyset$ we infer $\overset{(1)}{\bigcap} (\bigvee_{k\geq n} H_k) = \{0\}$. Applying Lemma 7 we obtain $H_n \cap H'_n = \{0\}$ and $\bigcap_n (\bigvee_{k\geq n} H_k) = \{0\}$.

Let us consider the regular factorization $\Theta_T = \Theta_{2,\alpha_n} \Theta_{1,\alpha_n}$ of Θ_T corresponding to the invariant subspace H_n. Since the equality $\Theta_{1,\alpha_n}(\lambda)\Theta_{1,\alpha_n}(\lambda)^{-1} = I$ $(\lambda \in \mathbf{D})$ holds with the bounded analytic function $\Theta_{1,\alpha_n}(\lambda)^{-1}$, it follows by a result of Teodorescu (cf. [20]) that there exists an invariant subspace H''_n of T such that H is the direct (non-necessarily orthogonal) sum of the subspaces H_n and $H''_n : H_n \dot{+} H''_n = H$. T is similar to the orthogonal sum $(T|H_n) \oplus (T|H''_n)$, hence $T|H''_n \in C_{11}$.

Let $P_{H''_n}$ denote the projection onto the subspace H''_n with respect to the decomposition $H = H_n \dot{+} H''_n$. The operator $P_{H''_n}|H'_n \in I(T|H'_n, T|H''_n)$ is obviously a quasi--affinity. Since $T|H'_n$ and $T|H''_n$ are quasi-similar to unitaries, it follows that they are quasi-similar to each other (cf. [15, Proposition II.3.4]). Hence **rank** $\Delta_{T|H'_n}(e^{it}) =$ $=$ **rank** $\Delta_{T|H''_n}(e^{it})$ a.e. (cf. [9, Corollary 1]), and so taking into consideration that $H'_n \in$ **Hyplat**$_1$ T we infer by [15, Theorem VII.5.2] that $H''_n \subset H'_n$. We conclude that $H_n + H'_n \supset H_n + H''_n = H$, consequently $H_n \dot{+} H'_n = H$.

We have proved that $\{H_n\}_n$ is an appropriate basic system.

b) Now we turn to the proof of necessity. If T is weakly similar to a unitary, then $T \in C_{11}$ and we can find a sequence $\{H_n\}_n$ of subspaces from **Hyplat**$_1$ T such that $\bigvee_n H_n = H$ and $T|H_n$ is similar to a unitary for every n. Let $\alpha_n \subset \mathbf{C}$ be a Borel set such that $q_T(\hat{\alpha}_n) = H_n$ $(n \in \mathbf{N})$. The equality $\bigvee_n H_n = H$ implies that $(\bigcup_n \alpha_n)^\wedge = \mathbf{C}^\wedge$, hence $\Theta_T(e^{it})$ is isometric a.e. on $\mathbf{C} \setminus (\bigcup_n \alpha_n)$. T being a C_{11}-contraction Θ_T is outer from both sides, and so $\Theta_T(e^{it})$ is a quasi-affinity a.e. on $\mathbf{C}$. We infer that $\Theta_T(e^{it})$ is unitary a.e. on $\mathbf{C} \setminus (\bigcup_n \alpha_n)$.

On the other hand, for a.e. $e^{it} \in \alpha_n$ we have $\| \Theta_T(e^{it})^{-1} \| = \| \Theta_{T|H_n}(e^{it})^{-1} \|$, and the latter norm is finite on account of Lemma 6. Summing up $\| \Theta_T(e^{it})^{-1} \|$ is finite a.e. on $\mathbf{C}$, and the proof is completed.

Now we list some immediate consequences of this theorem.

COROLLARY 8. (i) *If* T *is a contraction weakly similar to a unitary, then so* *are its adjoint* T^* *and its restriction* $T|M$ *to any subspace* $M \in \mathbf{Lat}_1 T$.

(ii) *If the contractions* T_1 *and* T_2 *are weakly similar to unitaries, then so is their orthogonal sum* $T_1 \oplus T_2$.

(iii) *If* $T \in L(H)$ *is a* C_{11}*-contraction and there is a system* $\{H_\alpha\}_{\alpha\in A} \subset \mathbf{Hyplat}_1 T$ *such that* $\bigvee_{\alpha\in A} H_\alpha = H$ *and* $T|H_\alpha$ *is weakly similar to a unitary for every* $\alpha \in A$*, then* T *is also weakly similar to a unitary.*

(iv) *If* T_1 *and* T_2 *are similar* C_{11}*-contractions and* $\Theta_{T_1}(e^{it})$ *is invertible a.e. on* $\mathbf{C}$*, then so is* $\Theta_{T_2}(e^{it})$ *too.*

PROOF. On account of Theorem 4, the first part of (i) and (ii) follow from the facts that $\Theta_{T^*}(\lambda) = \Theta_T(\bar{\lambda})^*$ and $\Theta_{T_1 \oplus T_2}(\lambda) = \Theta_{T_1}(\lambda) \oplus \Theta_{T_2}(\lambda)$, respectively. In proving the second part of (i) we may assume that T is c.n.u. . Let $\Theta_T = \Theta_2\Theta_1$ be the regular factorization corresponding to the invariant subspace $M \in \mathbf{Lat}_1 T$. It is immediate that together with $\Theta_T(e^{it})$ the operator $\Theta_1(e^{it})$ is also bounded from below a.e. on $\mathbf{C}$. On the other hand, since $T|M \in C_{11}$ it follows that Θ_1 is outer from both sides. Hence $\Theta_1(e^{it})$ is quasi-affinity, and so invertible a.e. on $\mathbf{C}$. The proof of (iii) is the same as part b) in the proof of Theorem 4, taking into consideration that $H = \bigvee_{\alpha\in A'} H_\alpha$ for a countable subset $A' \subset A$, since H is separable; (iv) is evident.

2. THE BICOMMUTANT PROPERTY

Let $T \in L(H)$ be a contraction, weakly similar to a unitary, and $\{H_n\}_n$ a basic system in H such that $H_n \in \mathbf{Hyplat}_1 T$ and $T|H_n$ is similar to a unitary for every n.

If $M_n \in \mathbf{Lat}_1(T|H_n)$ $(n \in \mathbf{N})$, then $M = \bigvee_n M_n$ clearly belongs to $\mathbf{Lat}_1 T$. Moreover, if M_n, $N_n \in \mathbf{Lat}_1(T|H_n)$ $(n \in \mathbf{N})$ and $\bigvee_n M_n = \bigvee_n N_n$, i.e. $M_n \vee (\bigvee_{k\neq n} M_k) = N_n \vee (\bigvee_{k\neq n} N_k)$, where M_n, $N_n \subset H_n$ and $\bigvee_{k\neq n} M_k$, $\bigvee_{k\neq n} N_k \subset \bigvee_{k\neq n} H_k$ $(n \in \mathbf{N})$, then $M_n = N_n$ for every n. Conversely, let us consider a subspace $M \in \mathbf{Lat}_1 T$. It follows by [10, Lemma 5] that $M = (AH)^-$ with a suitable $A \in \{T\}'$. Since $A|H_n \in \{T|H_n\}'$, we infer that $M_n = (AH_n)^- \in$ $\in \mathbf{Lat}_1(T|H_n)$ $(n \in \mathbf{N})$. Moreover, we have $\bigvee_n M_n = \bigvee_n (AH_n)^- = (A(\bigvee_n H_n))^- = (AH)^- = M$. Summing up these facts we say that $\mathbf{Lat}_1 T$ can be decomposed into the direct sum of lattices $\mathbf{Lat}_1(T|H_n)$:

$$\mathbf{Lat}_1 T = \dot{+}_n \mathbf{Lat}_1(T|H_n).$$

We may assume that $\Theta_{T|H_n}(e^{it})$ is isometric on a set of positive measure, hence

$\mathbf{Lat}_1(T|H_n) = \mathbf{Lat}(T|H_n)$, for every n. Therefore, we conclude

$$\mathbf{Lat}_1\, T = \dot{+}_n\, \mathbf{Lat}(T|H_n),$$

and so $\mathbf{Lat}_1 T = \mathbf{Lat}\, T$ if and only if

$$\mathbf{Lat}\, T = \dot{+}_n\, \mathbf{Lat}(T|H_n).$$

In the following theorem extending results of Wu (cf. [21, Theorem 3.8], [22, Theorem 3] and [23, Theorem 3]), and applying his method, we give among other things a sufficient condition for the coincidence $\mathbf{Lat}\, T = \mathbf{Lat}_1\, T$.

THEOREM 9. *Let T be a contraction, weakly similar to a unitary such that $\Theta_T(e^{it})$ is isometric on a set α of positive measure, and the absolutely continuous unitary part of T is unitarily equivalent to an operator of the form $\oplus_n M_{\alpha_n}$ where $\mathbf{C} \supset \alpha_1 \supset \alpha_2 \supset \dots$ and $m((\mathbf{C} \setminus \alpha_1) \cap \alpha) > 0$. If Θ_T has a scalar multiple, then*

(i) $\mathbf{Lat}\, T = \mathbf{Lat}_1\, T$, *hence* $\mathbf{Lat}\, T = \mathbf{Lat}''T$,

(ii) $\mathbf{Alg}\, T = \{T\}''$, *and*

(iii) T *is reflexive.*

PROOF. (i) On account of [10, Lemmas 1, 2] we may assume that T is c.n.u. . Let $M \in \mathbf{Lat}\, T$ be an arbitrary subspace, and $\Theta_T = \Theta_2\Theta_1$ the corresponding regular factorization. By the assumption the set $\alpha = \{e^{it} : \Theta_T(e^{it})$ is an isometry$\}$ is of positive measure. On account of [15, Proposition VII 3.3.d] we infer that $\Theta_2(e^{it})$ is also isometric a.e. on α. Hence, it follows by [15, Lemma VII.6.1] that Θ_1 has a scalar multiple too. Consequently, we obtain that $T|M \in C_{11}$ (cf. [15, Theorem V.6.2 and Proposition VI.3.5]).

(ii) Let $T^{(n)}$ denote the orthogonal sum of n copies of T $(n \in \mathbf{N})$. Since $\Theta_{T^{(n)}}(e^{it}) = \Theta_T(e^{it})^{(n)}$ a.e., it follows that the assumptions of the theorem hold for the operator $T^{(n)}$ too $(n \in \mathbf{N})$. Hence we infer by (i) that

$$\mathbf{Lat}\, T^{(n)} = \mathbf{Lat}_1\, T^{(n)} = \mathbf{Lat}''\, T^{(n)}$$

for every n.

Let us consider now an arbitrary operator $S \in \{T\}''$. Then, for every n, $S^{(n)}$ clearly belongs to $\{T^{(n)}\}''$, and so

$$\mathbf{Lat}\, S^{(n)} \supset \mathbf{Lat}''\, T^{(n)} = \mathbf{Lat}\, T^{(n)}.$$

We conclude by [14, Theorem 7.1] that $S \in \mathbf{Alg}\, T$.

(iii) The reflexivity of T follows from (ii) by a result of Takahashi (cf. [18]), which implies that the bicommutant of every C_{11}-contraction is reflexive.

In what follows we show that relations (i) and (ii) of Theorem 9 do not hold if we omit the assumption that Θ_T has a scalar multiple. Consequently, the bicommutant property $\mathbf{Alg}\, T = \{T\}''$ is not a quasi-similarity invariant in the class of contractions weakly similar to unitaries.

We need the following propositions concerning the bicommutant of contractions, weakly similar to unitaries. The proof of the first one is the same as the one of [11, Theorem 7].

PROPOSITION 10. *If T is a contraction, weakly similar to a unitary, then*

$$\mathbf{Hyplat}_1\, T = \{(\mathbf{ran}\, A)^- : A \in \{T\}''\}.$$

PROPOSITION 11. *Let T be a C_{11}-contraction. If* $\mathbf{ker}\, A_1 = \mathbf{ker}\, A_2$ *for operators* $A_1, A_2 \in \{T\}''$, *then* $(\mathbf{ran}\, A_1)^- = (\mathbf{ran}\, A_2)^-$. *In particular, if* $A \in \{T\}''$ *is injective, then it is a quasi-affinity.*

PROOF. Let us assume that $\mathbf{ker}\, A_1 = \mathbf{ker}\, A_2 = M$ for the operators A_1, $A_2 \in \{T\}''$. It is clear that $A_i|M^\perp \in I(P_{M^\perp} T|M^\perp, T|(\mathbf{ran}\, A_i)^-)$ is a quasi-affinity $(i = 1, 2)$. Since $P_{M^\perp} T|M^\perp \in C_{\cdot 1}$ and $T|(\mathbf{ran}\, A_i)^- \in C_{1\cdot}$, it follows that the operator $T|(\mathbf{ran}\, A_i)^-$ belongs to C_{11} and is quasi-similar to $P_{M^\perp} T|M^\perp$ $(i = 1, 2)$. Therefore, we infer that $(\mathbf{ran}\, A_i)^- \in \mathbf{Hyplat}_1\, T$ for $i = 1, 2$, and the operators $T|(\mathbf{ran}\, A_1)^-$ and $T|(\mathbf{ran}\, A_2)^-$ are quasi-similar. Now [11, Theorem 3] yields that $(\mathbf{ran}\, A_1)^- = (\mathbf{ran}\, A_2)^-$.

COROLLARY 12. *If* $T \in L(H)$ $(H \neq \{0\})$ *is a c.n.u. contraction, weakly similar to a unitary, then*

$$\{T\}'' \neq H^\infty(T) \ (:= \{u(T) : u \in H^\infty\}).$$

PROOF. Let U be a unitary operator, quasi-similar to T. It is known that $\mathbf{Hyplat}_1\, T$ is isomorphic with $\mathbf{Hyplat}\, U$ (cf. e.g. [11]). Hence there is a subspace $M \in \mathbf{Hyplat}_1\, T$ such that $\{0\} \neq M \neq H$. It follows by Proposition 10 that M is of the form $M = (AH)^-$, where $A \in \{T\}''$.

Let us consider now an arbitrary function $u \in H^\infty$. If $u \equiv 0$, then $u(T)H = \{0\}$. On the other hand, if $u \not\equiv 0$, then taking into account $T \in C_{11}$ we infer that $u(T)$ is injective (cf. [15, Proposition III.4.1]), hence in virtue of Proposition 11 $(u(T)H)^- = H$. Consequently, A does not belong to $H^\infty(T)$.

In connection with the following proposition we remind that a subset S of **D** is dominating in **D** if $\|u\|_\infty = \sup_{\lambda \in S} |u(\lambda)|$ for every $u \in H^\infty$ (cf. e.g. [4]).

PROPOSITION 13. *Let* T *be a c.n.u. contraction, weakly similar to a unitary. If* $\sigma(T)$ *is dominating in* **D**, *then*

(i) $H^\infty(T) = \mathbf{Alg}\, T \neq \{T\}''$,

(ii) $\mathbf{Lat}\, T \neq \mathbf{Lat}_1\, T$, *and*

(iii) T *is reflexive.*

PROOF. Since T is a C_{11}-contraction, it follows that $\sigma(T) = \sigma_e(T)$. Hence, by the results of [2] we know that $H^\infty(T) = \mathbf{Alg}\, T$ is reflexive. Then Corollary 12 implies $\mathbf{Alg}\, T \neq \{T\}''$. In order to verify (ii) it is enough to show that $\mathbf{Alg\,Lat}_1\, T = \{T\}''$. However this relation, being true for every C_{11}-contraction, follows from the proof of [18].

Indeed, let U be a unitary operator quasi-similar to T. It has been proved in [18] that intertwining quasi-affinities $X \in I(T,U)$ and $Y \in I(U,T)$ can be found such that XY is a positive operator, moreover an operator A belongs to the bicommutant $\{T\}''$ if and only if $XYHXAY \in \mathbf{Alg\,Hyplat}(XYHXY)$, for every injective, selfadjoint operator $H \in \{U\}'$.

Let $A \in \mathbf{Alg\,Lat}_1\, T$ be an arbitrary operator, and let us consider an injective, selfadjoint operator $H \in \{U\}'$ and a subspace $M \in \mathbf{Hyplat}(XYHXY)$. We have to check whether the relation $(XYHXAY)M \subset M$ holds. Since XYHXY commutes with U it follows that $\{XYHXY\}' \supset \{U\}''$, hence $M \in \mathbf{Lat}''\, U = \mathbf{Lat}_1\, U$. This implies that $(YM)^- \in \mathbf{Lat}_1\, T$ and so AYM is contained in $(YM)^-$, which immediately yields that $(XYHXAY)M \subset M$.

The following proposition shows that Theorem 9 is far from being true without the assumpution on the existence of a scalar multiple. In fact, in the quasi-similarity orbit of contractions considered we can find both ones with properties in Theorem 9 and ones with properties in Proposition 13.

PROPOSITION 14. *Let* T *be a c.n.u.* C_{11}*-contraction, and let us assume that* $\Theta_T(e^{it})$ *is isometric on a set of positive measure. Then there exist c.n.u.* C_{11} *-contractions* T_1 *and* T_2 *which are both weakly similar to unitaries and quasi-similar to* T *such that*

$$H^\infty(T_1) \neq \mathbf{Alg}\, T_1 = \{T_1\}'', \quad \mathbf{Lat}\, T_1 = \mathbf{Lat}_1\, T_1,$$

and

$$H^\infty(T_2) = \mathbf{Alg}\, T_2 \neq \{T_2\}'', \quad \mathbf{Lat}\, T_2 \neq \mathbf{Lat}_1\, T_2.$$

PROOF. T is quasi-similar to a unitary operator U of the form $U = \bigoplus_n M_{\alpha_n}$, where $\alpha_n = \{e^{it} \in \mathbf{C} : \mathbf{rank}\, \Delta_T(e^{it}) \geq n\}$ (cf. [9, Corollary 1]). Since $\Theta_T(e^{it})$ is isometric on a set of positive measure, it follows that $m(\mathbf{C} \setminus \alpha_1) > 0$.

Let T_1 be a c.n.u. contraction which is similar to U (cf.[10, Lemma 2]). Then T_1 is quasi-similar to T, hence $\mathbf{rank}\, \Delta_{T_1}(e^{it}) = \mathbf{rank}\, \Delta_T(e^{it})$ a.e., and so $\Theta_{T_1}(e^{it})$ is an isometry a.e. on the set $\mathbf{C} \setminus \alpha_1$ of positive measure. T_1 is clearly weakly similar to unitary and Θ_{T_1} has a scalar multiple. Consequently, applying Theorem 9 and Corollary 12 we obtain

$$H^{\infty}(T_1) \neq \mathbf{Alg}\, T_1 = \{T_1\}'', \quad \text{and} \quad \mathbf{Lat}\, T_1 = \mathbf{Lat}_1\, T_1.$$

On account of [3] there exists a c.n.u. contraction T_2' with cyclic vector such that T_2' is quasi-similar to M_{α_1} and $\sigma(T_2')$ is dominating in $\mathbf{D}$. Moreover, let T_2'' be a c.n.u. contraction being similar to $\bigoplus_{n \geq 2} M_{\alpha_n}$, and let T_2 be the orthogonal sum: $T_2 = T_2' \oplus T_2''$. Then T_2 is weakly similar to a unitary, quasi-similar to T and $\sigma(T_2)$ is dominating. Therefore, Proposition 13 yields that

$$H^{\infty}(T_2) = \mathbf{Alg}\, T_2 \neq \{T_2\}'' \quad \text{and} \quad \mathbf{Lat}\, T_2 \neq \mathbf{Lat}_1\, T_2.$$

REMARK. In a subsequent paper we are going to deal with contractions T, weakly similar to unitaries such that $\Theta_T(e^{it})$ is not isometric a.e. on the unit circle $\mathbf{C}$.

3. CYCLIC CONTRACTIONS

In this final section we characterize the case when the subspace lattices $\mathbf{Lat}_1\, T$ and $\mathbf{Hyplat}_1\, T$ of a C_{11}-contraction T coincide.

THEOREM 15. *For every* C_{11}*-contraction* $T \in L(H)$ *the following statements are equivalent:*

(i) T *has a cyclic vector ;*

(ii) *If* $L, M \in \mathbf{Lat}_1\, T$ *and* $T|L$ *is quasi-similar to* $T|M$, *then* $L = M$;

(iii) *If* $L, M \in \mathbf{Lat}_1\, T$ *and* $T|L$ *is similar to* $T|M$ *then* $L = M$;

(iv) $\mathbf{Lat}_1\, T = \mathbf{Hyplat}_1\, T$;

(v) $\{T\}'' = \{T\}'$;

(vi) $\mathbf{Lat}''\, T = \mathbf{Hyplat}\, T$.

We note that the implication (i)$\Longrightarrow$(v) follows from the general theorem of [17] and that the equivalence (i)$\Longleftrightarrow$(ii) was proved in [10, Corollary 1]. For the sake of readers' convenience we give here a complete proof of this theorem.

PROOF. (i)$\Longrightarrow$(ii) Let U be a unitary operator quasi-similar to T. Since T has a cyclic vector, so does U too, hence U possesses property (ii). Let $X \in I(T,U)$ be a quasi-affinity. On account of [10, Theorem 1] the mapping

$$\phi_X : \mathbf{Lat}_1 T \to \mathbf{Lat}_1 U, \quad \phi_X : L \to (XL)^-$$

is a lattice-isomorphism such that $T|L$ is quasi-similar to $U|(XL)^-$ for every $L \in \mathbf{Lat}_1 T$. This yields that T also has property (ii).

(ii)$\Longrightarrow$(iii) trivial.

(iii)$\Longrightarrow$(iv) Let $L \in \mathbf{Lat}_1 T$ be arbitrary. Since every operator $A \in \{T\}'$ can be written in the form $A = \lambda I + (A - \lambda I)$, where $|\lambda| > \|A\|$ and $(A - \lambda I) \in \{T\}'$, it is clear that the subspace

$$M = \bigvee \{AL : A \in \{T\}' \text{ is invertible}\}$$

belongs to $\mathbf{Hyplat}_1 T$ (cf. also [10, Lemma 5]). However, $T|L$ being similar to $T|AL$ for an invertible A, it follows by the assumption that $AL = L$, and so $M = L$.

(iv)$\Longrightarrow$(v) This follows from the facts that for every C_{11}-contraction T we have

$$\mathbf{Alg\,Lat}_1 T = \{T\}'' \quad \text{and} \quad \mathbf{Alg\,Hyplat}_1 T = \{T\}'.$$

In connection with the first relation we refer to the proof of Proposition 13, while the proof of the second equality follows the outline of the proof of [16, Proposition 4.1]. For the sake of completeness we give the details.

Let $A \in \mathbf{Alg\,Hyplat}_1 T$ be an arbitrary operator. Let us consider a unitary operator U, quasi-similar to T, and intertwining quasi-affinities $X \in I(T,U)$ and $Y \in I(U,T)$. For an arbitrary subspace $M \in \mathbf{Hyplat}\, U = \mathbf{Hyplat}_1 U$, $(YM)^-$ is contained in the C_{11}-hyperinvariant subspace $N = \bigvee\{CYM : C \in \{T\}'\} \in \mathbf{Hyplat}_1 T$ (cf. [10, Lemma 5 and Proposition 1]). Therefore A leaves invariant N, and so we infer that $XAYM \subset XAN \subset \subset XN \subset \bigvee\{XCYM : C \in \{T\}'\} \subset M$, using the fact that XCY commutes with U if C commutes with T. Hence XAY belongs to $\mathbf{Alg\,Hyplat}\, U$, and since $\{U\}'$ is reflexive (cf. [14, Chapter 7]), it follows that $XAY \in \{U\}'$. Then, taking into account that X and Y are quasi-affinities, the commuting relation $X(AT)Y = (XAY)U = U(XAY) = X(TA)Y$ implies that $AT = TA$.

(v)$\Longrightarrow$(vi) trivial.

(vi)$\Longrightarrow$(i) It is immediate that $\mathbf{Lat}_1 T = \mathbf{Hyplat}_1 T$. Then for any injection $A \in \{T\}'$ we get that $(\mathbf{ran}\, A)^- \in \mathbf{Lat}_1 T = \mathbf{Hyplat}_1 T$, and T is quasi-similar to $T|(\mathbf{ran}\, A)^-$.

Hence, it follows by [11, Theorem 3] that $(\mathbf{ran}\,A)^- = H$.

Let U be a unitary operator, quasi-similar to T, and $X \in I(T,U)$ a quasi-affinity. On account of [10, Theorems 1, 2] we know that the mappings

$$\phi_X : \mathbf{Lat}_1\, T \to \mathbf{Lat}_1\, U, \quad \phi_X : L \to (XL)^-$$

and

$$q_X : \mathbf{Hyplat}_1\, T \to \mathbf{Hyplat}_1\, U, \quad q_X : L \to (XL)^-$$

are lattice-isomorphisms. This yields that $\mathbf{Lat}_1\, U = \mathbf{Hyplat}_1\, U$. Consequently, U and so T also have cyclic vectors.

REFERENCES

1. **Apostol, C.** : Operators quasi-similar to a normal operator, *Proc. Amer. Math. Soc.* **53**(1975), 104-106.
2. **Bercovici, H. ; Foiaș, C. ; Langsam, J. ; Pearcy, C.** : BCP-operators are reflexive, *Michigan Math. J.* **29**(1982), 371-379.
3. **Bercovici, H. ; Kérchy, L.** : On the spectra of C_{11}-contractions, *Proc. Amer. Math. Soc.* to appear.
4. **Brown, S. ; Chevreau, B. ; Pearcy, C.** : Contractions with rich spectrum have invariant subspaces, *J. Operator Theory* **1**(1979), 123-136.
5. **Dunford, N. ; Schwartz, J.** : *Linear operators. Part II*, New York–London, 1963.
6. **Foiaș, C.** : Spectral maximal spaces and decomposable operators in Banach spaces, *Archiv. Math.* **14**(1963), 341-349.
7. **Halmos, P. R.** : Ten problems in Hilbert space, *Bull. Amer. Math. Soc.* **76**(1970), 887-933.
8. **Jafarian, A.** : The weak contractions of Sz.-Nagy and Foiaș are decomposable, *Rev. Roumaine Math.Pures Appl.* **22**(1977), 489-497.
9. **Kérchy, L.** : On the commutant of C_{11}-contractions, *Acta Sci. Math. (Szeged)* **43**(1981), 15-26.
10. **Kérchy, L.** : On invariant subspace lattices of C_{11}-contractions, *Acta Sci. Math. (Szeged)* **43**(1981), 281-293.
11. **Kérchy, L.** : Subspace lattices connected with C_{11}-contractions, in *Anniversary Volume on Approximation Theory and Functional Analysis* (eds. P.L. Butzer, R.L. Stens, B. Sz.-Nagy), Birkhäuser Verlag, Basel–Stuttgart, 1984, pp. 89-98.
12. **Kérchy, L.** : Approximation and quasisimilarity, to appear in *Acta Sci. Math. (Szeged).*
13. **Lange, R.** : On weak contractions, *Bull. London Math. Soc.* **13**(1981), 69-72.
14. **Radjavi, H. ; Rosenthal, P.** : *Invariant subspaces*, Springer-Verlag, New York, 1973.
15. **Sz.-Nagy, B. ; Foiaș, C.** : *Harmonic analysis of operators on Hilbert space*, North Holland–Akádémiai Kiadó, Amsterdam–Budapest, 1970.

16. **Bercovici, H. ; Foiaş, C. ; Sz.-Nagy, B.** : Reflexive and hyper-reflexive operators of class C_0, *Acta Sci. Math. (Szeged)* **43**(1981), 5-13.

17. **Sz.-Nagy, B. ; Foiaş, C.** : Vecteurs cycliques et commutativité des commutants, *Acta Sci. Math. (Szeged)* **32**(1971), 177-183.

18. **Takahashi, K.** : Double commutants of operators quasi-similar to normal operators, preprint.

19. **Teodorescu, R. I.** : Factorisations régulières et sous-espaces hyperinvariants, *Acta Sci. Math. (Szeged)* **40**(1978), 389-396.

20. **Teodorescu, R. I.** : Sur les décompositions directes des contractions de l'espace de Hilbert, *J. Functional Analysis* **18**(1975), 414-428.

21. **Wu, P. Y.** : Bi-invariant subspaces of weak contractions, *J. Operator Theory* **1** (1979), 261-272.

22. **Wu, P. Y.** : C_{11}-contractions are reflexive, *Proc. Amer. Math. Soc.* **77**(1979), 68-72.

23. **Wu, P. Y.** : C_{11}-contractions are reflexive. II, *Proc. Amer. Math. Soc.* **82**(1981), 226-230.

L.Kérchy

Bolyai Institute, University of Szeged,
Aradi vértanuk tere 1, 6720 Szeged,
Hungary.

Operator Theory:
Advances and Applications, Vol.17

A CHARACTERIZATION OF GENERALIZED ZEROS OF NEGATIVE TYPE OF FUNCTIONS OF THE CLASS N_κ

Heinz Langer

1. INTRODUCTION

Recall ([1], [2], [3]) that N_κ denotes the set of all complex valued functions Q which are meromorphic in the open upper half plane $\mathbf{C}^+$ and such that the kernel N_Q:

(1.1) $$N_Q(z,\zeta) := (Q(z) - \overline{Q(\zeta)})/(z - \overline{\zeta})$$

for $z,\zeta \in D_Q$ has κ negative squares (here D_Q ($\subset \mathbf{C}^+$) denotes the domain of holomorphy of Q). This means that for arbitrary $n \in \mathbf{Z}$ and $z_1, z_2, \ldots, z_n \in D_Q$ the matrix $(N_Q(z_i,z_j))_1^n$ has at most κ negative eigenvalues and for at least one choice of n, $z_1, \ldots, z_n$ it has exactly κ negative eigenvalues. The class N_0 coincides with the Nevanlinna class of all functions which are holomorphic in $\mathbf{C}^+$ and map $\mathbf{C}^+$ into $\mathbf{C}^+ \cup \mathbf{R}$. The following two examples of functions of the class N_1 were considered in [2], [4], respectively:

(1.2) $$w(z) := \alpha - z + \int_{-\infty}^{\infty} ((t - z)^{-1} - t(1 + t^2)^{-1})d\sigma_0(t),$$
$$v(z) := \alpha + (1/z) + \int_{-\infty}^{\infty} ((t - z)^{-1} - t(1 + t^2)^{-1})d\sigma_1(t),$$

where $\alpha \in \mathbf{R}$ and σ_0, σ_1 are nondecreasing functions on $\mathbf{R}$ such that

$$\int_{-\infty}^{\infty} (1 + t^2)^{-1} d\sigma_j(t) < \infty, \quad j = 0,1, \quad \sigma_1(0+) = \sigma_1(0-).$$

The function $Q \in N_\kappa$ can be extended to the lower half plane $\mathbf{C}^-$ putting

$$Q(\overline{z}) := \overline{Q(z)} \qquad (z \in D_Q),$$

and possibly it can also be extended to some intervals of the closed real axis $\mathbf{R} \cup \{\infty\}$ by holomorphy (see [1]). This extended function $\tilde{Q}$ has still the property that the kernel $N_{\tilde{Q}}$ of (1.1) with $z,\zeta \in D_{\tilde{Q}}$ and $N_{\tilde{Q}}(z,\overline{z}) := \tilde{Q}'(z)$ has κ negative squares. We often suppose that the function Q has been extended in this (unique) way and denote the extension also by Q.

Let $Q \in N_\kappa$. The following definitions were given in [2]. The point $z_0 \in \mathbf{R} \cup \{\infty\}$ is called a *generalized zero (pole) of negative type of* Q *of multiplicity* $\pi(z_0)$ if for each sufficiently small neighbourhood U of z_0 there exists a $\delta_U > 0$ such that $0 < \alpha < \delta_U$ ($\delta_U < \alpha < \infty$, respectively) implies that the equation $Q(z) = -i\alpha$ has $\pi(z_0)$ solutions in $U \cap \mathbf{C}^+$. Here for $z_0 = \infty$ "sufficiently small" means that U is contained in the complement of a sufficiently large disc around zero. If $z_0 \in \mathbf{C}^+$ is a zero (pole) of Q the zero multiplicity (pole multiplicity) of z_0 is defined as the order of the zero (pole, respectively) z_0. It is well known that these multiplicities can be characterized in the same way as the multiplicities of generalized zeros or poles have been defined above. Recall (see [2]) that for $Q \in N_\kappa$ the total multiplicity of the zeros (poles) in $\mathbf{C}^+$ and of the generalized zeros (poles, respectively) of negative type on $\mathbf{R} \cup \{\infty\}$ equals κ.

If $Q \in N_\kappa$ then also the function $\hat{Q}$:

$$\hat{Q}(z) := -(Q(z))^{-1} \qquad (z \in D_Q,\ Q(z) \neq 0)$$

belongs to N_κ. It follows immediately from the definitions that the poles (zeros) or generalized poles (zeros) of negative type of Q are the zeros (poles) or generalized zeros (poles, respectively) of negative type of $\hat{Q}$.

The poles in $\mathbf{C}^+$ or generalized poles of negative type on $\mathbf{R} \cup \{\infty\}$ of $Q \in N_\kappa$ can be characterized as follows. The function Q admits an operator representation

$$Q(z) = \alpha - \tfrac{1}{2}(z_1 - \bar{z}_1)[(U + \lambda(z)I)(U - \lambda(z)I)^{-1}u, u] \tag{1.3}$$

with a cyclic π - unitary operator U in some π_κ - space Π_κ and a generating element $u \in \Pi_\kappa$, $z_1 \in \mathbf{C}^+ \cap D_Q$, $\alpha \in \mathbf{R}$ and

$$\lambda(z) := (z - \bar{z}_1)(z - z_1)^{-1}$$

(see [1], [2]). Then $z_0 \in \mathbf{C}^+$ ($z_0 \in \mathbf{R} \cup \{\infty\}$) is a pole (a generalized pole of negative type) of Q of multiplicity $\pi(z_0)$ if and only if $\lambda(z_0)$ is an eigenvalue of algebraic multiplicity $\pi(z_0)$ of $U|L_0$, where L_0 denotes a κ - dimensional nonpositive subspace of Π_κ which is invariant under U and such that $|\sigma(U|L_0)| \geq 1$.

It is the aim of this note to characterize the generalized zeros of negative type of $Q \in N_\kappa$ by the boundary behavior and the spectral measure of Q. For the special functions w and v above these characterizations were given in [2] and [4].

2. AUXILIARY STATEMENTS

For the convenience of the reader we start with two statements about functions $Q \in N_\kappa$ which were proved earlier (see [1], [3]).

(1) Suppose that the function $Q \in N_\kappa$ has a pole (generalized pole of negative type) at $z_0 \in \mathbf{C}^+$ ($z_0 \in \mathbf{R} \cup \{\infty\}$) of multiplicity $\pi(z_0)$ and let U be a sufficiently small neighbourhood of z_0. Then Q can be decomposed as $Q = Q_0 + Q_1$ such that $Q_0 \in N_{\pi(z_0)}$, z_0 is a pole (generalized pole of negative type, respectively) of multiplicity $\pi(z_0)$ of Q_0 and Q_0 is holomorphic outside of $U \cup U^*$, $Q_1 \in N_{\kappa-\pi(z_0)}$ and Q_1 is holomorphic in $U \cup U^*$, where $U^* := \{\bar{z} : z \in U\}$.

(2) If the function $Q \in N_\kappa$ has a generalized pole of negative type of multiplicity κ at $z_0 = 0$ and is holomorphic at ∞ then it admits a unique representation

$$Q(z) = z^{-2\rho} \int_\Delta (t - z)^{-1} d\sigma(t) + a_0 + a_1 z^{-1} + \ldots + a_\beta z^{-\beta} \tag{2.1}$$

where $a_j \in \mathbf{R}$ ($j = 0,1,\ldots,\beta$), $a_\beta \neq 0$ (otherwise the sum disappears), σ is a bounded nondecreasing function on some compact interval $\Delta \subset \mathbf{R}$, $\sigma(0+) = \sigma(0-)$, $\int_\Delta t^{-2} d\sigma(t) = \infty$, $\rho \in \mathbf{Z}$ and

$$\kappa = \max(\rho\,, [(\beta + 1)/2]) - \delta \quad ^{\dagger)}$$

with

$$\delta := \begin{cases} 1 & \text{if } \beta > 2\rho,\ \beta \text{ odd},\ a_\beta < 0, \\ 0 & \text{otherwise.} \end{cases} \tag{2.2}$$

The representation (2.1) of Q with the above properties is called *irreducible*.

In the following, if $z_0 \in \mathbf{R}$ then $\widehat{\lim_{z\to z_0}}$, $\widehat{\lim_{z,z'\to z_0}}$ or $z \hat{\to} z_0$ denote the nontangential limit from the upper half plane.

LEMMA 2.1. *If* $Q \in N_\kappa$ *and* $z_0 \in \mathbf{R}$ *is a generalized pole (zero) of negative type of* Q *then the following relations hold:*

$$\widehat{\lim_{z\to z_0}} |Q(z)| = \infty \qquad (\widehat{\lim_{z\to z_0}} |Q(z)| = 0, \textit{ respectively}).$$

PROOF. Here we prove only the weaker statements

$$\lim_{y\downarrow 0} |Q(z_0 + iy)| = \infty \qquad (\lim_{y\downarrow 0} |Q(z_0 + iy)| = 0, \text{ respectively}), \tag{2.3}$$

their extension to nontangential limits will be given in Theorem 3.1.

In order to prove the first relation in (2.3) we decompose Q according to (1). Then it is sufficient to prove the statement for Q_0 instead of Q, and we can also assume that $z_0 = 0$. Thus we can suppose that Q admits a representation (2.1).

If $\rho = 0$ then $\beta \geq 1$ and the statement follows from the relation

---

$^{\dagger)}$ [a] denotes the greatest integer $\leq$ a.

$$(2.4) \qquad \lim_{y\downarrow 0} y \int_{\Delta} (t - iy)^{-1} d\sigma(t) = 0$$

(observe that σ is continuous at $t = 0$). If $\rho \geq 1$ we multiply $Q(z)$ by $z^{2\rho-2}$ and consider the function R:

$$R(z) := z^{-2}\int(t - z)^{-1}d\sigma(t) + \sum_{\nu=1}^{\beta'} b_\nu z^{-\nu},$$

$\beta' := \beta - 2\rho + 2$ if $\beta > 2\rho - 2$, $b_\nu = a_{\nu+2\rho-2}$ $(\nu = 1, \ldots, \beta')$, and the sum is zero if $\beta \leq 2\rho - 2$. If $\beta' \leq 2$ we have

$$|\mathrm{Im}\, R(iy)| \geq (1/y^2)\int y d\sigma(t)/(t^2 + y^2) - |b_1|/y = (1/y)(\int d\sigma(t)/(t^2 + y^2) - |b_1|),$$

and the statement follows as $\int t^{-2} d\sigma(t) = \infty$. If $\beta' > 2$ we consider the function $z^2R(z)$:

$$|z^2R(z)| \geq (1/|z|^{\beta'-2})(|b_{\beta'}| - \ldots - |b_1||z|^{\beta'-1} - |z|^{\beta'-2}|\int(t - z)^{-1}d\sigma(t)|),$$

and the statement follows if we observe (2.4) and $|b_{\beta'}| > 0$. Thus the first relation in (2.3) is proved. The second relation in (2.3) follows immediately if we apply the first relation to $\hat{Q} = -Q^{-1}$, observing that z_0 is a generalized pole of negative type of $\hat{Q}$ if it is a generalized zero of negative type of Q.

REMARK 2.1. If $\rho = 0$ and $\beta > 1$, or if $\rho \geq 1$ the reasoning in the proof of Lemma 2.1 gives actually the stronger statement

$$\lim_{y\downarrow 0} y|Q(iy)| = \infty .$$

COROLLARY 2.1. *If $Q \in N_\kappa$, then the set of generalized zeros of negative type of Q and the set of generalized poles of negative type of Q are disjoint.*

3. THE MAIN RESULTS

THEOREM 3.1. *If $Q \in N_\kappa$, then $z_0 \in \mathbf{R}$ is a generalized zero of negative type of Q if and only if*

(3.1) $\qquad \widehat{\lim_{z\to z_0}} (Q(z))/(z - z_0) \qquad$ *exists and is finite and nonpositive.*

PROOF. Let $z_0 \in \mathbf{R}$ be a generalized zero of negative type of Q. Again, we prove the relation (3.1) first in the weaker form with $\lim_{y\downarrow 0}$, $z = z_0 + iy$, instead of $\widehat{\lim_{z\to z_o}}$. Consider $\hat{Q} := -Q^{-1}$. Then z_0 is a generalized pole of negative type of $\hat{Q}$. We decompose $\hat{Q}$ according to (1) with respect to $z = z_0$: $\hat{Q} = \hat{Q}_0 + \hat{Q}_1$, and suppose again that $z_0 = 0$. Then we have $\lim_{y\downarrow 0} |\hat{Q}_0(iy)| = \infty$,

$$(3.2)\qquad Q(z)/z = -(1/z)(\hat{Q}_o(z) + \hat{Q}_1(z))^{-1}$$

and

$$Q(iy)/iy = -(1/iy)(\hat{Q}_o(iy) + \hat{Q}_1(iy))^{-1} = -(1 + o(1))/(iy\hat{Q}_o(iy)) \qquad (y\downarrow 0),$$

$$\lim_{y\downarrow 0} Q(iy)/(iy) = -\lim_{y\downarrow 0} (iy\hat{Q}_o(iy))^{-1}.$$

Now consider the irreducible representation (2.1) of $\hat{Q}_o$:

$$(3.3)\qquad \hat{Q}_o(z) = z^{-2\rho}\int(t - z)^{-1}d\sigma(t) + a_o + a_1 z^{-1} + \ldots + a_\beta z^{-\beta}.$$

If $\rho = 0$, $\beta = 1$, then $a_1 > 0$ (otherwise $\hat{Q}_o$ would belong to the class N_o which is impossible) and we find

$$(3.4)\qquad \lim_{y\downarrow 0} (iy)^{-1}Q(iy) = -a_1 < 0.$$

If $\rho = 0$ and $\beta > 1$, or if $\rho \geq 1$ we find now easily (comp. the proof of Lemma 2.1 and Remark 2.1)

$$(3.5)\qquad \lim_{y\downarrow 0} (iy)^{-1}Q(iy) = 0.$$

The relations (3.4) and (3.5) also imply that

$$\lim_{y\downarrow 0} (2iy)^{-1}(Q(iy) - \overline{Q(iy)})$$

exists and is finite and nonpositive.

Now we extend the relations (3.4) and (3.5) to $\widehat{\lim_{z\to 0}}$. To this end we first observe that $z_o = 0$ is not a generalized pole of Q (see Corollary 2.1), hence we can choose a bounded interval Δ around $z_o = 0$ such that $\bar{\Delta}$ does not contain any generalized pole of negative type of Q. Then Q admits a decomposition

$$(3.6)\qquad Q = \tilde{Q}_o + \tilde{Q}_1$$

with $\tilde{Q}_o \in N_o$, $\tilde{Q}_1 \in N_\kappa$ and such that $\tilde{Q}_1$ is holomorphic on Δ and $\tilde{Q}_o$ is holomorphic outside $\bar{\Delta}$ (see [1], [3]). The function $\tilde{Q}_o$ can be chosen to have the representation

$$\tilde{Q}_o(z) = \int_\Delta (t - z)^{-1}d\sigma(t),$$

where σ is the restriction of the spectral measure of Q to the interval Δ. Further,

$$(2iy)^{-1}(\tilde{Q}_o(iy) - \overline{\tilde{Q}_o(iy)}) = (2iy)^{-1}(Q(iy) - \overline{Q(iy)}) - (2iy)^{-1}(\tilde{Q}_1(iy) - \overline{\tilde{Q}_1(iy)}),$$

and if $y \downarrow 0$ both quotients on the right hand side have a limit: For Q this was shown above, for $\hat{Q}_1$ it follows from the holomorphy at $z_0 = 0$. Hence

$$\lim_{y\downarrow 0} (2iy)^{-1}(\tilde{Q}_0(iy) - \overline{\tilde{Q}_0(iy)}) = \lim_{y\downarrow 0} \int_\Delta (t^2 + y^2)^{-1} d\sigma(t)$$

exists, that is

$$\int_\Delta t^{-2} d\sigma(t) < \infty .$$

Using this fact and the decomposition (3.6) of Q it is easy to see that $\widehat{\lim_{z\to 0}} z^{-1}Q(z)$ exists and is finite. Indeed, it is sufficient to show that

$$\widehat{\lim_{z\to 0}} z^{-1}(\tilde{Q}_0(z) - \tilde{Q}_0(0)), \qquad \tilde{Q}_0(0) := \int_\Delta t^{-1} d\sigma(t)$$

exists. If z is such that

$$\pi/2 - \gamma \leq \arg z \leq \pi/2 + \gamma$$

for some γ, $0 < \gamma < \pi/2$, we have for $t \in \mathbf{R}$

$$| t(t - z)^{-1} | \leq (\cos\gamma)^{-1} ,$$

and

$$\widehat{\lim_{z\to 0}} z^{-1}(\tilde{Q}_0(z) - \tilde{Q}_0(0)) - \int_\Delta t^{-2} d\sigma(t) = \widehat{\lim_{z\to 0}} \int_\Delta (t(t - z)^{-1} - 1)t^{-2} d\sigma(t) = 0.$$

Conversely, let $z_0 \in \mathbf{R}$ be such that (3.1) holds and assume that z_0 is not a generalized zero of negative type of Q. Then it is not a generalized pole of negative type of $\hat{Q} = -Q^{-1}$, and we choose a decomposition

$$\hat{Q} = \hat{Q}_0 + \hat{Q}_1 ,$$

such that $\hat{Q}_0 \in N_0$, $\hat{Q}_1 \in N_\kappa$ and $\hat{Q}_1$ is holomorphic at z_0 , $\hat{Q}_0$ is holomorphic outside of a neighbourhood of z_0. As $\widehat{\lim_{z\to z_0}} | \hat{Q}(z) | = \infty$ it follows that $\widehat{\lim_{z\to z_0}} | \hat{Q}_0(z) | = \infty$ and

$$\widehat{\lim_{z\to z_0}} (z - z_0)^{-1}Q(z) = -\widehat{\lim_{z\to z_0}} (z - z_0)^{-1}(\hat{Q}_0(z)(1 + \hat{Q}_1(z)\hat{Q}_0(z)^{-1})^{-1}) =$$

$$= -\widehat{\lim_{z\to z_0}} (z - z_0)^{-1}\hat{Q}_0(z) = (\hat{\sigma}(z_0+) - \hat{\sigma}(z_0-))^{-1} ,$$

where $\hat{\sigma}$ is the spectral measure of $\hat{Q}$. The last expression is either positive or ∞ , a contradiction to the hypothesis (3.1). The theorem is proved.

REMARK 3.1. Evidently, the relation (3.1) implies that

$$\widehat{\lim_{z \to z_o}} Q(z) = 0.$$

Thus, a generalized zero of negative type of Q is characterized by the properties that the nontangential limit of Q at z_o is zero and, additionally, the nontangential derivative (or derivative in Caratheodory's sense) exists in z_o and is nonpositive. We mention that this derivative is negative if and only if $\lambda(z_o)$ is an eigenvalue of negative (or elliptic) type of the π - unitary operator U in a representation (1.3) of $\hat{Q} = -Q^{-1}$, that is if and only if in case $z_o = 0$ the representation (3.3) reduces to

$$\hat{Q}_o(z) = \int (t - z)^{-1} d\sigma(t) + a_o + a_1 z^{-1}$$

with $a_1 > 0$.

REMARK 3.2. If $z_o \in \mathbf{R}$ is a generalized zero of negative type of $Q \in N_\kappa$ it can also be characterized by the following two properties:

(i) $\widehat{\lim_{z \to z_o}} Q(z) = 0$;

(ii) $\widehat{\lim_{z,z' \to z_o}} (z - \bar{z}')^{-1}(Q(z) - \overline{Q(z')})$ exists and is finite and nonpositive.

If we introduce the spectral measure σ of Q then the existence of the limit in (ii) is equivalent to the relation

$$\int_\Delta (t - z_o)^{-2} d\sigma(t) < \infty,$$

where Δ is a sufficiently small interval around z_o.

REMARK 3.3. The statement of Theorem 3.1 is closely related to the following simple observation. If $Q \in N_o$ then also $\hat{Q} = -Q^{-1} \in N_o$. Denote the spectral measure of Q and $\hat{Q}$ by σ and $\hat{\sigma}$, respectively, e.g.

$$Q(z) = \alpha + \beta z + \int_{-\infty}^{\infty} ((t - z)^{-1} - t(1 + t^2)^{-1}) d\sigma(t)$$

($\alpha \in \mathbf{R}$, $\beta \geq 0$, σ nondecreasing, $\int (1 + t^2)^{-1} d\sigma(t) < \infty$). Then for arbitrary $z_o \in \mathbf{R}$ the relation

$$\int_{-\infty}^{\infty} (t - z_o)^{-2} d\sigma(t) = (\hat{\sigma}(z_o+) - \hat{\sigma}(z_o-))^{-1}$$

holds, where possibly both sides are ∞. As a consequence, if $\hat{\sigma}$ has a jump at z_o then

$$(3.7)\qquad \widehat{\lim_{z\to z_0}} (z - z_0)^{-1}Q(z) = (\hat{\sigma}(z_0+) - \hat{\sigma}(z_0 -))^{-1} > 0.$$

A particular case of this statement is the following fact which can also easily be shown directly. If A is a selfadjoint operator in a Hilbert space H with $z_0 \in \sigma_p(A)$ and $u \in H$ is not orthogonal to the eigenspace of A at z_0, then the function $Q_A(z) := -((A - zI)^{-1}u,u)$ has the properties

$$(3.8)\qquad \widehat{\lim_{z\to z_0}} Q_A(z) = 0,\quad \widehat{\lim_{z\to z_0}} (z - z_0)^{-1}Q_A(z) = \|Pu\|^{-2} > 0,$$

where P denotes the orthogonal projection onto the eigenspace of A at z_0. If (3.7) or (3.8) holds it would be natural to call z_0 a generalized zero of positive type of Q or Q_A, respectively.

In the following theorem the multiplicity of a generalized zero z_0 of negative type of $Q \in N_\kappa$ is characterized by the nontangential derivatives of higher order of Q at z_0.

THEOREM 3.2. *If* $z_0 \in \mathbf{R}$ *is a generalized zero of negative type of* $Q \in N_\kappa$ *of multiplicity* π_0 *then*

$$\widehat{\lim_{z\to z_0}} (z - z_0)^{-(2\pi_0-1)}Q(z) \text{ exists and is finite and nonpositive}$$

and

$$\widehat{\lim_{z\to z_0}} (z - z_0)^{-(2\pi_0+1)}Q(z) \text{ exists and is positive } (\leq \infty).$$

PROOF. We assume again that $z_0 = 0$. Then π_0 is the multiplicity of the generalized pole $z_o = 0$ of negative type of $\hat{Q} = -Q^{-1}$ or (with the same notation as in the proof of Theorem 3.1) of $\hat{Q}_0$. We use again the representation (3.3) of $\hat{Q}_0$ and obtain

$$z^{-(2\pi_0-1)}Q(z) = -[z^{2\pi_0-1}\hat{Q}_0(z)(1 + o(1))]^{-1} =$$
$$= -[z^{2\pi_0-1}(z^{-2\rho}\int(t - z)^{-1}d\sigma(t) + a_0 + \ldots + a_\beta z^{-\beta})(1 + o(1))]^{-1}$$

if $z \hat{\to} 0$. According to (2) we have

$$\pi_0 = \max(\rho, [(\beta + 1)/2]) - \delta$$

with δ given by (2.2). If $\rho \geq [(\beta + 1)/2]$, then $\beta \leq 2\rho$ and hence $\pi_0 = \rho\ (\geq 1)$. It follows that

$$z^{-(2\rho-1)}Q(z) = -[z^{2\rho-1}(z^{-2\rho}\int(t - z)^{-1}d\sigma(t) + a'_0 + \ldots + a'_{2\rho}z^{2\rho})\cdot(1 + o(1))]^{-1} =$$
$$= -[(z^{-1}\int(t - z)^{-1}d\sigma(t) + o(1) + a'_{2\rho-1} + a'_{2\rho}z^{-1})(1 + o(1))]^{-1}\qquad (z \hat{\to} 0)$$

where $a'_{2\rho-1}$, $a'_{2\rho}$ possibly vanish. It follows from Theorem 3.1 and the last statement in Remark 3.1 that

$$z^{-1}\int(t-z)^{-1}d\sigma(t) + a'_{2\rho-1} + a'_{2\rho}z^{-1} = z(z^{-2}\int(t-z)^{-1}d\sigma(t) + a'_{2\rho-1}z^{-1} + a'_{2\rho}z^{-2})$$

tends to ∞ if $z \hat{\to} 0$, hence

$$\widehat{\lim_{z\to 0}}\, z^{-(2\rho-1)}Q(z) = 0.$$

Moreover,

$$\widehat{\lim_{z\to 0}}\, z^{-(2\rho+1)}Q(z) = -\widehat{\lim_{z\to 0}}\, [z^{2\rho+1}(z^{-2\rho}\int(t-z)^{-1}d\sigma(t) + a'_o + \ldots + a'_{2\rho}z^{-2\rho})]^{-1} = \infty,$$

as the expression in the square brackets tends to zero if $z \hat{\to} 0$.

If $\rho < [(\beta+1)/2]$ then $\rho < k$ for $\beta = 2k$ and $\rho \leq k$ for $\beta = 2k+1$. In the first case we have $\pi_o = k$ and

$$|z^{2k-1}(z^{-2\rho}\int(t-z)^{-1}d\sigma(t) + a_o + \ldots + a_{2k}z^{-2k})| \to \infty \text{ if } z \hat{\to} 0,$$

$$z^{2k+1}(z^{-2\rho}\int(t-z)^{-1}d\sigma(t) + a_o + \ldots + a_{2k}z^{-2k}) \to 0 \text{ if } z \hat{\to} 0.$$

In the second case

$$\pi_o = \begin{cases} k & \text{if } a_\beta < 0, \\ k+1 & \text{if } a_\beta > 0 \end{cases}$$

and if $z \hat{\to} 0$:

$$|z^{2k-1}(z^{-2\rho}\int(t-z)^{-1}d\sigma(t) + a_o + \ldots + a_{2k+1}z^{-(2k+1)})| \to \infty,$$

$$z^{2k+1}(z^{-2\rho}\int(t-z)^{-1}d\sigma(t) + a_o + \ldots + a_{2k+1}z^{-(2k+1)}) \to a_{2k+1},$$

$$z^{2k+3}(z^{-2\rho}\int(t-z)^{-1}d\sigma(t) + a_o + \ldots + a_{2k+1}z^{-(2k+1)}) \to 0.$$

The theorem is proved.

In a similar way ∞ can be characterized as a generalized zero of negative type by the boundary behavior of the function $Q \in N_\kappa$. To this end we have only to observe that ∞ is a generalized zero of negative type of multiplicity $\pi(\infty)$ of $Q \in N_\kappa$ if and only if $z_o = 0$ is a generalized zero of negative type of the function Q_1:

$$Q_1(z) := Q(-z^{-1})$$

of the same multiplicity. Now Theorem 3.1 implies:

THEOREM 3.1'. *If* $Q \in N_\kappa$, *then* ∞ *is a generalized zero of negative type of* Q *if and only if*

$$\widehat{\lim_{z\to\infty}} zQ(z) \text{ exists and is finite and nonnegative.}$$

Here $\widehat{\lim_{z\to\infty}}$ means that $|z| \to \infty$ with $\operatorname{Im} z |\operatorname{Re} z|^{-1} \geq \gamma$ for some $\gamma > 0$. Theorem 3.2 can be transferred in a similar way. This is left to the reader.

4. AN EXAMPLE

Let $b_j \in \mathbf{R}$ $(j = 0,1,\dots,n)$, $b_n \neq 0$, $n \geq 1$, σ a nondecreasing function on $\mathbf{R}$, $\int_{-\infty}^{\infty} (1 - t^2)^{-1} d\sigma(t) < \infty$. We consider the function Q :

$$(4.1) \qquad Q(z) := \sum_{j=0}^{n} b_j z^j + \int_{-\infty}^{\infty} ((t - z)^{-1} - t(1 + t^2)^{-1}) d\sigma(t) .$$

It is easy to see that the function Q belongs to the class N_κ with

$$\kappa = \begin{cases} [n/2] + 1 & \text{if } n \text{ is odd and } b_n < 0 , \\ [n/2] & \text{otherwise} . \end{cases}$$

Evidently, the function w in (1.2) is of the form (4.1). Theorem 3.1 implies that $z_0 \in \mathbf{R}$ is a generalized zero of negative type of Q if and only if

$$\int_{-\infty}^{\infty} (t - z_0)^{-2} d\sigma(t) < \infty , \quad \sum_{j=0}^{n} b_j z_0^j + \int_{-\infty}^{\infty} ((t - z_0)^{-1} - t(1 + t^2)^{-1}) d\sigma(t) = 0$$

and

$$\sum_{j=0}^{n} j b_j z_0^{j-1} + \int_{-\infty}^{\infty} (t - z_0)^{-2} d\sigma(t) \leq 0 .$$

Consider again the function $\hat{Q} := -Q^{-1}$. Then $\hat{Q} \in N_\kappa$, and it admits an operator representation

$$(4.2) \qquad (Q(z)^{-1} =) \hat{Q}(z) = [(\hat{A} - zI)^{-1} u , u]$$

with a π - selfadjoint operator A in some π_κ - space Π_κ and $u \in \Pi_\kappa$ (see [1]). Here it is possible to describe Π_κ and $\hat{A}$ in more explicit terms.

We suppose first that $n \geq 2$ and $b_n > 0$. It is no loss of generality to assume that $b_n = 1$. Consider the space

$$\mathbf{C}^n \oplus L^2_\sigma ,$$

and define in this space an (indefinite) π_κ - scalar product $[x,y] := (Gx,y)$ by means of the Gram operator

$$G = \begin{bmatrix} G_{11} & G_{12} \\ G_{21} & 1 \end{bmatrix}$$

with

$$G_{11} := \begin{bmatrix} 0 & 0 & \dots & 0 & 1 \\ 0 & 0 & \dots & 1 & -b_{n-1} \\ & & \vdots & & \vdots \\ 0 & 1 & \dots & -b_3 & -b_2 \\ 1 & -b_{n-1} & \dots & -b_2 & -b_1 + \int t^2(1+t^2)^{-2} d\sigma(t) \end{bmatrix},$$

$$G_{12} := (0 \quad 0 \dots 0 \quad (\cdot, -t(1+t^2)^{-1})^T,$$

$$G_{21} := (0 \quad 0 \dots 0 \quad -t(1+t^2)^{-1}).$$

Then the operator $\hat{A}$:

$$\hat{A} := \begin{bmatrix} \hat{A}_{11} & \hat{A}_{12} \\ \hat{A}_{21} & t\cdot \end{bmatrix}$$

with

$$\hat{A}_{11} := \begin{bmatrix} -b_{n-1} & -b_{n-2} & \dots & -b_1 & -b_o \\ 1 & 0 & \dots & 0 & 0 \\ \vdots & \vdots & & \vdots & \\ 0 & 0 & \dots & 1 & 0 \end{bmatrix}$$

$$\hat{A}_{12} := ((\cdot, 1) \quad 0 \dots 0)^T,$$

$$\hat{A}_{21} := (0 \quad 0 \dots 0 \quad t(1+t^2)^{-1} \quad (1+t^2)^{-1})$$

is π - selfadjoint in this π_κ - space and for

$$u := (1 \quad 0 \dots 0 \quad 0)^T$$

the relation (4.2) holds.

If $n \geq 2$ and $b_n < 0$, we suppose without loss of generality that $b_n = -1$. Then the signs in the first line of $\hat{A}_{11}$ and of $\hat{A}_{12}$ and in the second diagonal of G_{11} have to be changed.

Finally, if $n = 1$, $b_1 = -1$, we consider in the space $\mathbf{C} \oplus L^2_\sigma$ the operator

$$\hat{A} := \begin{bmatrix} b_o & (\cdot, 1) \\ -(b_o t + 1)(1+t^2)^{-1} & t\cdot - t(1+t^2)^{-1}(\cdot, 1) \end{bmatrix}.$$

It is π-selfadjoint in the π_1 - space with the indefinite scalar product defined by means of the Gram operator

$$G := \begin{bmatrix} \int t^2(1+t^2)^{-2} d\sigma(t) - 1 & (\cdot, t(1+t^2)^{-1}) \\ t(1+t^2)^{-1} & 1 \end{bmatrix}$$

and it is easy to check that with $u := (1 \ \ - t(1+t^2)^{-1})^T$ the relation (4.2) holds.

In each of these cases the generalized zeros of negative type of Q and the zeros of Q in $\mathbf{C}^+$ form the spectrum $\sigma(\hat{A}\,|L_o)$, and the multiplicity of such a zero or generalized zero coincides with its algebraic multiplicity as an eigenvalue of $\hat{A}\,|L_o$; here L_o is a κ - dimensional nonpositive invariant subspace of $\hat{A}$ such that $\mathrm{Im}\,\sigma(\hat{A}\,|L_o) \geq 0$.

REFERENCES

1. **Kreĭn, M.G. ; Langer, H. :** Über einige Fortsetzungsprobleme, die eng mit der Theorie hermitescher Operatoren in Räume Π_κ zusammenhängen. I. Einige Funktionenklassen und ihre Darstellungen, *Math. Nachr.* **77**(1977), 187-236.

2. **Kreĭn, M.G. ; Langer, H. :** Some propositions on analytic matrix functions related to the theory of operators in the space Π_κ, *Acta Sci. Math. (Szeged)* **43** (1981), 181-205.

3. **Daho, K. ; Langer, H. :** Matrix functions of the class N_κ, *Math. Nachr.* (to appear).

4. **Kreĭn, M.G. ; Šmul'jan, Ju.L. :** On Wiener-Hopf equations whose kernels admit an integral representation by means of exponents (Russian), *Izv. Akad. Nauk Armjan. SSR Ser. Mat.* **17**(1982), 307-327.

Heinz Langer
Sektion Mathematik
Technische Universität Dresden
8027 Dresden
D D R

Operator Theory:
Advances and Applications, Vol.17

SKEW-SYMMETRIC OPERATORS AND ISOMETRIES ON A REAL BANACH SPACE WITH A HYPERORTHOGONAL BASIS

Peter Legiša

A Schauder basis $\{e_i ; i \in \mathbf{N}\}$ of a Banach space X is *hyperorthogonal* if

$$\|\sum a_i e_i\| = \|\sum |a_i| e_i\|$$

for arbitrary scalars a_i $(i \in \mathbf{N})$. In this case the following is true (see [6]):

1. PROPOSITION. *If* $x = \sum a_i e_i$ *and* $|b_i| \leq |a_i|$ *for all* $i \in \mathbf{N}$, *then* $\sum b_i e_i$ *converges and the norm of the sum is smaller or equal to* $\|x\|$.

Let X be a real Banach space with a hyperorthogonal basis $\{e_i\}$. We embed X into a complex Banach space Y with a hyperorthogonal basis in a natural way.

Let Y be the set of all complex sequences (a_i) for which the series $\sum |a_i| e_i$ converges in X. Then, by Proposition 1, Y is a linear space over $\mathbf{C}$. Define a norm in Y by $\|(a_i)\| = \|\sum |a_i| e_i\|$. Then it is not hard to see that Y becomes a Banach space. The sequences $f_i = (\delta_{ij})$ $(j \in \mathbf{N})$ form a basis of Y. For any $a = (a_i) \in Y$

$$\|a - \sum_{i=1}^{n} a_i f_i\| = \|\sum_{i=n+1}^{\infty} |a_i| e_i\| \to 0 \quad (n \to \infty).$$

Obviously, $\{f_i\}$ is a hyperorthogonal basis of Y. The map

$$x = \sum a_i e_i \to (a_i) = \sum a_i f_i \in Y$$

is an isometric embedding of X into Y. Thus we will assume that $X \subset Y$.

Let from now on $\{e_i\}$ be a normalized hyperorthogonal basis in X, i.e. $\|e_i\| = 1$ for all i. Then we can assume that $\{e_i\}$ is a normalized hyperorthogonal basis in Y. It follows that for any $y \in Y$,

$$y = \sum a_i e_i = \sum (\mathbf{Re}\, a_i) e_i + i \sum (\mathbf{Im}\, a_i) e_i = y_1 + i y_2$$

where $y_1, y_2 \in X$. We call Y the *complexification* of X.

The properties of complex Banach spaces with a hyperorthogonal basis are well known (see [1], [2], [3], [4], [5], [7], [8]). We will use the results from [4].

2. DEFINITION. An operator A on a Banach space is *skew-symmetric* iff $\|\exp(tA)\| = 1$ for all $t \in \mathbf{R}$. An operator H on a complex Banach space is *hermitian* iff iH is skew-symmetric.

Let $A(Y)$ be the linear span of all hermitian operators on Y. Theorem 2.2 in [4] states that $A(Y)$ is a type I W^*-algebra (for the natural involution which preserves hermitian operators). There exists a family $\{G_z ; z \in Z\}$ of mutually orthogonal central projections in $A(Y)$, such that $Y_z = G_z Y$ is a Hilbert subspace in Y for all z and $\sum G_z = I_Y$ in the strong operator topology.

If $N_z = \{i \in \mathbf{N} ; G_z e_i = e_i\}$, then $\{e_i ; i \in N_z\}$ is an orthonormal basis of the Hilbert subspace Y_z. Let

$$X_z = \{\sum a_i e_i \in X ; i \in N_z\}.$$

Obviously X_z is a real Hilbert space and $\{e_i ; i \in N_z\}$ is an orthogonal basis. For any $y \in Y_z$, $y = y_1 + iy_2$ where $y_1, y_2 \in X_z$ and

$$\|y\|^2 = \sum |a_i|^2 = \sum(\mathbf{Re}\, a_i)^2 + \sum(\mathbf{Im}\, a_i)^2 = \|y_1\|^2 + \|y_2\|^2.$$

Now $G_z X = X_z$. Let $G'_z = G_z|X$. Then $I_X = \sum G'_z$ $(z \in Z)$ in the strong operator topology. Thus X is a direct sum of real Hilbert spaces: $X = \sum X_z$ $(z \in Z)$.

3. PROPOSITION. *Every bounded linear operator* $A : X_z \to X_z$ *can be extended to an operator* $E(A) : Y_z \to Y_z$ *by* $E(A)(y_1 + iy_2) = Ay_1 + iAy_2$. *The map* $A \to E(A)$ *is a homomorphism such that* $\|A\| = \|E(A)\|$. *Also, if* A *is skew-symmetric (isometry),* $E(A)$ *is skew-symmetric (isometry).*

We denote by $L(X)$ the algebra of all bounded linear operators on X.

In [4, Theorem 2.2] we prove that the map

$$A \to \{G_z A G_z | Y_z ; z \in Z\}$$

is a $*$-isomorphism of the algebra $A(Y)$ on the ℓ^∞-direct sum of the algebras $L(Y_z)$.

4. PROPOSITION. *For every* $z \in Z$ *let* A_z *be the bounded linear operator on* X_z, *such that* $\sup \|A_z\| = M < \infty$. *There exists one and only one bounded linear operator* A *on* X, *such that* $A(X_z) \subset X_z$ *and* $A|X_z = A_z$. *Also,* $\|A\| = M$. *If all* A_z *are skew-symmetric (isometries), then* A *is also skew-symmetric (isometry).*

PROOF. $E(A_z)$ is a bounded linear operator on Y_z, such that $\|E(A_z)\| = \|A_z\|$.

By the above remark there exists one and only one operator $B \in L(Y)$ such that $B(Y_z) \subset Y_z$ and $B|Y_z = E(A_z)$. Also, $\|B\| = \sup\|E(A_z)\| = \sup\|A_z\| = M$. Let $A = B|X$. Then obviously $\|A\| \leq \|B\| = M$. On the other hand, $\|A\| \geq \sup\|A_z\|$. Thus $\|A\| = M$. The rest follows from Proposition 3.

5. PROPOSITION. *If* V *is a surjective linear isometry on* X, $B \in L(X)$, *then* $\|VB\| = \|BV\| = \|B\|$.

6. PROPOSITION. *An operator* $A \in L(X)$ *is skew-symmetric iff* $\|1 + tA\| = 1 + o(t)$ *as* $t \to 0$.

(The proof uses the facts that

$$\mathbf{exp}(tA) = 1 + tA + (t^2/2)A^2 + \ldots = \lim_{n \to \infty} (1 + (t/n)A)^n \ ).$$

7. PROPOSITION. *If* A,B *are skew-symmetric, so are* A + B, sA $(s \in \mathbf{R})$.

PROOF. $\|\mathbf{exp}(tA)\mathbf{exp}(tB)\| = 1$ (Proposition 5) for all $t \in \mathbf{R}$, or $\|I + t(A + B) + t^2(\ldots)\| = 1$, so $\|1 + t(A + B)\| = 1 + o(t)$ $(t \to 0)$.

We denote by $S(X)$ the real linear space (in fact, a Lie algebra) of all skew-symmetric operators on X.

In [4, Theorem 2.2] we prove that one-parameter groups of isometries on Y preserve the subspaces Y_z.

We define a relation $\backsim$ in $\mathbf{N}$ by $i \backsim j$ iff $|a_i|^2 + |a_j|^2 = |b_i|^2 + |b_j|^2$ and $|a_k| = |b_k|$ for $k \neq i,j$ implies $\|\sum a_i e_i\| = \|\sum b_i e_i\|$. It is obvious that $i \backsim j$ relative to the space X iff $i \backsim j$ relative to the space Y. Now we prove that $i \backsim j$ iff there exists $z \in Z$, such that $e_i, e_j \in Y_z$ (iff there exists $z \in Z$ such that $e_i, e_j \in X_z$). In fact, if $i \backsim j$, then define V_t by

$$V_t(\sum a_k e_k) = (a_i \mathbf{cos}\, t - a_j \mathbf{sin}\, t)e_i + (a_i \mathbf{sin}\, t + a_j \mathbf{cos}\, t)e_j + \sum_{k \neq i,j} a_k e_k.$$

Then V_t is a one parameter group of isometries. By the above remark, e_i and $V_{\pi/2} e_i = e_j$ lie in the same Y_z. The converse is clear.

8. THEOREM. *If* $A \in L(X)$ *is skew-symmetric, then* $AX_z \subset X_z$.

PROOF. We will use the method from [5]. If V is an isometry on X, then $\|\mathbf{exp}(tVAV^{-1})\| = \|V\mathbf{exp}(tA)V^{-1}\| = 1$ and so VAV^{-1} is skew-symmetric as well as $B = (A + VAV^{-1})/2$. Let (a_{ij}) be the matrix of the operator A corresponding to the basis

$\{e_i\}$. Let $V : X \to X$ be the isometry defined by $Ve_i = \lambda_i e_i$ where $\lambda_i = \pm 1$. Now

$$VAV^{-1}e_j = \lambda_j VAe_j = \lambda_j V(\sum_i a_{ij}e_i) = \lambda_j \sum_i a_{ij}Ve_i = \sum(\lambda_i\lambda_j a_{ij})e_i.$$

So B has a matrix (b_{ij}) with $b_{ij} = a_{ij}(1 + \lambda_i\lambda_j)/2$.

If $\lambda_1 = 1$ and $\lambda_i = -1$ $(i \neq 1)$, then $b_{1i} = b_{i1} = 0$ $(i \neq 1)$, $b_{11} = a_{11}$, and $b_{ij} = a_{ij}$ $(i, j > 1)$. Since $\mathbf{exp}(tB)e_1 = \mathbf{exp}(ta_{11})e_1$ and $\|\mathbf{exp}(tB)e_1\| = \|e_1\| = 1$, it follows that $a_{11} = 0$. In the same way we prove that $a_{ii} = 0$ for all i. We can also prove that the operator C with entries $c_{ij} = a_{ij}$ $(i, j \geq n)$ and $c_{ij} = 0$ otherwise is skew-symmetric.

Let now $\lambda_1 = \lambda_2 = 1$, $\lambda_i = -1$ $(i \neq 1, 2)$. By repeated use of the above argument we can construct a skew-symmetric operator B such that $b_{12} = a_{12}$, $b_{21} = a_{21}$, and all other entries 0. Thus $Be_1 = a_{21}e_2$, $Be_2 = a_{12}e_1$, $Be_i = 0$ $(2 < i)$.

Suppose now that $a_{12} \neq 0$. Then $B^2e_2 = a_{12}a_{21}e_2 = de_2$. If $d = 0$, $\|\mathbf{exp}(tB)e_2\| = \|e_2 + ta_{12}e_1\| > 1$ for $|t|$ large enough – a contradiction. If $d = k^2 > 0$, then

$$\mathbf{exp}(tB)e_2 = (a_{12}/k)(\mathbf{sh}\, kt)e_1 + (\mathbf{ch}\, kt)e_2$$

and we arrive to a contradiction as well. Thus we see that $d = -k^2 < 0$. Then $a_{12} = k$ and $a_{21} = -k$.

Let now $x = \sum a_i e_i \in X$ and $\varepsilon > 0$. Take n such that if $y = \sum a_i e_i$ $(i \geq n + 1)$, then $\|y\| < \varepsilon$. Then

$$\mathbf{exp}(tB)x = a_1\mathbf{exp}(tB)e_1 + a_2\,\mathbf{exp}(tB)e_2 + \sum_{i=3}^{n} a_i\mathbf{exp}(tB)e_i + \mathbf{exp}(tB)y =$$

$$= (a_1\mathbf{cos}\, kt - a_2\mathbf{sin}\, kt)e_1 + (a_1\mathbf{sin}\, kt + a_2\mathbf{cos}\, kt)e_2 + \sum_{i=3}^{\infty} a_i e_i +$$

$$+ \mathbf{exp}(tB)y - y = x_t + \mathbf{exp}(tB)y - y.$$

Since $\|y - \mathbf{exp}(tB)y\| \leq 2\|y\| < 2\varepsilon$, we see that

$$\|x\| - 2\varepsilon \leq \|x_t\| \leq \|x\| + 2\varepsilon.$$

Thus $\|x_t\| = \|x\|$ for all $t \in \mathbf{R}$ and thus $1 \backsim 2$. In the same way we can prove that $a_{ij} \neq 0$ implies $i \backsim j$, or equivalently, $i \not\backsim j$ implies $a_{ij} = 0$. This proves that $AX_z \subset X_z$ for all z.

In this case $A|X_z$ is skew-symmetric on the real Hilbert subspace X_z. We have thus characterized all the skew-symmetric operators (and, consequently, all one-parameter groups of isometries on X). Summary:

The space $S(X)$ of all skew-symmetric operators on X is the ℓ^{∞}-direct sum of the spaces $S(X_z)$ $(z \in \mathbf{Z})$.

9. PROPOSITION. *If $V : X \to X$ is a surjective linear isometry on X, then $\{Ve_i\}$ is also a normalized hyperorthogonal basis of X.*

From now on let V be a given surjective linear isometry on X. Let $X = \sum X'_w$ $(w \in Z')$ be the decomposition of X into a direct sum of Hilbert subspaces relative to the new basis $\{f_i\}$, $f_i = Ve_i$. Then $i \backsim j$ iff $a_i^2 + a_j^2 = b_i^2 + b_j^2$ and $|a_k| = |b_k|$ for $k \neq i,j$ imply $\| \sum a_k f_k \| = \| \sum b_k f_k \|$. Thus we may assume that $Z' = Z$.

10. PROPOSITION. *Let $L = \{x \in X \,;\, Ax = 0$ for all $A \in S(X)\}$. Then L is the closed linear span of all those e_i which lie in the one-dimensional subspaces X_z (equivalently, of all those f_i which lie in the one-dimensional subspaces X'_w).*

PROOF. If $A \in S(X)$, then $Ae_i = 0$ for all such i. Now let $x = \sum a_i e_i$ where $a_i \neq 0$, $e_i, e_j \in X_z$, $i \neq j$. Define A by $Ae_i = e_j$, $Ae_j = -e_i$, and $Ae_k = 0$ for $k \neq i,j$. Then A is skew-symmetric and $Ax \neq 0$.

11. COROLLARY. $V(L) \subset L$.

12. PROPOSITION. *If* **dim** $X'_w \geq 2$, *then* $X'_w = X_z$ *for some* $z \in Z$.

PROOF. We know that $S(X)$ is the ℓ^{∞}-direct sum of the Lie algebras $S(X'_w)$ $(w \in Z)$. Let $M \subset S(X)$ be all operators A such that $A|X'_w = 0$ for $z \neq w$. Then $M|X'_w = S(X'_w)$. Now any X_z is an invariant subspace for M (since it is invariant for $S(X)$). Thus $MX_z \subset X_z$. On the other hand, $MX_z \subset X'_w$. If $MX_z \neq \{0\}$ (which is true for at least one z), then $X'_w \cap X_z = X'_w$ (it is invariant for all skew-symmetric operators on the real Hilbert space X'_w). It follows that $X'_w \subset X_z$. Because of symmetry, $X'_w = X_z$.

13. COROLLARY. *V maps every subspace X_z with* **dim** $X_z \geq 2$ *onto itself or onto some other subspace X_w of the same dimension.*

14. COROLLARY. *The connected component of the identity in the group of isometries of X contains only those V which preserve all the subspace X_z with* **dim** $X_z \geq 2$.

PROOF. Let $e_i \in X_z$, **dim** $X_z \geq 2$. Let $f(V) = \| G'_z Ve_i \|$. If $VX_z \subset X_z$, $G'_z Ve_i = Ve_i$

and thus $f(V) = 1$. Otherwise $Ve_i \in X_w$ for $w \neq z$ and $G'_z Ve_i = 0$, so $f(V) = 0$.

SUMMARY. Let $V : X \to X$ be a surjective linear isometry. If **dim** $X_z \geq 2$, then $VX_z = X_w$ for some $w \in Z$. Also, V preserves the closed linear span of the one-dimensional subspaces X_z.

REFERENCES

1. **Fleming, R.J. ; Jamison, J.E.** : Isometries on certain Banach spaces, *J. London Math. Soc.* **9**(1974), 121-127.
2. **Fleming, R.J. ; Jamison, J.E.** : Hermitian and adjoint abelian operators on certain Banach spaces, *Pacific J. Math.* **52**(1974), 67-84.
3. **Kalton, N.J. ; Wood, G.V.** : Orthonormal systems in Banach spaces and their applications, *Math. Proc. Cambridge Phil. Soc.* **79**(1976), 493-510.
4. **Legiša, P.** : W^*-algebras on Banach spaces, *Studia Math.* **72**(1982), 97-107.
5. **Partington, J.R.** : Hermitian operators for absolute norms and absolute direct sums, *Linear Algebra Appl.* **23**(1979), 275-280.
6. **Singer, I.** : *Bases in Banach spaces. I*, Grundl. der Math. Wiss. Bd. **154**, Springer Verlag, Berlin–Heidelberg–New York, 1970.
7. **Schneider, H. ; Turner, R.E.L.** : Matrices hermitian for an absolute norm, *Linear Multilin. Algebra* **1**(1973), 9-31.
8. **Vidav, I.** : The group of isometries and the structure of a finite dimensional normed space, *Linear Algebra Appl.* **14**(1976), 227-236.

Peter Legiša

Department of Mathematics
Institute of Mathematics, Physics and Mechanics
Jadranska 19, 61000 Ljubljana
Yugoslavia.

Operator Theory:
Advances and Applications, Vol.17

ON THE OPERATOR EQUATION $ax + (ax)^* - \lambda x = b$

Bojan Magajna

1. THE SPECTRUM

Let A be either the algebra $B(H)$ of all bounded operators on some Hilbert space H or the Calkin algebra $C(H)$ and let T be the **R** - linear operator defined on A by

$$Tx = ax + (ax)^* \tag{1}$$

where $a \in A$ is some fixed element. In this note various parts of the spectrum $\sigma(T)$ will be characterized.

Here the spectrum $\sigma(T)$ is defined to be the set of all $\lambda \in \mathbf{C}$ such that the operator $T - \lambda I$ is not invertible as a bounded **R** - linear operator. Similarly the left spectrum $\sigma_\ell(T)$, the right spectrum $\sigma_r(T)$, the approximate point spectrum $\sigma_\pi(T)$ and the defect spectrum $\sigma_\delta(T)$ are defined by the analogy with bounded **C** - linear operators.

The equation $ax + (ax)^* = b$ has been studied for matrices in [5], and in [8] the problem of eigenvalues of the operator T (when defined on matrix algebras) has been treated. I am indebted to Professor M. Omladič for bringing the article [5] to my attention.

Our characterization of the spectrum $\sigma(T)$ is the following:

THEOREM. *Let* T *be the operator defined on* A *by* (1) *and for each subset* $F \subset \mathbf{C}$ *put*

$$\hat{F} = \{\lambda \in \mathbf{C} : \exists\, \alpha, \beta \in F,\ |\lambda|^2 = \bar{\alpha}\lambda + \beta\bar{\lambda}\}.$$

Then:

(i) $\sigma(T) = \hat{\sigma}(a)$;

(ii) $\sigma_\ell(T) = \hat{\sigma}_\ell(a) = \sigma_\pi(T)$;

(iii) $\sigma_r(T) = \hat{\sigma}_r(a) = \sigma_\delta(T)$.

REMARK. For each $\alpha \in \mathbf{C}$ denote by C_α the circle with center α and radius $|\alpha|$ and for every $\alpha, \beta \in \mathbf{C}$ such that $|\alpha| \neq |\beta|$

$$z_{\alpha,\beta} = (|\beta|^2 - |\alpha|^2)/(\bar{\beta} - \bar{\alpha}) .$$

Then it can be easily verified that

$$\hat{F} = (\bigcup_{\alpha \in F} C_\alpha) \cup \{z_{\alpha,\beta} : \alpha, \beta \in F, \ |\alpha| \neq |\beta|\}.$$

Denote by $A_{\mathbf{R}}$ the real vector space of all self-adjoint elements in A.

PROOF OF THE THEOREM. Let us identify A with $A_{\mathbf{R}} \oplus A_{\mathbf{R}}$. Then T can be represented by the matrix

$$\begin{bmatrix} R & S \\ 0 & 0 \end{bmatrix}$$

where R and S are $\mathbf{R}$ - linear operators on $A_{\mathbf{R}}$ given by

(2) $$Rx = ax + xa^*, \quad Sx = i(ax - xa^*) .$$

Then for each $\lambda = \mu + \nu i$, $\mu, \nu \in \mathbf{R}$, the operator $T - \lambda I$ is represented by the matrix

$$\begin{bmatrix} R - \mu I & S + \nu I \\ -\nu I & -\mu I \end{bmatrix} .$$

This matrix is invertible (respectively left invertible, right invertible, bounded from below or onto) if and only if the same holds for its determinant D defined by $D = (-\mu)(R - \mu I) + \nu(S + \nu I)$. (To see this, suppose first $\lambda \neq 0$. Then it is not hard to find an invertible 2×2 matrix U such that the matrix $V = (T - \lambda I)U$ has the form

$$V = \begin{bmatrix} D & D_1 \\ 0 & 1 \end{bmatrix}$$

where D_1 is some $\mathbf{R}$ - linear operator on $A_{\mathbf{R}}$ (see [3], Problem 71 for a similar computation). Now observe that V is left invertible (respectively right invertible, bounded from below or onto) if and only if the same holds for D. In the case $\lambda = 0$ just note that T is neither one - to - one (since $T(ia^*) = 0$) nor onto (since $\mathbf{im}\, T \subset A_{\mathbf{R}}$).) Now we insert the expressions (2) for R and S into the expression for D and thus we get

(3) $$Dx = (|\lambda|^2 - \bar{\lambda}a)x - x(\lambda a^*), \quad x \in A_{\mathbf{R}} .$$

The operator D can be extended by (3) to an operator $\tilde{D}$ on A such that $\tilde{D}x^* = (\tilde{D}x)^*$ for every $x \in A$. Not that $\tilde{D}$ is a generalized derivation, that is, it has the form $\tilde{D}x = bx - xc$ where $b = |\lambda|^2 - \bar{\lambda}a$ and $c = \lambda a^*$. For the spectrum of $\tilde{D}$ the following relations are known: $\sigma(\tilde{D}) = \sigma(b) - \sigma(c)$, $\sigma_\ell(\tilde{D}) = \sigma_\ell(b) - \sigma_r(c) = \sigma_\pi(\tilde{D})$, $\sigma_r(\tilde{D}) = \sigma_r(b) - \sigma_\ell(c) = \sigma_\delta(\tilde{D})$ (see [7], [6], [2], [1], or [4] p.50). Thus $\tilde{D}$ is invertible (respectively left invertible, etc.) if and only if $\sigma(b) \cap \sigma(c) = \emptyset$ (respectively $\sigma_\ell(b) \cap \sigma_r(c) = \emptyset$, etc.). It is evident from the expressions for b and c that these last conditions are equivlent to $\lambda \notin \hat{\sigma}(a)$ (respectively $\lambda \notin \hat{\sigma}_\ell(a)$, etc.). Now the theorem follows from the following simple *observation*: If $L : A \to A$ is any $\mathbf{C}$ - linear operator such that $Lx^* = (Lx)^*$ for every $x \in A$ then L is invertible (respectively left invertible, right invertible, bounded from below, onto or one - to - one) if and only if the restriction $L|A_{\mathbf{R}}$ is invertible (respectively left invertible, right invertible, bounded from below, onto or one - to - one.) ■

The following proposition is a generalization of Theorem 2^* from [8].

PROPOSITION 1. *Denote by* R *the restriction* $T|A_{\mathbf{R}}$ *where* T *is defined by* (1). *Then* $\sigma(R) \cap \mathbf{R} = (\sigma(a) + \overline{\sigma(a)}) \cap \mathbf{R}$ *and the same equalities hold for the left, the right, the approximate point, and the defect spectrum.*

PROOF. For every $\lambda \in \mathbf{R}$ let L_λ be the operator defined on A by $L_\lambda x =$ $= ax + xa^* - \lambda x$. Then $(L_\lambda x)^* = L_\lambda x^*$ for every $x \in A$ and $L_\lambda | A_{\mathbf{R}} = R - \lambda I$, hence the proposition follows from the observation at the end of the proof of the above theorem and from well-known results about the spectrum of the generalized derivations (that have been already used in the proof of the above theorem). ■

2. THE EQUATION $ax + (ax)^* = b$

As it has been already observed the operator T (defined by (1)) is never one-to-one (since $ia^* \in \mathbf{ker}\, T$) nor is it ever onto. The question, when $\mathbf{im}\, T$ is equal to $A_{\mathbf{R}}$, has an easy answer at least for the algebra $A = B(H)$. In this section we will restrict our attention to the algebra $B(H)$.

PROPOSITION 2. *Let* $a, b \in B(H)$ *and* $b = b^*$. *Then the necessary condition for the solvability of the equation*

(4) $$ax + x^* a^* = b$$

is $b(\mathbf{ker}\, a^*) \subset \mathbf{im}\, a$. *If* $\mathbf{im}\, a$ *is closed then this condition is also sufficient.*

PROOF. Necessity of the condition is obvious. In order to prove sufficiency denote by p the orthogonal projection onto $\mathbf{ker}\, a^*$ and let x be any solution of the equation

(5) $$ax = \tfrac{1}{2}(1 - p)b(1 + p)\,.$$

That this equation has a solution follows from $\mathbf{im}(\tfrac{1}{2}(1 - p)b(1 + p)) \subset \mathbf{im}(1 - p) = \mathbf{im}\, a$ (since $\mathbf{im}\, a$ was assumed to be closed). Here we have used the familiar fact that the equation of the form $ax = c$, $a, c \in B(H)$, has a solution x in $B(H)$ if and only if **im** $c \subset$ **im** a (see [3], Problem 59). Now from the hypothesis $b($**ker** $a^*) \subset$ **im** a we have $pbp = 0$ and this suffices that every solution of (5) is also a solution of (4). ■

COROLLARY 1. *If the operator* b *is positive and invertible then the equation* (4) *has a solution if and only if the operator* a *is onto.*

PROOF. If a is onto then (4) has a solution by the above proposition. Conversely, if x is a solution of (4) then the operator h defined by $ax = \frac{1}{2}b + ih$ is self--adjoint. But, since the operator $c = \frac{1}{2}b$ is positive and invertible, the operator $c + ih$ must be invertible (because $c + ih = c^{\frac{1}{2}}(1 + ic^{-\frac{1}{2}}hc^{-\frac{1}{2}})c^{\frac{1}{2}}$), hence **im** $a \supset$ **im**$(ax) =$ $=$ **im**$(c + ih) = H$. ■

From Proposition 2 and Corollary 1 we get:

COROLLARY 2. *For the operator* T *defined on the algebra* $A = B(H)$ *the equality* **im** $T = A_{\mathbf{R}}$ *holds if and only if the operator* a *is onto.*

PROBLEM. Find a necessary and sufficient condition for solvability of the equation (4) when the range of the operator a is not closed.

Note that the equation (4) is equivalent to the equation $ax = \frac{1}{2}b + iy$ in two unknowns x and y where y is self-adjoint. Further this last equation is equivalent to the condition **im**$(\frac{1}{2}b + iy) \subset$ **im** a and since **im** a = **im**$(aa^*)^{\frac{1}{2}}$ we see that it would be enough to solve the above problem when the operator a is positive.

REFERENCES

1. **Fialkow, L.A.** : A note on norm ideals and the operator $X \to AX - XB$, *Israel J. Math.* **32** (1979), 331-348.
2. **Fialkow, L.A.** : A note on the operator $X \to AX - XB$, *Trans. Amer. Math. Soc.* **243** (1978), 147-168.
3. **Halmos, P.R.** : *A Hilbert space problem book*, Springer, New York, 1982.
4. **Herrero, D.A.** : *Approximation of Hilbert spaces operators*, Pitman, Boston, 1982.
5. **Lancaster, P. ; Rozsa, P.** : On the matrix equation $AX + X^*A^* = C$, *Siam J. Alg. Disc. Meth.* **4** (1983), 432-436.
6. **Lumer, G. ; Rosenblum, M.** : Linear operator equations, *Proc. Amer. Math. Soc.* **10** (1959), 32-41.
7. **Rosenblum, M.** : On the operator equation $BX - XA = Q$, *Duke Math. J.* **23** (1956), 263-269.
8. **Taussky, O. ; Wielandt, H.** : On the matrix function $AX + X'A'$, *Arch. Rat. Mech. Anal.* **9** (1962), 93-96.

Bojan Magajna
Department of Mathematics, University of Ljubljana, Jadranska 19, Ljubljana 61000 Yugoslavia.

Operator Theory:
Advances and Applications, Vol.17

ON THE INVERSE PROBLEM OF A DISSIPATIVE SCATTERING THEORY. I

Hagen Neidhardt

1. INTRODUCTION

Let L be a selfadjoint operator on the separable Hilbert space L. By H we denote a maximal dissipative operator,

(1.1) $$\mathrm{Im}(Hf, f) \leq 0,$$

$f \in \mathbf{dom}\, H$, on a separable Hilbert space G. The operator H generates a one-parameter contraction semigroup T(t),

(1.2) $$T(t) = e^{-itH},$$

$t \geq 0$. By J we denote a bounded operator from L into G. The wave operator $W_+(H^*,L;J)$ is defined by

(1.3) $$W_+(H^*,L;J) = \underset{t \to +\infty}{\text{s-lim}}\; e^{itH^*} J e^{-itL} P_{ac}(L),$$

where $P_{ac}(L)$ denotes the projection onto the absolutely continuous subspace L_{ac} of the selfadjoint operator L. Similarly we introduce the wave operator $W_-(H,L;J)$,

(1.4) $$W_-(H,L;J) = \underset{t \to +\infty}{\text{s-lim}}\; e^{-itH} J e^{itL} P_{ac}(L).$$

In the following we assume in addition to the existence of the wave operators $W_+(H^*,L;J)$ and $W_-(H,L;J)$ the completeness of these operators. To explain the completeness we recall that for every one-parameter contraction semigroup T(t) on G, there is a larger Hilbert space H, $G \subset H$, and a unitary group $U(t) = e^{-itK}$, $t \in \mathbf{R}^1$, such that

(1.5) $$e^{-itH} = P^{H}_{G} e^{-itK} \restriction G,$$

$t \geq 0$, holds. The group U(t) is called a unitary dilation of the contraction semigroup T(t), $t \geq 0$. It is well-known that the unitary dilation U(t) is defined up to unitary equivalence if

(1.6) $$H = \bigvee_{t\in\mathbf{R}^1} e^{-itK} G$$

is valid. If (1.6) is valid, then U(t) is called a minimal unitary dilation of T(t). In the following we consider only minimal dilations.

The limits $P^{\pm}_{ac}(K)$,

(1.7) $$P^{\pm}_{ac}(K) = \operatorname*{s-lim}_{t\to\pm\infty} e^{itK} P^{H}_{G} e^{-itK} P_{ac}(K)$$

exist and define projections on H. By $P_{ac}(K)$ we denote the projection onto the absolutely continuous subspace of the selfadjoint operator K. The subspaces $H^{+}_{ac}(K)$ and $H^{-}_{ac}(K)$,

(1.8) $$H^{\pm}_{ac}(K) = P^{\pm}_{ac}(K)H$$

are called the absolutely continuous absorption and radiation subspaces of the minimal unitary dilation U(t), respectively.

Now it is possible to consider the wave operators $W_{\pm}(K,L;P^{H}_{G} J)$,

(1.9) $$W_{\pm}(K,L;P^{H}_{G}J) = \operatorname*{s-lim}_{t\to\pm\infty} e^{itK}P^{H}_{G} Je^{-itL}P_{ac}(L).$$

Assuming in addition to the existence of $W_{+}(H^{*},L;J)$ and $W_{-}(H,L;J)$ the existence of $W_{\pm}(K,L;P^{H}_{G} J)$ we find

(1.10) $$P^{H}_{G} W_{+}(K,L;P^{H}_{G} J) = W_{+}(H^{*},L;J)$$

and

(1.11) $$P^{H}_{G} W_{-}(K,L;P^{H}_{G} J) = W_{-}(H,L;J).$$

A simple calculation proves the inclusions

(1.12) $$R(W_{\pm}(K,L;P^{H}_{G} J)) \subset H^{\pm}_{ac}(K).$$

On account of (1.12) the following definition makes sense.

DEFINITION 1.1. The wave operators $W_{+}(H^{*},L;J)$ and $W_{-}(H,L;J)$ are *complete* , if the wave operators $W_{\pm}(K,L,P^{H}_{G} J)$ exist and are partial isometries from L_{ac} onto $H^{\pm}_{ac}(K)$.

Using the notion of a complete wave operator we are able to introduce the notion of a scattering system.

DEFINITION 1.2. The triplet $A = \{H,L;J\}$ is called a *complete scattering system* if the wave operators $W_+(H^*,L;J)$ and $W_-(H,L;J)$ exist and are complete.

For every complete scattering system it is possible to introduce the scattering operator $S(A)$,

$$S(A) = W_+^*(H^*,L;J)W_-(H,L;J). \tag{1.13}$$

Fixing the pair $\{L;J\}$ and changing the operator H in the class of all maximal dissipative operators on G such that the triplet $A = \{H,L;J\}$ is a complete scattering system we can introduce the set $S(L;J)$ of all possible scattering operators of the pair $\{L;J\}$.

For the sake of simplicity we restrict ourself in the following to the case that H belongs to the class C_{11}. This means the maximal dissipative operator H generates a contraction semigroup $T(t)$ of class C_{11}. In such a way allowing to vary the operator H only in the class C_{11} we obtain a subset of $S(L;J)$ which will be denoted by $S_0(L;J)$.

The aim of the present paper is to characterize the set $S_0(L;J)$ in a necessary and sufficient manner. We use a method which allows to obtain from a given $S \in S_0(L;J)$ a maximal dissipative operator H of class C_{11} such that the scattering system $A = \{H,L;J\}$ is complete and S has the representation (1.13).

2. DIRECT PROBLEM

In this chapter we obtain some necessary properties of the scattering operator of a complete scattering system. To describe these properties we introduce the set of fictive scattering operators.

DEFINITION 2.1. The bounded operator S on L belongs to the *set of fictive scattering operators* $FS(L)$ if the conditions

$$\|S\| \leq 1, \tag{2.1}$$

$$e^{-itL} S = S e^{-itL}, \quad t \in \mathbf{R}^1, \tag{2.2}$$

$$(R(S))^- = (R(S^*))^- = L_{ac} \tag{2.3}$$

are fulfilled.

The relation between the sets $S_0(L;J)$ and $FS(L)$ is clarified by the following:

PROPOSITION 2.2. *If* $A = \{H,L;J\}$ *is a complete scattering system and* H *belongs to the class* C_{11}, *then we have*

$$S(\mathcal{A}) \in FS(L). \tag{2.4}$$

Hence

$$S_0(L;J) \subset FS(L). \tag{2.5}$$

PROOF. Using the properties of a minimal unitary dilation we prove the representation

$$S(\mathcal{A}) = W_+^*(K,L;P_G^H J)\, W_-(K,L,P_G^H J). \tag{2.6}$$

On account of (2.6) and the completeness of the wave operators $W_\pm(K,L;P_G^H J)$ we obtain (2.1) and (2.2). The property (2.3) is a consequence of the representation (2.6), the completeness and the property that H belongs to the class C_{11}. ■

A surprising observation is the fact the completeness of the scattering system $\mathcal{A} = \{H,L;J\}$ has also some consequences concerning the bounded operator J.

DEFINITION 2.3. Let J be a bounded operator from $\mathcal{L}$ into $\mathcal{G}$. If there is a partial isometry $F : \mathcal{L} \to \mathcal{G}$, $F^*F = P_{ac}(L)$, such that

$$\operatorname*{s-lim}_{t \to \pm\infty} (F - J)\, e^{-itL} P_{ac}(L) = 0 \tag{2.7}$$

holds, then the operator J is called *asymptotically partially isometric* with respect to $L \restriction \mathcal{L}_{ac} = L_{ac}$. We use the notation $J \underset{L_{ac}}{\backsim} F$.

Definition 2.3 is in accordance with Definitions 71 and 72 of [1, p.125].

With the help of Definition 2.3 the following proposition is obtained.

PROPOSITION 2.4. *If* $\mathcal{A} = \{H,L;J\}$ *is a complete scattering system, then there is a partial isometry* $F: \mathcal{L} \to \mathcal{G}$, $F^*F = P_{ac}(L)$, *such that* $J \underset{L_{ac}}{\backsim} F$.

PROOF. On account of Theorem 4 of [1, p.127] there is a projection $Q_+ \le P_{ac}(L)$ satisfying

$$\operatorname*{s-lim}_{t \to +\infty} e^{itL} Q_+ e^{-itL} P_{ac}(L) = P_{ac}(L) \tag{2.8}$$

and

$$\operatorname*{s-lim}_{t \to -\infty} e^{itL} Q_+ e^{-itL} P_{ac}(L) = 0. \tag{2.9}$$

We introduce the operator $\tilde{J}: L \to G$,

(2.10) $$\tilde{J} = W_+(H^*,L;J)\,Q_+ + W_-(H,L;J)(1 - Q_+).$$

Using (2.8), (2.9) and the completeness of the wave operators $W_+(H^*,L;J)$ and $W_-(H,L;J)$ we obtain

(2.11) $$\underset{t \to \pm\infty}{\text{s-lim}} \; (\tilde{J} - J)\, e^{-itL} P_{ac}(L) = 0.$$

Moreover applying again the completeness we find

(2.12) $$\underset{t \to \pm\infty}{\text{s-lim}} \; e^{itL} \tilde{J}^* \tilde{J}\, e^{-itL} P_{ac}(L) = P_{ac}(L).$$

Now using the considerations of Theorem 12 of [1, p.238] we find a partial isometry F such that $\tilde{J} \underset{L_{ac}}{\sim} F$. But (2.11) implies $J \underset{L_{ac}}{\sim} F$. ■

REMARK 2.5. The assumption that H belongs to the set C_{11} is not necessary to prove $J \underset{L_{ac}}{\sim} F$. Moreover the completeness can be replaced by the assumption that the wave operators $W_\pm(K,L;P_G^H J)$ exist.

REMARK 2.6. Proposition 2.4 generalizes Theorem 12 of [1, p.238] to the case of dissipative operators.

3. INVERSE PROBLEM

The aim of this chapter is to show that under the assumption $J \underset{L_{ac}}{\sim} F$ every fictive scattering operator can be regarded as a scattering operator of a complete scattering system. Consequently the inclusion (2.5) can be replaced by an equality.

To show this we prove first of all a technical lemma which will be useful in the sequel.

LEMMA 3.1. *Let* B *be an absolutely continuous selfadjoint operator on the separable Hilbert space* B*. By* A *we denote a non-negative selfadjoint operator and in general unbounded operator on* B *commuting in the resolvent sense with* B, i.e.

(3.1) $$(B - i)^{-1}(A - i)^{-1} = (A - i)^{-1}(B - i)^{-1}.$$

Then there is a non-negative and densely defined closed quadratic form $\mu_0(\cdot,\cdot)$ *on* B *with the following properties*

(3.2) $$e^{-itB}\,\mathbf{dom}\,\mu_0 \subset \mathbf{dom}\,\mu_0, \quad t \geq 0,$$

(3.3) $$0 \leq \mu_0(e^{-itB}f, e^{-itB}f) \leq \mu_0(f, f), \quad t \geq 0,$$

(3.4) $$\lim_{t\to\infty} \mu_0(e^{-itB}f, e^{-itB}f) = 0, \quad f \in \mathbf{dom}\,\mu_0,$$

(3.5) $$D = \{f \in \bigcap_{t\geq 0} e^{-itB}\,\mathbf{dom}\,\mu_0 : \sup_{t\geq 0} \mu_0(e^{itB}f, e^{itB}f) < +\infty\} = D(A)$$

and

(3.6) $$\lim_{t\to\infty} \mu_0(e^{itB}f, e^{itB}f) = (Af, Af), \quad f \in D(A).$$

PROOF. By $E_A(\cdot)$ we denote the spectral resolution of the selfadjoint operator A. We introduce the Hilbert space $B_0 = L_2(\mathbf{R}^1, b_0)$, where b_0 is a separable and infinite dimensional Hilbert space. By B_0 we denote the generator of the right translation group, i.e.

(3.7) $$(e^{-itB_0}f)(x) = f(x - t),$$

$f \in B_0 = L_2(\mathbf{R}^1, b_0)$. Because the selfadjoint operator B is absolutely continuous there is an isometry $F_0 : B \to B_0$, $F_0^* F_0 = 1$, such that

(3.8) $$e^{-itB_0}F_0 = F_0 e^{-itB},$$

$t \in \mathbf{R}^1$. By P_- we denote the projection

(3.9) $$(P_- f)(x) = \chi_{\mathbf{R}^1_-}(x) f(x), \quad f \in B_0 = L_2(\mathbf{R}^1, b_0).$$

We define the quadratic form $\mu_0(\cdot,\cdot)$ by

(3.10) $$\mathbf{dom}\,\mu_0 = \{f \in B : \sum_{n=1}^{\infty} \|P_- F_0 A E_A([n-1,n))f\|^2 < +\infty\}$$

and

(3.11) $$\mu_0(f, g) = \sum_{n=1}^{\infty} (P_- F_0 A E_A([n-1,n))f, F_0 A E_A([n-1,n))g),$$

$f,g \in \mathbf{dom}\,\mu_0$. Because of

(3.12) $$\mu_0(f, f) = \sum_{n=1}^{\infty} \|P_- F_0 A E_A([n-1,n))f\|^2 \leq \|Af\|^2,$$

$f \in D(A)$, the set $\mathbf{dom}\,\mu_0$ is dense in B. Obviously the form $\mu_0(\cdot,\cdot)$ is positive. It remains to show that $\mu_0(\cdot,\cdot)$ is closed.

To this end we introduce the family $\{\mu_o^{(N)}(\cdot,\cdot)\}_{N=1}^{\infty}$ of non-negative closed quadratic forms,

$$\mu_o^{(N)}(f,g) = \sum_{n=1}^{N} (P_-F_oAE_A([n-1,n))f, F_oAE_A([n-1,n))g), \tag{3.13}$$

f,g $\in$ **dom** $\mu_o^{(N)} = B$, N = 1, 2, ... The family $\{\mu_o^{(N)}(\cdot,\cdot)\}$ is non-decrasing and satisfies the estimation

$$\mu_o^{(N)}(f,f) \leq \|Af\|^2, \tag{3.14}$$

f $\in D(A)$, N = 1, 2, ... On account of Theorem 3.1 of [5] the quadratic form $\mu_o^{(\infty)}(\cdot,\cdot)$,

$$\mathbf{dom}\, \mu_o^{(\infty)} = \{f \in B : \sup_N \mu_o^N(f,f) < +\infty\}, \tag{3.15}$$

$$\mu_o^{(\infty)} = \lim_{N\to\infty} \mu_o^N(f,f), \tag{3.16}$$

f $\in$ **dom** $\mu_o^{(N)}$, is densely defined and closed. But by a simple calculation we obtain

$$\mu_o(\cdot,\cdot) = \mu_o^{(\infty)}(\cdot,\cdot). \tag{3.17}$$

Consequently the quadratic form $\mu_o(\cdot,\cdot)$ is closed.

Because of the inequality

$$e^{itB_o}P_-e^{-itB_o} \leq P_- \tag{3.18}$$

$t \geq 0$, and (3.1) we obtain (3.2) and (3.3). Using the inequality

$$\|P_-e^{-itB_o}F_oAE_A([n-1,n))f\|^2 \leq \|P_-F_oAE_A([n-1,n))f\|^2, \tag{3.19}$$

$t \geq 0$, n = 1, 2, ..., and

$$\operatorname{s-lim}_{t\to+\infty} e^{itB_o}P_-e^{-itB_o} = 0 \tag{3.20}$$

we prove (3.4).

Let f $\in D$. Then we have

$$\mu_o^{(N)}(e^{itB}f, e^{itB}f) \leq \sup_{t\geq 0}(e^{itB}f, e^{itB}f), \tag{3.21}$$

$t \geq 0$, N = 1, 2, ... Taking into account

$$\operatorname{s-lim}_{t\to+\infty} e^{-itB_o}P_-e^{itB_o} = 1 \tag{3.22}$$

we obtain

(3.23) $$\| AE_A([0,N))f \|^2 \leq \sup_{t\geq 0} \mu_0(e^{itB}f, e^{itB}f)$$

which implies $f \in D(A)$. Consequently we have

(3.24) $$D \subset D(A),$$

Because of

(3.25) $$e^{itB} D(A) \subset D(A),$$

$t \in \mathbf{R}^1$, and (3.12) we get

(3.26) $$D(A) \subset D.$$

On account of (3.24) and (3.26) we obtain (3.5). Moreover we have

(3.27) $$\mu_0^{(N)}(e^{itB}f, e^{itB}f) \leq \mu_0(e^{itB}f, e^{itB}f),$$

$f \in D(A)$, $t \geq 0$, $N = 1, 2, \ldots$. The inequalities (3.14) and (3.27) yield

(3.28) $$\| AE_A([0,N))f \|^2 \leq \lim_{t\to\infty} \mu_0(e^{itB}f, e^{itB}f) \leq \| Af \|^2,$$

$f \in D(A)$, $N = 1, 2, \ldots$, which proves (3.6). ∎

COROLLARY 3.2. *The set* $D(A) \subset \mathbf{dom}\, \mu_0$ *is a core of the quadratic form* $\mu_0(\cdot,\cdot)$.

PROOF. Denoting by $\tilde{\mu}_0(\cdot,\cdot)$ the restriction of the quadratic form $\mu_0(\cdot,\cdot)$ to the set $D(A)$ we must show that the closure $\tilde{\mu}_0^-(\cdot,\cdot)$ coincides with $\mu_0(\cdot,\cdot)$. To this end we introduce the sequence $\{f_N\}_{N=1}^{\infty}$,

(3.29) $$f_N = E_A([0,N))f,$$

$f \in \mathbf{dom}\, \mu_0$. We have $\text{s-}\lim_{t\to+\infty} f_N = f$ and $f_N \in D(A)$, $N = 1, 2, \ldots$. We find

(3.30) $$\tilde{\mu}_0(f_N - f_{N'}, f_N - f_{N'}) = \sum_{n=N'+1}^{N} \| P_-F_0AE_A([n-1,n))f \|^2,$$

$N' < N$. Consequently for every $\varepsilon > 0$ there is a N_0 such that

(3.31) $$\tilde{\mu}_0(f_N - f_{N'}, f_N - f_{N'}) < \varepsilon,$$

is valid for every $N_0 < N' < N$. Hence f belongs to $\mathbf{dom}\, \tilde{\mu}_0^-$ and

(3.32) $$\tilde{\mu}_0^-(f, f) = \mu_0(f, f)$$

holds. ∎

In the following we use the polar decomposition of the operator S^*, i.e.

(3.33) $$S^* = \mathbf{sign}\, S^* \, |S^*|.$$

Because of $S \in FS(L)$ the operator $\mathbf{sign}\, S^*$ is a unitary operator on the absolutely continuous subspace of L such that

(3.34) $$e^{-itL}(\mathbf{sign}\, S^*) = (\mathbf{sign}\, S^*)e^{-itL},$$

$t \in \mathbf{R}^1$, holds. The operator $|S^*| = (SS^*)^{1/2}$ is a positive selfadjoint operator on L satisfying

(3.35) $$e^{-itL}|S^*| = |S^*|e^{-itL},$$

$t \in \mathbf{R}^1$, and

(3.36) $$\mathbf{ker}\, |S^*| = L \ominus L_{ac}.$$

Consequently the inverse operator $|S^*|^{-1}$ exists on L_{ac} and fulfils

(3.37) $$|S^*|^{-1} \geq P_{ac}(L).$$

LEMMA 3.3. *If* $S \in FS(L)$, *then there is a non-negative and densely defined quadratic form* $\gamma_o(\cdot,\cdot)$ *on the absolutely continuous subspace* L_{ac} *of* L *with the following properties:*

(3.38) $$e^{-itL}\,\mathbf{dom}\,\gamma_o \subset \mathbf{dom}\,\gamma_o \subset L_{ac}, \quad t \geq 0,$$

(3.39) $$\|f\|^2 \leq \gamma_o(e^{-itL}f, e^{-itL}f) \leq \gamma_o(f,f), \quad t \geq 0,$$

(3.40) $$\lim_{t\to\infty} \gamma_o(e^{-itL}f, e^{-itL}f) = \|f\|^2, \quad t \geq 0, \quad f \in \mathbf{dom}\,\gamma_o,$$

(3.41) $$\{f \in \bigcap_{t\geq 0} e^{-itL}\,\mathbf{dom}\,\gamma_o : \sup_{t\geq 0} \gamma_o(e^{itL}f, e^{itL}f) < +\infty\} = R(|S^*|),$$

(3.42) $$\lim_{t\to\infty} \gamma_o(e^{itL}f, e^{itL}f) = (|S^*|^{-1}f, |S^*|^{-1}f), \quad f \in R(|S^*|).$$

PROOF. In order to apply Lemma 3.1 we introduce the notations $B = L_{ac}$ and $B = L\ L_{ac}$. By A we denote the operator

(3.43) $$A = \sqrt{1 - |S^*|^2}\, |S^*|^{-1}$$

which is well defined on $B = L_{ac}$. A simple calculation shows

(3.44) $$D(A) = R(|S^*|) \subset B = L_{ac}.$$

Because of (3.35) the operator A commutes with $B = L \restriction L_{ac}$ in the resolvent sense. On account of Lemma 3.1 there is a non-negative and densely defined closed quadratic form $\mu_o(\cdot,\cdot)$ on B obeying (3.2)–(3.6). Defining $\gamma_o(\cdot,\cdot)$ by

(3.45) $$\gamma_o(f, g) = (f, g) + \mu_o(f, g),$$

$f,g \in \mathbf{dom}\, \gamma_o = \mathbf{dom}\, \mu_o$, we prove Lemma 3.3. ■

COROLLARY 3.4. *The set* $R(|S^*|) \subset \mathbf{dom}\, \gamma_o$ *is a core of the quadratic form* $\gamma_o(\cdot,\cdot)$.

PROOF. Corollary 3.4 follows from Corollary 3.2. ■

Because the form $\gamma_o(\cdot,\cdot)$ is closed there is a positive selfadjoint operator C on L_{ac} such that the representation

(3.46) $$\gamma_o(f, g) = (C^{\frac{1}{2}} f, C^{\frac{1}{2}} g),$$

$f,g \in \mathbf{dom}\, C^{\frac{1}{2}} = \mathbf{dom}\, \gamma_o$, holds. On account of (3.45) we have

(3.47) $$C \geq P_{ac}(L).$$

Consequently we find

(3.48) $$R(C^{\frac{1}{2}}) = L_{ac}.$$

Because of

(3.49) $$R(|S^*|) \subset D(C^{\frac{1}{2}})$$

the restriction $C^{\frac{1}{2}} \restriction R(|S^*|)$ makes sense.

COROLLARY 3.5. *The relation*

(3.50) $$(C^{\frac{1}{2}} \restriction R(|S^*|))^{-} = L_{ac}$$

holds.

PROOF. Because of Corollary 3.4 for every $f \in D(C^{\frac{1}{2}}) = \mathbf{dom}\, \gamma_o$ there is a sequence of elements $\{f_N\}_{N=1}^{\infty}$ of $R(|S^*|)$ converging to f such that $\{C^{\frac{1}{2}} f_N\}_{N=1}^{\infty}$ is a Cauchy sequence. The operator $C^{\frac{1}{2}}$ is closed. Hence we obtain

(3.51) $$\operatorname*{s-lim}_{t\to\infty} C^{\frac{1}{2}} f_N = C^{\frac{1}{2}} f.$$

Taking into account (3.51) we prove (3.50). ■

Because of (3.47) the inverse operator $C^{-1/2}$ exists on L_{ac} and is a contraction.

COROLLARY 3.6. *The operator* $C^{-\frac{1}{2}}$ *fulfils the following properties:*

$$(R(C^{-\frac{1}{2}}))^{-} = L_{ac}, \tag{3.52}$$

$$e^{-itL}C^{-1}e^{itL} \leq C^{-1}, \quad t \geq 0, \tag{3.53}$$

$$\underset{t\to+\infty}{\text{s-lim}}\ (1 - C^{-\frac{1}{2}})e^{-itL}P_{ac}(L) = 0, \tag{3.54}$$

$$\underset{t\to-\infty}{\text{s-lim}}\ (C^{-\frac{1}{2}} - |S^*|)e^{-itL}P_{ac}(L) = 0. \tag{3.55}$$

PROOF. The relation (3.52) follows from $R(C^{-\frac{1}{2}}) = \mathbf{dom}\,\gamma_0$. The inequality (3.53) is a consequence of (3.38) and (3.39).

To prove (3.54) we use the estimation

$$\begin{aligned}&((1 - C^{-1})e^{-itL}f, e^{-itL}f) \leq \\ &\leq (C^{\frac{1}{2}}e^{-itL}f, C^{\frac{1}{2}}e^{-itL}f) - (f, f),\end{aligned} \tag{3.56}$$

$f \in D(C^{\frac{1}{2}}) = \mathbf{dom}\gamma_0$, $t \geq 0$. Because of (3.40) we obtain

$$\underset{t\to+\infty}{\text{s-lim}}\ (1 - C^{-1})e^{-itL}P_{ac}(L) = 0. \tag{3.57}$$

But (3.57) implies (3.54).

To prove (3.55) we introduce a family of non-negative and densely defined closed quadratic forms $\{\gamma_t(\cdot,\cdot)\}_{t\geq 0}$,

$$\gamma_t(f, g) = \gamma_0(e^{itL}f, e^{itL}g), \tag{3.58}$$

$f,g \in \mathbf{dom}\,\gamma_t = e^{-itL}\,\mathbf{dom}\,\gamma_0$, $t \geq 0$. The sequence $\{\gamma_t(\cdot,\cdot)\}_{t\geq 0}$ is non-decreasing, i.e.

$$\mathbf{dom}\,\gamma_{t'} \subset \mathbf{dom}\,\gamma_t, \tag{3.59}$$

$$\gamma_t(f, f) \leq \gamma_{t'}(f, f), \tag{3.60}$$

$f \in \mathbf{dom}\,\gamma_{t'}$, $0 \leq t \leq t'$. On account of Theorem 3.1 of [5] the form $\gamma_\infty(\cdot,\cdot)$,

(3.61) $$\mathbf{dom}\,\gamma_\infty = \{f \in \bigcap_{t\geq 0} \mathbf{dom}\,\gamma_t : \sup_{t\geq 0} \gamma_t(f, f) < +\infty\},$$

(3.62) $$\gamma_\infty(f, f) = \lim_{t\to\infty} \gamma_t(f, f),$$

$f \in \mathbf{dom}\,\gamma_\infty$, is closed. Moreover taking into account (3.41) and (3.42) we find

(3.63) $$\mathbf{dom}\,\gamma_\infty = R(|S^*|)$$

and

(3.64) $$\gamma_\infty(f, g) = (|S^*|^{-1} f, |S^*|^{-1} g),$$

$f,g \in \mathbf{dom}\,\gamma_\infty$. Using again Theorem 3.1 of [5] we get

(3.65) $$\operatorname*{s-lim}_{t\to+\infty} e^{-itL} C^{-1} e^{itL} P_{ac}(L) = |S^*|^2.$$

Proposition 27 of [1, p.137] completes the proof. ∎

Now we solve the inverse problem.

THEOREM 3.7. *Let* $S \in FS(L)$. *If there is a partial isometry* F, $F^*F = P_{ac}(L)$, *obeying* $J \underset{L_{ac}}{\sim} F$, *then there is a maximal dissipative operator* H *of class* C_{11} *such that* $A = \{H,L;J\}$ *is a complete scattering system and* S *has the representation* (1.13).

PROOF. We prove Theorem 3.7 in several steps.

1. The operator $\mathbf{sign}\,S^*$ is a unitary operator on L_{ac} obeying (3.34). Because of Theorem 15 of [1, p.241] there is a partial isometry $R : L \to L$ satisfying

(3.66) $$R^*R = RR^* = P_{ac}(L),$$

(3.67) $$\operatorname*{s-lim}_{t\to+\infty} (R - 1)\, e^{-itL} P_{ac}(L) = 0,$$

(3.68) $$\operatorname*{s-lim}_{t\to-\infty} (R - \mathbf{sign}\,S^*)\, e^{-itL} P_{ac}(L) = 0.$$

Moreover we have

(3.69) $$\operatorname*{s-lim}_{t\to+\infty} (R^* - 1)\, e^{-itL} P_{ac}(L) = 0$$

and

(3.70) $$\operatorname*{s-lim}_{t\to-\infty} (R^* - (\mathbf{sign}\,S^*)^*)\, e^{-itL} P_{ac}(L) = 0.$$

2. We introduce the operator $W : L_{ac} \to G$,

(3.71) $$W = FRC^{-\frac{1}{2}},$$

where C is defined by (3.46). Because of (3.52) we have

(3.72) $$(R(W))^{-} = R(F) = \hat{G} \subset G.$$

On account of (3.53) the relation

(3.73) $$\hat{T}^*(t)\, W f = W e^{itL} f,$$

$f \in L$, $t \geq 0$, defines an one-parameter contraction semi-group on $\hat{G}$. Using (3.72) and

(3.74) $$\lim_{t \to \infty} \| W^* \hat{T}(t) f \| = \| W^* f \| \neq 0,$$

$f \in \hat{G}$, we find that the maximal dissipative operator $\hat{H}$, $\hat{T}(t) = e^{-it\hat{H}}$, $t \geq 0$, belongs to the class $C_{1\cdot}$.

Now we show that $\hat{H}$ belongs to the class C_{11}. To this end we introduce the operators $\tilde{D} : \hat{G} \to L_{ac}$,

(3.75) $$\tilde{D} f = |S^*|\, W^{-1} f,$$

$f \in D(\tilde{D}) = R(W)$, and E:

(3.76) $$E = \text{s-}\lim_{t \to +\infty} e^{-it\hat{H}} e^{it\hat{H}^*}.$$

Using (3.71) and (3.55) we obtain

(3.77) $$W^* E W = \text{s-}\lim_{t \to +\infty} e^{-itL} W^* W e^{itL} = |S^*|^2.$$

Hence we find

(3.78) $$\|\tilde{D} f\|^2 = \|E f\|^2 \leq \|f\|^2$$

$f \in D(\tilde{D}) = R(W)$. Consequently the operator $\tilde{D}$ is closable. The closure D is a contraction. We have

(3.79) $$\ker D = \ker E.$$

Because of (3.49) and (3.71) we get

(3.80) $$D^* f = (W^*)^{-1} |S^*| f = FRC^{\frac{1}{2}} |S^*| f,$$

$f \in L_{ac}$. But (3.80) and Corollary 3.5 yield

(3.81) $$(R(D^*))^{-} = \hat{G}$$

which implies

(3.82) $$\ker E = \ker D = \{0\}.$$

3. Now we prove the existence and completeness of the wave operator $W_+(\hat{H}^*,L;J)$. Because of (3.67) and (3.54) we get

(3.83) $$\underset{t\to+\infty}{\text{s-lim}}\ (W - F)e^{-itL} P_{ac}(L) = 0.$$

Hence the wave operator $W_+(\hat{H}^*,L;F)$ exists and the equality

(3.84) $$W = W_+(\hat{H}^*,L;F)$$

holds. The minimal unitary dilation of $\hat{T}(t)$ we denote by $\hat{U}(t) = e^{-it\hat{K}}$ which is defined on $\hat{\mathcal{H}}$, $\hat{G}\subset\hat{\mathcal{H}}$. Theorem 4.4 of [2, p.36] yields the existence of $W_+(K,L;P^{\hat{\mathcal{H}}}_{\hat{G}}F)$.

It remains to show the completeness of $W_+(\hat{H}^*,L;F)$. For this purpose we establish

(3.85) $$\underset{t\to+\infty}{\text{s-lim}}\ (W^* - F^*)e^{-it\hat{H}} = 0.$$

We have

(3.86) $$W^* = \underset{t\to+\infty}{\text{w-lim}}\ e^{itL} F^* e^{-it\hat{H}}.$$

The weak limit can be transformed into the strong limit if we show

(3.87) $$\lim_{t\to\infty} \|F^* e^{-it\hat{H}} f\|^2 = \|W^* f\|^2,$$

$f \in \hat{G}$. We have

(3.88) $$e^{-itL} R(W^*)\subset R(W^*)$$

and

(3.89) $$e^{-it\hat{H}} (W^*)^{-1} f = (W^*)^{-1} e^{-itL} f,$$

$f \in R(W^*)$, $t \geq 0$. Taking into account (3.40) we get

(3.90) $$\lim_{t\to\infty} \|F^* e^{-it\hat{H}}(W^*)^{-1} g\|^2 = \lim_{t\to\infty} \|C^{\frac{1}{2}} e^{-itL} g\|^2 = \|g\|^2,$$

$g \in R(W^*)$, which shows (3.87). Hence the wave operator $W_+(\hat{H}^*,L;F)$ is complete.

4. In this section we establish the existence and completeness of the wave operator $W_-(\hat{H},L;F)$. Because of (3.55) and (3.68) we have

(3.91) $$\underset{t\to -\infty}{\text{s-lim}}\ e^{itL} W^* F e^{-itL} = S.$$

Hence we get the existence of $\tilde{W}_-(\hat{H},L;F)$,

(3.92) $$\tilde{W}_-(\hat{H},L;F) = \underset{t\to +\infty}{\text{w-lim}}\ e^{-it\hat{H}} F e^{itL} P_{ac}(L)$$

and the representation

(3.93) $$S = W_+^*(\hat{H}^*,L;F)\tilde{W}_-(\hat{H},L;F).$$

Because of $S \in FS(L)$ we obtain

(3.94) $$(R(\tilde{W}_-^*(\hat{H},L;F)))^- = L_{ac}.$$

Theorem 4.1 of [2, p.34] implies the existence of $\tilde{W}_-(\hat{K},L;P^{\hat{H}}_{\hat{G}} F)$,

(3.95) $$\tilde{W}_-(K,L;P^{\hat{H}}_{\hat{G}} F) = \underset{t\to -\infty}{\text{w-lim}}\ e^{it\hat{K}} P^{\hat{H}}_{\hat{G}} F e^{-itL} P_{ac}(L).$$

By a simple calculation we find

(3.96) $$\lim_{t\to\infty} e^{-itL} F^* e^{it\hat{H}^*} Wg = \lim_{t\to\infty} e^{-itL} RC^{-\frac{1}{2}} e^{itL} g = S^* g,$$

$g \in L_{ac}$. Therefore the limit $\underset{t\to +\infty}{\text{s-lim}}\ e^{-itL} F^* e^{it\hat{H}^*}$ exists. Moreover the limit $\underset{t\to +\infty}{\text{s-lim}}\ e^{-itL} F^* P^{\hat{H}}_{\hat{G}} e^{it\hat{K}}$ exists, too, and the representation

(3.97) $$\tilde{W}_-^*(\hat{K},L;P^{\hat{H}}_{\hat{G}} F) = \underset{t\to -\infty}{\text{s-lim}}\ e^{itL} F^* P^{\hat{H}}_{\hat{G}} e^{-it\hat{K}}$$

is valid. Consequently the operator $\tilde{W}_-^*(\tilde{K},L;P^{\hat{H}}_{\hat{G}} F)$ is a partial isometry fulfilling

(3.98) $$\tilde{W}_-(\hat{K},L;P^{\hat{H}}_{\hat{G}} F)W_-^*(\hat{K},L;P^{\hat{H}}_{\hat{G}} F) = P^-_{ac}(\hat{K})$$

Because of

(3.99) $$\tilde{W}_-^*(\hat{K},L;P^{\hat{H}}_{\hat{G}} F) \restriction \hat{G} = \tilde{W}_-^*(\hat{H},L;F)$$

and (3.94) we find

(3.100) $$R(\tilde{W}_-^*(\hat{K},L;P^{\hat{H}}_{\hat{G}} F)) = L_{ac}$$

or

(3.101) $$\tilde{W}_-^*(\hat{K},L;P^{\hat{H}}_{\hat{G}} F)\tilde{W}_-(\hat{K},L;P^{\hat{H}}_{\hat{G}} F) = P_{ac}(L).$$

But (3.101) implies the existence and completeness of $W_{-}(\hat{H},L;F)$.

5. In accordance with the decomposition $G = \hat{G} \oplus (G \ominus \hat{G})$ we define the operator H by

$$H = \hat{H} \oplus 0. \tag{3.102}$$

The operator H is maximal dissipative and belongs to the class C_{11}. Using (2.7) we obtain the existence and completeness of the wave operators $W_{+}(H,L;J)$ and $W_{-}(H,L;J)$. The representation (1.13) is obvious. ■

REMARK 3.8. If S is a partial isometry satisfying $S^{*}S = SS^{*} = P_{ac}(L)$ the problem was solved in [6]. See also [1]. In this special case the operator H can be chosen selfadjoint.

REMARK 3.9. Theorem 3.7 shows that for every selfadjoint operator L there exists a maximal dissipative operator H of class C_{11} such that the absolutely continuous absorption part $K \restriction H^{+}_{ac}(K)$ and radiation part $K \restriction H^{-}_{ac}(K)$ are unitarily equivalent to the absolutely continuous part $L \restriction L_{ac}$ of L.

REMARK 3.10. A generalization of the results of this paper was obtained in [3, 4].

REFERENCES

1. **Baumgärtel, H. ; Wollenberg, M. :** *Mathematical scattering theory,* Akademie Verlag, Berlin, 1983.
2. **Neidhardt, H. :** Scattering theory of contraction semigroups, Institut für Mathematik, AdW der DDR, Report R-Math-05/81.
3. **Neidhardt, H. :** On the representation of intertwining operators, Institut für Mathematik, AdW der DDR, Preprint P-Math-17/84.
4. **Neidhardt, H. :** On the inverse problem of a dissipative scattering theory. II, to appear.
5. **Simon, B. :** A canonical decomposition for quadratic forms with applications to monotone convergence theorems, *J. Functional Analysis* **28**(1978), 377-385.
6. **Wollenberg, M. :** Wave algebras and scattering theory. I, Zentralinstitut für Mathematik und Mechanik, AdW der DDR, Report R-10/79.

H. Neidhardt
Zentralinstitut für Mathematik und Mechanik
Akademie der Wissenschaften der DDR
Mohrenstrasse 39, Berlin 1080
DDR.

Operator Theory:
Advances and Applications, Vol.17

SOME SPECTRAL PROPERTIES OF AN OPERATOR

Matjaž Omladič *)

0. INTRODUCTION

In this note we study some spectral properties of a class of operators. Denote by I the closed interval $[0,1]$ and by L^p the usual Banach space of (equivalence classes of) measurable functions on I with integrable p-th power for $p \in [1,\infty)$, while L^∞ stands for the space of essentially bounded functions. We use the ordinary Lebesgue measure in all these definitions. Furthermore, denote by C the Banach space of continuous functions on I with the supremum norm. Note that the space C can be regarded as a closed subspace of L^∞. In what follows, the symbol X will stand either for some of the spaces L^p with $p \in [1,\infty]$, or for the space C, unless stated otherwise.

Let us now define an operator T on X. For any function $x \in X$ put

$$(Tx)(t) = t\,x(t) - \int_0^t x(s)\,ds, \quad \text{for } t \in I. \tag{1}$$

Note that T is always bounded, in fact $\|Tx\|_p \leq 2\|x\|_p$ for every $x \in L^p$ with $p \in [1,\infty]$, while $C \subset L^\infty$ is invariant under T. In what follows, the symbol T will always stand for the operator defined by (1), unless stated otherwise.

An operator much like our T was studied by J.R. Ringrose (see Section 6 of [16]) and a similar one by Sh. Kantorovitz (see Examples 1.3(c) and 4.4 of [11], Example 3.2 of [12] and Example 3.2 of [13]; for a deep study of linear combinations of a Volterra operator and a multiplication in an abstract setting see also [14]). Some of the results presented here were given in Section 4.2 of the author's Ph. D. thesis [15].

1. DECOMPOSABILITY AND WELL-BOUNDEDNESS

1.1. PROPOSITION. *For any* $\lambda \in \mathbf{C} \setminus I$ *and any* $x \in X$ *put*

$$(R_\lambda x)(t) = (\lambda - t)^{-1} x(t) - \int_0^t (\lambda - s)^{-2} x(s)\,ds,$$

*) This research was supported by the Research Council of Slovenia.

for $t \in I$. *Then* R_λ *is a bounded operator on* X *and* $(\lambda - T)R_\lambda = R_\lambda(\lambda - T) = I$.

PROOF. An easy computation.

1.2. PROPOSITION. *For any* $r \in I$, $r \neq 1$, *the function defined for* $t \in I$ *by*

$$e_r(t) = \begin{cases} 0 & \text{for } t \leq r \\ 1 & \text{for } t > r \end{cases}$$

is a member of L_p *for every* $p \in [1,\infty]$ *and* $Te_r = re_r$. *These are (up to linear independence) the only eigenfunctions of* T *in any space* X.

PROOF. For any possible eigenfunction $e_r \in X$ with eigenvalue $r \in I$ we have

$$(t - r)e_r(t) = \int_0^t e_r(s)ds, \quad \text{for } t \in I.$$

The proposition follows.

1.3. PROPOSITION. $\sigma(T) = I$.

PROOF. By Proposition 1.1 and 1.2 we only need to consider the case $X = C$. But for any $r \in I$, $r \neq 1$, and for any positive integer n define for $t \in I$

$$e_r^n(t) = \begin{cases} 0 & \text{if} \quad t \leq r - n^{-1} \\ n2^{-1}(t - r + n^{-1}) & \text{if } r - n^{-1} \leq t \leq r + n^{-1} \\ 1 & \text{if } r + n^{-1} \leq t \end{cases}.$$

After a short computation we get $\|e_r^n\|_\infty = 1$ and $\|(r - T)e_r^n\|_\infty = (4n)^{-1}$.

Let G be some open set in $\mathbf{C}$ which contains the interval I as a subset and let $C^\infty(G)$ be the algebra of all complex valued infinitely differentiable functions on G. For any $f \in C^\infty(G)$ define for $x \in X$

$$(U(f)x)(t) = f(t)x(t) - \int_0^t f'(s)x(s)ds, \tag{2}$$

for $t \in I$. Then, U(f) is clearly a bounded operator on X which coincides with T in case $f(t) = t$.

1.4. PROPOSITION. (a) *The operator* T *is a generalized scalar operator and* U *is* *its spectral distribution in the sense of C. Foiaş* [8].

(b) *The operator* T *is decomposable in the sense of C. Foiaş* [9].

PROOF. (a) A straightforward verification of the Foiaş axioms. (b) follows from the assertion (a) say, by Theorem 1.16 of Chapter 3 in [5].

It turns out that U is even a regular spectral distribution of the operator T, but we will not need this result.

It is sometimes convenient to have the spectral maximal subspaces of a decomposable operator. A rather nice way to get them for the operator T goes over the spectral capacities in the sense of C. Apostol [2] (see also [10], [1], [21], etc.). For any closed subset F of the complex plane $\mathbf{C}$ denote by E(F) the set of those $x \in X$ for which it holds that for every closed interval $J \subset I \setminus F$ the function x is constant (almost everywhere) on J, this constant being zero, if $0 \in J$. Note that E(F) is always a closed subspace of X, invariant under the operator T, and the mapping E, given by $F \mapsto E(F)$, satisfies the axioms of a spectral capacity.

1.5. PROPOSITION. *For any closed set* $F \subset \mathbf{C}$ *we have*

$$X_T(F) = E(F).$$

PROOF. The only thing that remains to see is that $\sigma(T\,|\,E(F)) \subset F$ holds for closed sets F. The proposition will then follow by a well-known result of C. Foiaş (see [21]). To get this inclusion take any $x \in E(F)$ and choose some closed interval $J = [a,b]$ such that $J \subset I \setminus F$. By Proposition 1.1 we get for $\lambda \in \mathbf{C} \setminus I$ the local resolvent function of the vector x. To see that $R_\lambda x$ can be analytically extended at least over the interval (a,b) (respectively [a,b) in case a = 0, respectively (a,b] in case b = 1), consider at first the cases a = 0 and b = 1. In the general case, write x = u + v, where u is equal to zero on [0,b] and v is constant on [a,1].

Recall that an operator A on a Banach space is unconditionally decomposable in the sense of C. Apostol [4], if it is decomposable and for any system $\{F_k\}_{k=1}^m$ of disjoint closed subsets of $\mathbf{C}$ we have

$$\sup \Big\| \sum_{k=1}^{m} \alpha_k x_k \Big\| \leq M \Big\| \sum_{k=1}^{m} x_k \Big\|,$$

where $x_k \in X_A(F_k)$ are fixed, the supremum being taken over all $\alpha_k \in \mathbf{C}$ with $|\alpha_k| = 1$, while the finite constant M depends only on the operator A.

1.6. PROPOSITION. *The operator* T *is not unconditionally decomposable.*

PROOF. Let us first discuss the case $X \neq C$. Take any positive integer m and put $\lambda_k = (k-1)m^{-1}$, $F_k = \{\lambda_k\}$ and $x_k = (-1)^k e_{\lambda_k}$, for $k = 1, 2, \ldots, m$. Then the sequence of norms

$$\|\sum_{k=1}^{m} (-1)^k x_k\|_p = (m^{-1} \sum_{k=1}^{m} k^p)^{1/p}$$

diverges as m increases, while the sequence of norms $\|\sum_{k=1}^{m} x_k\|_p$ converges to $2^{-1/p}$. The case $X = C$ follows in a similar way, using the approximation eigenfunctions $e^n_{\lambda_k}$ (those functions were introduced in the proof of Proposition 1.3) with the integers n chosen large enough, instead of the functions e_{λ_k}.

1.7. PROPOSITION. *The operator* T *is not spectral in the sense of N. Dunford.*

PROOF. This is a simple corollary of the previous proposition, taking in account a result of C. Apostol [4] by which every spectral operator is unconditionally decomposable.

Next, we would like to see whether our operator T is well-bounded in the sense of D.R. Smart (see [18], see also [16], [17], [20], etc). Recall that an operator A is well--bounded if there is a finite positive constant M such that for every polynomial p

$$\|p(A)\| \leq M(|p(0)| + V(p)),$$

holds, where V(p) denotes the total variation of p on the interval I. Note that the interval I can be interchanged by any other closed bounded interval. For a generalization where I is replaced by a segment on some nice curve in the plane, see [16]. Note that we have as an immediate consequence of this definition that $\sigma(A) \subset I$.

1.8. PROPOSITION. *If either* $X = L^\infty$ *or* $X = C$, *then* T *is well-bounded.*

PROOF. Use (2) and Proposition 1.4(a) to get

$$(p(T)x)(t) = p(t)x(t) - \int_0^t p'(s)x(s)\,ds,$$

for $t \in I$. Since

$$|\int_0^t p'(s)x(s)\,ds| \leq V(p)\|x\|_\infty,$$

we have

$$\| p(T)x \|_{\infty} \leq 2(|p(0)| + V(p)) \| x \|_{\infty}$$

which implies the proposition.

1.9. PROPOSITION. *If neither* $X = L^{\infty}$ *nor* $X = C$*, then* T *is not well-bounded.*

PROOF. Fix $p \in [1, \infty)$, choose $m \in (0, p^{-1})$, and take some positive integer k. Define the polynomial

$$p_k^m(t) = \sum_{j=0}^{k} (-1)^j \binom{m}{j} (1 - t)^j$$

and differentiate it

$$p_k^{m\prime}(t) = m \sum_{j=1}^{k} \binom{j-m-1}{j-1} (1 - t)^{j-1}$$

to get a strictly positive decreasing function of $t \in I$.

Hence $|p_k^m(0)| + V(p_k^m) = p_k^m(1) = 1$. Set $x(t) = t^{-m}$ for $t \in I$ to get

$$\| x \|_p = [\int_0^1 t^{-mp}\, dt]^{1/p} = (1 - mp)^{-1/p},$$

therefore $x \in L^p$. Clearly it holds for the function defined by $z_k^m(t) = p_k^m(t)x(t)$, for $t \in I$, that

(3)
$$\| z_k^m \|_p \leq \sup | p_k^m | \, \| x \|_p = \| x \|_p .$$

On the other hand, define

$$y_k^m(t) = \int_0^t p_k^{m\prime}(s)\, x(s)\, ds$$

and choose any positive $\varepsilon < 1$ to get for $t \in I$, $t \geq \varepsilon$,

$$|y_k^m(t)| = y_k^m(t) \geq \sum_{j=1}^{\infty} \int_{\varepsilon(j+1)^{-1}}^{\varepsilon j^{-1}} p_k^{m\prime}(s) s^{-m}\, ds \geq (1 - m)^{-1} \varepsilon^{1-m} \sum_{j=1}^{\infty} p_k^{m\prime}(\varepsilon j^{-1})(j^{m-1} - (j + 1)^{m-1}).$$

Consequently,

$$\| y_k^m \|_p \geq [\int_{\varepsilon}^{1} |y_k^m(t)|^p\, dt]^{1/p} \geq \inf_{t \in [\varepsilon, 1]} |y_k^m(t)| (1 - \varepsilon)^{1/p} \geq$$

$$\geq (1 - \varepsilon)^{1/p} (1 - m)^{-1} \varepsilon^{1-m} \sum_{j=1}^{\infty} p_k^{m\prime}(\varepsilon j^{-1})(j^{m-1} - (j + 1)^{m-1}).$$

As $k \to \infty$, the positive terms of the series on the right side increase. Actually, $p_k^{m'}(\varepsilon j^{-1})$ increases to $(\varepsilon j^{-1})^{m-1}$ and therefore

$$\liminf_{k\to\infty} \|y_k^m\|_p \geq (1-\varepsilon)^{1/p}(1-m)^{-1} \sum_{j=1}^{\infty} 1 - (j(j+1)^{-1})^{1-m}.$$

But, the series on the right side is obviously divergent, thus $\|y_k^m\|_p \to \infty$ as $k \to \infty$. Since

$$\|p_k^m(T)x\|_p = \|z_k^m - y_k^m\|_p \geq \|y_k^m\|_p - \|z_k^m\|_p \geq \|y_k^m\|_p - \|x\|_p,$$

where we have used (3), we get the desired result.

2. THE INNER AND THE OUTER SPECTRALNESS

We shall introduce two notions, just for the purposes limited to this note.

2.1. DEFINITION. Let Y be any non-trivial complex Banach space and A a bounded operator on Y. If there is a Banach space Y_0, a bounded injective homomorphism $V_0 : Y_0 \to Y$ with dense range and a spectral operator A_0 on Y_0 such that $V_0A_0 = AV_0$, then A_0 will be called an *inner spectral representation* of the operator A.

2.2. PROPOSITION. *If* $X \neq L^\infty$, *then* T *has an inner spectral representation.*

PROOF. Assume at first that $X \neq C$. Denote by X_0 the set of all functions $x \in X$ with bounded variation. We will assume that the functions are left continuous on I and right continuous at the point $t = 0$, while $x(0) = 0$. Define a norm on X_0 by $\|x\|_0 = V(x)$, the total variation of the function. The space X_0 is a Banach space and by

$$\|x\|_p \leq \|x\|_\infty \leq V(x) = \|x\|_0, \quad \text{for } x \in X_0,$$

we see that the natural embedding $V_0 : X_0 \to X$ is bounded. It is clear that V_0 is injective and has dense range. Furthermore, for any $x \in X_0$ we have by (1)

$$(Tx)(t) = \int_0^t s\,dx(s), \quad \text{for } t \in I,$$

using the formula of partial integration for the Riemann-Stieltjes integral. Next, identify X_0 with the space of all regular measures x on I for which $x(\{0\}) = x(\{1\}) = 0$. Note that T_0 is on X_0 necessarily defined by $(T_0x)(A) = \int_A s\,dx(s)$, for Borel subsets $A \subset I$. This operator is obviously spectral of the scalar type. In the case $X = C$, define first

X_o, V_o and T_o as above and then take the space $\mathbf{C} + X_o$ instead of X_o, where $\mathbf{C}$ stays for complex constants, take the homomorphism $(\alpha,x) \mapsto \alpha + V_o x \in X$ instead of V_o and the operator $(\alpha,x) \mapsto (0, T_o x)$ instead of T_o, to get the result.

2.3. PROPOSITION. *If* $X = L^\infty$, *then* T *has no inner spectral representation.*

PROOF. For any $t \in I$ let F_t be a functional of norm 1 on L^∞ which has the value x(t) on every $x \in L^\infty$, continuous at least on some neighbourhood of the point t. Furthermore, let LIM be some Banach limit (see [6], page 73 for the definition and the existence of Banach limits), and define the bounded sequence

$$a_n = \begin{cases} -1 & \text{if n divides by 3} \\ 1 & \text{otherwise} \end{cases}$$

for n = 1, 2, . . . , and put for any $x \in L^\infty$

$$\langle x | F \rangle = \mathrm{LIM}\, (3a_n - 1)\langle x | F_{1/n} \rangle.$$

Note that F is a bounded functional on L^∞ which is non-trivial, since for the function $x_o \in L^\infty$ which equals a_n on $(2(2n+1)^{-1},\ 2(2n-1)^{-1}) \cap I$, for n = 1, 2, . . ., we have $\langle x_o | F \rangle = 8/3$. Suppose now that T has inner spectral representation T_o acting on some Banach space X_o being continuously and densely embedded in the space X by a homomorphism V_o such that $V_o T_o = T V_o$. Furthermore, denote by P_o the resolution of the identity of the operator T_o and choose some decreasing sequence of positive reals r_m such that $r_m \to 0$. For any $x \in \mathbf{Im}\, V_o$ denote $u_m = V_o P_o([0, r_m]) V_o^{-1} x$ and $v_m = x - u_m$. Then $v_m \in X_T([r_m, 1])$ is necessarily trivial on $[0, r_m]$ and hence $\langle v_m | F \rangle = 0$. On the other hand, the sequence u_m converges necessarily to a constant function in the norm of L^∞, but the continuous functional F obviously annihilates constants. Hence the sequence $\langle u_m | F \rangle$ converges to zero which implies $\langle x | F \rangle = 0$. Thus $\mathbf{Im}\, V_o$ can not be dense in L^∞.

The technique of going to either some smaller or some bigger space with the aim of getting a nicer operator is often used. In this connection we would first like to call attention to the notion of quasisimilarity (see [19], and also [3]). A bounded homomorphism V from a Banach space X into a Banach space Y is usually called quasi--invertible if it is injective and has dense range. A quasi-invertible homomorphism is sometimes also called quasiaffinity and this notion has often been used in the case when X = Y is a Hilbert space. For operators T and S acting on some Hilbert space X we say that T is a quasiaffine transform of S if there is a quasiaffinity V on X such that TV = VS. The operators T and S are said quasisimilar if they are quasiaffine transforms of one another. We could use this terminology if in our definition of inner spectral representations all Banach spaces were replaced by Hilbert spaces. Having inner spectral representations would, under this assumptions, become actually the same as being a quasiaffine transform of a spectral operator. Let us give now a spectral notion, similar to the inner spectral repressentation.

2.4. DEFINITION. Let T be an operator on a Banach space X. If there is a Banach space X^o and a bounded quasi-invertible homomorphism $V^o : X \to X^o$ such that for a spectral operator T^o on X^o the relation $T^oV^o = V^oT$ holds, then T^o will be said an *outer spectral representation* of T.

Actually, having an outer spectral representation means something like having a quasiaffine transform which is a spectral operator.

Return now again to our operator T to see that an operator without any outer spectral representations can have some inner spectral representations.

2.5. PROPOSITION. *If neither* $X = L^\infty$ *nor* $X = C$, *then* T *has no outer spectral representation..*

PROOF. Suppose on the contrary that the operator T has some outer spectral representation. Choose $r \in (0,1)$, then $e_r \in X_T([0,r])$ and for $s \in I \setminus [0,r] = J$, we have for the local spectrum $\sigma_T(e_s) \subset J$, but $e_s \to e_r$ in the norm of X as $s \to r$, for $s \in J$. If X^o, V^o and T^o are as in the definition of outer spectral representations and P^o is the resolution of the identity of the operator T^o, then $P^o([0, r])V^oe_s = 0$ for $s \in J$, but it converges to $P^o([0,r])V^oe_r = V^oe_r$, as $s \to r$, contradicting the injectivity of V^o.

It turns out that in the cases $X = L^\infty$ and $X = C$, the operator T again has no outer spectral representation, but the proof is more complicated and will be omitted. One may think from our examples that the operators having inner spectral representations are decomposable; however this is not true in general.

2.6. EXAMPLE. *There is an operator with both inner and outer spectral representations which is not decomposable.*

PROOF. We give here an example which seems to be due to I. Colojoară and C. Foiaş (see [5]). Let T_k be the operator, represented in the orthonormal basis of the k-

-dimensional Hilbert space with the matrix, having entries 1 on the first upper diagonal and zeros everywhere else. Moreover, let T be the direct sum of operators T_k for k = = 1, 2, ... Then T is not decomposable (see [5]) and is quasisimilar to a quasinilpotent operator (see [7]), hence it has both inner and outer spectral representa-tions.

A spectral operator has, of course, both inner and outer spectral representa-tions, but the converse is not true. Example 2.6 shows that there exists an operator with both inner and outer spectral representations, which is not even decomposable. This proves that the classes of operators with inner and outer spectral representations are too large for a proper study.

Acknowledgements. The author would like to express his warmest thanks to his teacher Professor Ivan Vidav. The author is also indebted to the referee for his valuable remarks on this note.

REFERENCES

1. **Albrectht, E.J. ; Vasilescu, F.-H.** : On spectral capacities, *Rev. Roumaine Math. Pures Appl.* **19**(1974), 701-705.
2. **Apostol, C.** : Spectral decomposition and functional calculus, *Rev. Roumaine Math. Pures Appl.* **13**(1968), 1481-1528.
3. **Apostol, C.** : Operators quasi-similar to a normal operator, *Proc. Amer. Math. Soc.* **53**(1975), 104-106.
4. **Apostol, C.** : The spectral flavour of Scott Brown's techniques, *J. Operator Theory* **6**(1981), 3-12.
5. **Colojoarǎ, I. ; Foiaş, C.** : *Theory of generalized spectral operators*, Gordon and Breach, New York, 1968.
6. **Dunford, N. ; Schwartz, J. T.** : *Linear operators. Parts I, II, III*, Wiley--Interscience, New York, 1958, 1963, 1971.
7. **Fialkov, L.** : A note on quasisimilarity. II, *Pacific J. Math.* **70**(1977), 151-162.
8. **Foiaş, C.** : Une application des distributions vectorielles a la theorie spectrale, *Bull. Sci. Math. (2)* **84**(1960), 147-158.
9. **Foiaş, C.** : Spectral maximal spaces and decomposable operators in Banach space, *Arch. Math.* **14**(1963), 341-349.
10. **Foiaş, C.**: Spectral capacities and decomposable operators, *Rev. Roumaine Math. Pures Appl.* **13**(1968), 1539-1545.
11. **Kantorovitz, Sh.** : Classification of operators by means of their operational calculus, *Trans. Amer. Math. Soc.* **115**(1965), 194-224.
12. **Kantorovitz, Sh.** : A Jordan decomposition for operators in Banach space, *Trans. Amer. Math. Soc.* **120**(1965), 526-550.

13. **Kantorovitz, Sh.** : The semi-simplicity manifold of arbitrary operators, *Trans. Amer. Math. Soc.* **123**(1966), 241-252.

14. **Kantorovitz, Sh.** : *Spectral theory of Banach space operators*, Lecture Notes Math. **1012**, Springer, Berlin, 1983.

15. **Omladič, M.** : *Spectral invariant subspaces of bounded operators* (Slovene), Ph. D. thesis, E. K. University of Ljubljana, 1980.

16. **Ringrose, J. R.** : On well-bounded operators. II, *Proc. London Math. Soc. (3)* **13** (1963), 613-638.

17. **Sine, R. C.** : Spectral decompositions of a class of operators, *Pacific J. Math.* **14**(1964), 333-352.

18. **Smart, D. R.** : Conditionally convergent spectral expansions, *J. Austral. Math. Soc.* **1**(1959/60), 319-333.

19. **Sz.-Nagy, B. ; Foiaş, C.** : *Harmonic analysis of operators on Hilbert space*, North-Holland Publ. Co., Amsterdam, 1970.

20. **Turner, J. K.** : On well-bounded and decomposable operators, *Proc. London Math. Soc. (3)* **37**(1978), 521-544.

21. **Vasilescu, F.-H.** : An application of Taylor's functional calculus, *Rev. Roumaine Math. Pures Appl.* **19**(1974), 1165-1167.

Matjaž Omladič

Department of Mathematics
E.K.University of Ljubljana
Ljubljana
Yugoslavia.

Operator Theory:
Advances and Applications, Vol.17

HYPONORMAL OPERATORS AND EIGENDISTRIBUTIONS

Mihai Putinar

INTRODUCTION

This paper deals with two-dimensional models for hyponormal operators. We link the canonical model described in a previous paper [12] to more familiar function spaces and then we relate it to other functional realizations of some special classes of hyponormal operators. The paper is centered around the space of globally defined eigendistributions of the adjoint of a hyponormal operator.

A hyponormal operator on a Hilbert space H is by definition a linear bounded operator $T \in L(H)$ with the property $TT^* \leq T^*T$. The generic element of this class of operators was described by Daoxing Xia [16] as a combination between multiplication operators with bounded measurable functions and the Hilbert transform, on the Hilbert space $L^2(a,b)$, where (a,b) is an interval of the real line. Xia's model and the Cartesian decomposition of an operator into real and imaginary part were the principal methods in the theory of hyponormal operators. Recently, several authors (Xia [17], Clancey [4], Pincus-Xia-Xia [11]) have refered to hyponormal operators with one-dimensional self-commutator, and related objects to them, in complex coordinate terms. We adopt in this paper the same point of view. Clancey's report [4] was the motivation of the present paper, while the proofs below continues the technique developed in [12]. Let us recall the main construction from [12].

Let Ω be a bounded domain with smooth boundary of the complex plane $\mathbf{C}$, and let $T \in L(H)$ be a hyponormal operator. For every smooth H-valued function $f \in E(\overline{\Omega}, H)$, the Cauchy-Pompeiu formula gives rise to the inequality

$$\|(I - P)f\|_{2,\Omega} \leq C(\|(\bar{z} - T^*)\bar{\partial}f\|_{2,\Omega} + \|(\bar{z} - T^*)\bar{\partial}^2 f\|_{2,\Omega}), \tag{1}$$

where C is a positive constant depending only on Ω and $P : L^2(\Omega,H) \to A^2(\Omega,h)$ denotes the Bergman projection. When the domain Ω contains the spectrum of T, $\Omega \supset \sigma(T)$, the linear map

$$V : H \to H^2(\Omega,H)/\overline{(z - T)H^2(\Omega,H)},$$

where Vh represents the class $1 \tilde{\otimes} h$ of the constant function h on Ω, is one-to-one and has closed range. Here $H^2(\Omega,H)$ stands for the Sobolev space of order 2 of H-valued functions. This fact is a consequence of the inequality (1) and of the Riesz-Dunford functional calculus. Then the operator $\tilde{T}$ induced by the multiplication with z (the complex coordinate) on the quotient space above is (generalized) scalar in the terminology of Colojoară-Foiaş [5] and it extends T, that is $VT = \tilde{T}V$. The existence of this natural scalar extension explains several spectral properties of hyponormal operators.

The present paper deals with the dual picture of the above construction. More precisely, the dual space of the quotient space where the scalar extension acts, is the following space of distributions

$$W_T^{-2}(H) = \{u \in H^{-2}(\Omega, H) | (\bar{z} - T^*)u = 0\},$$

and the surjective map

$$V' : W_T^{-2}(H) \to H$$

acts by the formula

$$V'(u) = (u,1) = \int u, \quad u \in W_T^{-2}(H).$$

In other words, the Hilbert space H is generated by the global eigendistributions of the operator T^*.

The space $W_T^{-2}(H)$ carries a Hilbert space norm which is independent of Ω, and this fact will be used in the sequel in order to define a canonical norm on the space of the scalar extension $\tilde{T}$.

The surjectivity of the operator V' reminds of the characteristic property of a class of operators studied and classified by Cowen and Douglas [6]. Although the case of a general hyponormal operator is more complicated, this analogy suggests a correspondence between a hyponormal operator T and an operator-valued distribution kernel K_T which plays the role of the generalized Bergman kernel of Curto and Salinas [7]. This object offers a functional description of the initial Hilbert space, in which T^* becomes the multiplication operator with $\bar{z}$. The kernel K_T has a certain redundancy, and we were able to eliminate it and to describe a determining part of K_T only in a few well understood cases.

The content is the following.

In the first section we recall some facts concerning vector-valued Sobolev spaces.

Since the natural scalar extension of a hyponormal operator relies on the multiplication operator with the complex coordinate on a Sobolev space, we collect in the second section a series of properties of this prototype operator.

The third section deals with the naturality problem for the Hilbert space structure introduced by various Sobolev space norms on the space of the natural scalar extension of a hyponormal operator. The list of the properties of the natural scalar extension is completed in the last part of this section with a spectral preserving theorem.

The fourth section is devoted to eigendistributions of cohyponormal operators. We prove that any cohyponormal operator possesses a global distribution resolvent in H^{-1}_{loc}, localized at an arbitrary vector. Then we associate in a natural way to a hyponormal operator T the distribution kernel $K_T \in H^{-2}(\mathbf{C}, L(W^{-2}_T(H))$, which is a complete unitary invariant of T and behaves well to analytic changes of coordinates.

Thanks to the recent work of Clancey [3], [4] we determine in the fifth section a generating part (a compression) of the kernel K_T, in the case of irreducible hyponormal operators with one-dimensional self-commutator. As a by-product we derive a concrete functional model for such operators, which diagonalizes T^* and is expressed only in terms of the principal function of T.

§1. PRELIMINARIES

In this section we recall some properties of the Sobolev spaces of vector-valued functions. Although most of the results listed below are more general, we concentrate on Sobolev spaces of order 2, on $\mathbf{R}^2$. A complete and optimal reference on that subject is Hörmander's book [9].

Let H be a complex Hilbert space and let $D(\mathbf{C},H)$ be the LF-space of smooth, compactly supported H-valued functions on the complex plane $\mathbf{C}$. Its topological dual is the space of H-valued distributions, denoted by $D'(\mathbf{C},H)$. We shall use the nondegenerate sesquilinear pairing

$$\langle \cdot , \cdot \rangle : D'(\mathbf{C},H) \times D(\mathbf{C},H) \rightarrow \mathbf{C},$$

which extends the L^2-scalar product:

$$\langle \phi , \psi \rangle_2 = \int \langle \phi(z) , \psi(z) \rangle_H \, d\mu(z); \quad \phi , \psi \in D(\mathbf{C},H).$$

Here, and throughout this paper, μ stands for the planar Lebesgue measure.

Let z denote as usually the complex coordinate on $\mathbf{C}$. One denotes : $\partial = \partial / \partial z$, $\bar{\partial} = \partial / \partial \bar{z}$ and $\Delta = 4 \partial \bar{\partial}$.

The Hilbert space completion of $D(\mathbf{C},H)$ with respect to the norm

$$\|\phi\|_{H^2} = \|(1 - \Delta)\phi\|_2$$

is the Sobolev space $H^2(\mathbf{C},H)$. Its dual via the above sesquilinear form is the Sobolev space of order -2, denoted $H^{-2}(\mathbf{C},H)$. The norm of this space can be described in terms of the Fourier transform as follows:

$$\|u\|_{H^{-2}}^2 = \int \|\hat{u}(\xi)\|^2 (1 + |\xi|^2)^{-2} d\mu(\xi).$$

We point out that for any $\phi \in D(\mathbf{C},H)$,

(2) $$\|(1 - \Delta)\phi\|_2^2 = \|\phi\|_2^2 + 8\|\bar{\partial}\phi\|_2^2 + 16\|\bar{\partial}^2\phi\|_2^2.$$

Let Ω be a complex domain with smooth boundary $\partial\Omega$. Then

$$H_0^2(\Omega,H) = \{f \in H^2(\mathbf{C},H) | \operatorname{supp}(f) \subset \bar{\Omega}\}$$

is a closed subspace of $H^2(\mathbf{C},H)$. By the Sobolev embedding theorem the Hilbert space $H_0^2(\Omega,H)$ is continuously contained in the Banach space $C(\bar{\Omega},H)$ of continuous functions on $\bar{\Omega}$, uniformly bounded in norm. The dual of $H_0^2(\Omega,H)$ with respect to the above pairing is denoted by $H^{-2}(\Omega,H)$ and it is a quotient Hilbert space of $H^{-2}(\mathbf{C}, H)$.

Conversely, if one denotes by $H_0^{-2}(\Omega,H)$ the closed subspace of $H^{-2}(\mathbf{C},H)$ of those distributions supported by $\bar{\Omega}$, then $H^2(\Omega,H)$ will denote its dual. It is convenient to identify $H^2(\Omega,H)$ with the orthogonal complement of $H_0^2(\mathbf{C}\setminus\bar{\Omega}, H)$:

$$H^2(\Omega,H) = H^2(\mathbf{C},H) \ominus H_0^2(\mathbf{C}\setminus\bar{\Omega}, H).$$

We point out that the space $H_0^2(\Omega,H)$ is continuously contained in $H^2(\Omega,H)$.

By the definition of the H^2-norm, the operator $1 - \Delta : H_0^2(\Omega,H) \to L^2(\Omega,H)$ is an isometry, which is not onto. The dual, in the sense of distributions, of the differential operator $1 - \Delta$ has the same expression, hence the operator

$$(1 - \Delta)^2 : H_0^2(\Omega,H) \to H^{-2}(\Omega,H)$$

is unitary. We should remark at this point that the space $H^{-2}(\Omega,H)$ is naturally contained in $D'(\Omega,H)$, but not in $H^{-2}(\mathbf{C},H)$. However, we may indentify $H^{-2}(\Omega,H)$ with the range of the operator $(1 - \Delta)^2 : H_0^2(\Omega,H) \to H^{-2}(\mathbf{C},H)$. In such a way $H^{-2}(\Omega,H)$ becomes a subspace of $H_0^{-2}(\Omega,H)$.

If we assume in addition that the domain Ω is bounded, then the space $H_0^2(\Omega,H)$ carries the following equivalent norms:

$$\|f\|_{H^2} \sim \|\Delta f\|_2 \sim \|\bar{\partial}^2 f\|_2, \quad f \in H_0^2(\Omega,H).$$

Some formulae in this paper will be at hand in the third norm rather than in the first one. Consequently we denote throughout this paper by $W_0^2(\Omega,H)$ the space $H_0^2(\Omega,H)$ endowed with the following Hilbert space norm:

$$\|f\|_{W^2} = \|\bar{\partial}^2 f\|_2 = \tfrac{1}{4}\|\Delta f\|_2.$$

Its isometric dual is denoted by $W^{-2}(\Omega,H)$, and it is endowed with the norm that makes the operator

$$(\partial\bar{\partial})^2 : W_0^2(\Omega,H) \to W^{-2}(\Omega,H)$$

unitary.

At the level of local spaces we state the following.

LEMMA 1.1. *A locally integrable function* f *on* Ω *belongs to* $H^2_{loc}(\Omega,H)$ *iff* ϕf *belongs to* $W_0^2(\Omega,H)$ *for every* $\phi \in D(\Omega)$.

The space $W_0^2(\Omega,H)$ is again continuously embedded in $C(\bar{\Omega},H)$ and consequently the Dirac measures $\delta_\lambda \otimes h$ belong to $W^{-2}(\Omega,H)$, where $\lambda \in \Omega$ and $h \in H$. Moreover, these and only these are the elements of $W^{-2}(\Omega,H)$ supported by a single point.

§2. A SCALAR SUBNORMAL OPERATOR

In a previous paper, the multiplication operator with the complex coordinate on a Sobolev space of order 2 was the prototype in the functional model associated there to a hyponormal operator, see [12]. We present in this section some of the properties of that operator, though we will not make use of all of them in the sequel. We restrict ourselves to the scalar case $\dim(H) = 1$, the higher dimensional case being completely similar.

Let $\Omega \subset \mathbf{C}$ be a bounded domain with smooth boundary and let M denote the multiplication operator with z on the Hilbert space $W_0^2(\Omega)$. As we already remarked in [12], the operator M is scalar of order 2, in the sense of Colojoară and Foias [5], with the spectral distribution

$$U : D(\mathbf{C}) \to L(W_0^2(\Omega)),$$

$$U(\phi)f = \phi f\,, \quad \phi \in D(\mathbf{C}),\ \ f \in W_0^2(\mathbf{C}).$$

The maximal spectral space associated to a closed subset F of $\mathbf{C}$ is

$$W_0^2(\Omega)_M(F) = \{f \in W_0^2(\Omega) \,|\, \mathrm{supp}(f) \subset F \cap \bar{\Omega}\}.$$

Let E_F denote the orthogonal projection of $W_0^2(\Omega)$ onto this space. Then E behaves like a spectral measure, with one exception – the countable additivity property. Indeed, let ω be a subdomain relatively compact in Ω, with smooth boundary.

LEMMA 2.1. *Let* $\{K_n\}$ *be an increasing compact exhaustion of* ω. *Then* $I - E_{\mathbf{C}\setminus\omega} \neq \text{s-lim}\, E_{K_n}$.

PROOF. The orthogonal projections $P_1 = E_{\mathbf{C}\setminus\omega}$ and $P_2 = \text{s-lim}\, E_{K_n}$ are complementary and for every $f \in W_0^2(\Omega)$ the continuous functions $P_1 f$ and $P_2 f$ vanish on $\partial\omega$. Therefore $P_1 + P_2 \neq I$, **q.e.d.**

The operator M is subnormal because the operator

$$\bar\partial^2 : W_0^2(\Omega) \to L^2(\Omega)$$

is an isometry which intertwines M with the multiplication operator with z, acting on $L^2(\Omega)$.

LEMMA 2.2. *The adjoint of* M *has the following expression*

$$(3) \qquad (M^* f)(z) = \bar z f(z) + (2/\pi) p_\Omega\left(\int (\chi(z) f(\zeta)/(\zeta - z))\, d\mu(\zeta)\right), \quad f \in W_0^2(\Omega),$$

where $\chi \in D(\Omega')$, $\chi \equiv 1$ *on* $\bar\Omega$, $\Omega \subset\subset \Omega' \subset\subset \mathbf{C}$ *are arbitrary and* p_Ω *denotes the orthogonal projection of* $W_0^2(\Omega')$*onto* $W_0^2(\Omega)$.

PROOF. The Cauchy transform is a linear bounded operator from $W_0^2(\Omega)$ into $H^2(\Omega')$, hence right side of (3) makes a good sense.

Let $\phi, \psi \in D(\Omega)$. Then one gets from the Cauchy-Pompeiu formula:

$$\langle \bar z\phi, \psi\rangle_{W^2} + \langle (2/\pi) p_\Omega\left(\chi\int \phi(\zeta)/(\zeta - z)\, d\mu(\zeta)\right), \psi\rangle_{W^2} =$$

$$= \langle \bar\partial^2(\bar z\phi), \bar\partial^2\psi\rangle_2 + \langle (2/\pi)\bar\partial^2 \int \phi(\zeta)/(\zeta - z)\, d\mu(\zeta), \bar\partial^2\psi\rangle_2 =$$

$$= \langle \bar z\bar\partial^2\phi + 2\bar\partial\phi, \bar\partial^2\psi\rangle_2 - \langle 2\bar\partial\phi, \bar\partial^2\psi\rangle_2 = \langle \bar z\bar\partial^2\phi, \bar\partial^2\psi\rangle_2 = \langle \phi, M\psi\rangle_{W^2}.$$

Because the space $D(\Omega)$ is dense in $W_0^2(\Omega)$, the formula (4) is proved. As the operator M^* is unique, its formula doesn't depend on the choices of the domain Ω' and of the function χ, **q.e.d.**

The operator M^* is still scalar, with the spectral distribution U^*. Its maximal spectral spaces are:

$$W_0^2(\Omega)_{M^*}(F) = \{f \in W_0^2(\Omega) \,|\, \mathbf{supp}(\Delta^2 f) \subset F\},$$

where F is a closed subset of the complex plane. Indeed, $f \in W_0^2(\Omega)_{M^*}(F)$ iff $U^*(\phi)f = 0$ for every $\phi \in \mathcal{D}(\mathbf{C} \setminus F)$. That is $\langle U^*(\phi)f , g\rangle = 0$ for every function $g \in W_0^2(\Omega)$. But we have

$$\langle U^*(\phi)f , g\rangle = \langle f , U(\phi)g\rangle = \langle \bar{\partial}^2 f , \bar{\partial}^2(\phi g)\rangle_2 =$$

$$\langle \partial^2 \bar{\partial}^2 f , \phi g\rangle = 4^{-1}\langle \Delta^2 f , \phi g\rangle,$$

and consequently $\Delta^2 f = 0$ on $\mathbf{C} \setminus F$.

The maximal spectral spaces of the operator M^* are not orthogonal for disjoint supports, as those of M, but one can estimate the angle between them, as follows. Let e_F denote the orthogonal projection of $W_0^2(\Omega)$ onto the space $W_0^2(\Omega)_{M^*}(F)$.

PROPOSITION 2.3. *Let* F,G *be two disjoint closed subsets of* $\bar{\Omega}$. *Then there is a positive constant* C, *such that*

$$\|e_F \cdot e_G\| \leq 1 - C[\mathbf{dist}(F,G)]^2,$$

provided that **dist**(F,G) *is small.*

For a proof of Proposition 2.3 and related results we refer the reader to Simon's book [14, §III.4].

The operator $\bar{M}$ – the dual of M on $W^{-2}(\Omega)$ – will be of a certain interest in the next sections. As a first applications of the duality between distributions and functions, we compute various spectra of the operator M. Although some of the equalities below are true for arbitrary scalar opertors, we prove them in our context.

PROPOSITION 2.4. *The operator* M *has the following spectra:*

$$\sigma(M) = \sigma_{ess}(M) = \sigma_{ap}(M) = \bar{\Omega},$$

$$\sigma_r(M) = \Omega,$$

$$\sigma_c(M) = \partial\Omega.$$

PROOF. The point spectrum of M is empty. Indeed, if $(M - \lambda)f = 0$ for a point $\lambda \in \bar{\Omega}$ and a function $f \in W_0^2(\Omega)$, then $\mathbf{supp}(f) \subset \{\lambda\}$ and, since f is continuous, f = 0.

Let us assume that $0 \in \Omega \setminus \sigma_r(M)$, that is the operator M has dense range in $W_0^2(\Omega)$. On the other hand,

$$\mathbf{Ran}(M)^{\perp} = \{u \in W^{-2}(\Omega) \mid \langle u, zf\rangle = 0,\ f \in W_0^2(\Omega)\} =$$

$$= \{u \in W^{-2}(\Omega) \mid \bar{z}u = 0\} = \mathbf{C} \cdot \delta,$$

a contradiction! Consequently $\Omega \subset \sigma_r(M)$.

Similarly $\partial\Omega \subset \sigma_c(M)$. The inclusions are in fact equalities, because $\bar{\Omega} = \sigma(M) = \sigma_r(M) \cup \sigma_c(M)$ and $\sigma_r(M) \cap \sigma_c(M) = \emptyset$.

Let us assume $0 \in \Omega \setminus \sigma_{ap}(M)$. Then the operator M has, by the above computation of $\mathbf{Ran}(M)^{\perp}$ a closed range of condimension 1 in $W_0^2(\Omega)$. Then every element $f \in W_0^2(\Omega)$, $f(0) = 0$, would factorize as $f = zg$, with $g \in W_0^2(\Omega)$. But $\bar{z}/z$ doesn't belong to $H^2_{loc}(\Omega)$, which contradicts the assumption that M has closed range. In conclusion, $\sigma_{ess}(M) = \sigma_{ap}(M) = \bar{\Omega}$, and the proof is complete.

The dual $\bar{M}$ of the operator M, on $W^{-2}(\Omega)$, coincides with the multiplication with $\bar{z}$. Its adjoint can be easily computed, as follows.

LEMMA 2.5. *The operator* $\bar{M}^* \in L(W^{-2}(\Omega))$ *is unitarily equivalent with* M *and it is represented by the formula*

(4) $$\bar{M}^* u = zu - (2/\pi)(1/\bar{z} * u), \quad u \in W^{-2}(\Omega).$$

Here the space $W^{-2}(\Omega)$ is embedded into $H_0^{-2}(\Omega)$ as the range of the operator $(\partial\bar{\partial})^2 : W_0^2(\Omega) \to H^{-2}(\mathbf{C})$. This makes possible the convolution $1/\bar{z} * u$.

PROOF. An element $u \in W^{-2}(\Omega)$ can be approximated with a smooth function of the form $(\partial\bar{\partial})^2\phi$, $\phi \in D(\Omega)$. Since $\partial(1/\bar{z}) = -\pi$, one gets

$$1/\bar{z} * (\partial\bar{\partial})^2\phi = \partial(1/\bar{z}) * (\partial\bar{\partial}^2\phi) = -\pi\partial\bar{\partial}^2\phi,$$

hence the right part of the equality (4) represents a bounded operator on W^{-2}.

The relation $(\partial\bar{\partial})^2 M^* = \bar{M}(\partial\bar{\partial})^2$ implies $\bar{M}^*(\partial\bar{\partial})^2 = (\partial\bar{\partial})^2 M$, therefore the operator $\bar{M}^*$ is unitarily equivalent with M. Moreover, the same equality shows that in order to prove (4) it is enough to check that

$$\langle z(\partial\bar{\partial})^2\phi, \psi\rangle - (2/\pi)\langle 1/\bar{z} * (\partial\bar{\partial})^2\phi, \psi\rangle = \langle(\partial\bar{\partial})^2 z\phi, \psi\rangle,$$

for any functions $\phi, \psi \in D(\Omega)$. But in view of the above computation we have

$$-(2/\pi)\langle 1/\bar{z} * (\partial\bar{\partial})^2\phi, \psi\rangle = 2\langle\partial\bar{\partial}^2\phi, \psi\rangle = 2\langle\partial\bar{\partial}^2\phi, \psi\rangle,$$

and the proof is complete.

§3. SCALAR EXTENSIONS OF HYPONORMAL OPERATORS

Let T be a hyponormal operator on the Hilbert space H. We describe in this section several Hilbert space structures on the space of the natural scalar extension of T.

Let Ω be a bounded domain in $\mathbf{C}$, with smooth boundary and which contains the spectrum of T, $\Omega \supset \sigma(T)$. Then the Fréchet quotient space

$$\mathcal{H} = H^2_{loc}(\Omega,H)/\overline{(z - T)H^2_{loc}(\Omega,H)}$$

contains the space H, as classes of constant functions, via the embedding $V : H \to \mathcal{H}$ $Vh = (1\otimes h)^{\sim}$. The scalar operator $z\otimes I$ commutes on $H^2_{loc}(\Omega,H)$, as well as its spectral distribution, with the operator $z - T$, hence it induces a scalar operator on $\mathcal{H}$, denoted by $\tilde{T}$. Moreover, $VT = \tilde{T}V$, see [12].

The Frechet space topology of $\mathcal{H}$ is compatible with various Hilbert space norms, as for instance that used in [12], by relation (2) and Lemma 1.1. Most of these Hilbert space norms depend on a choice, e.g. of the domain Ω. For example, every domain Ω endows the space $\mathcal{H}$ with a Hilbert space structure, by identifying $\mathcal{H}$ with the space

$$\mathcal{H}_0(\Omega) = W^2_0(\Omega,H) \ominus (z - T)W^2_0(\Omega,H).$$

The bad behaviour of these norms when comparing them for different domains can be easily illustrated by the following.

EXAMPLE. Hyponormal operators on finite dimensional spaces.

We assume that **dim**(H) is finite. In that case T is a normal operator. Let consider two domains with the property $\sigma(T) \subset \Omega' \subset \Omega$. Then the operator $A : \mathcal{H}_0(\Omega') \to \mathcal{H}_0(\Omega)$ induced by the natural extension map $W^2_0(\Omega') \subset W^2_0(\Omega)$ is not, in general, unitary. Notice that the operator A is invertible.

Indeed, let us assume that $f \in \mathcal{H}_0(\Omega)$ satisfies $\|f\| = \|A^{-1}f\|$. Then the function f vanishes on $\Omega \setminus \Omega'$, because the space $W^2_0(\Omega')$ is isometrically contained in $W^2_0(\Omega)$. As the spectrum $\sigma(T)$ is finite and the space $\mathcal{H}_0(\Omega)$ contains only continuous functions, it is not possible that A would be unitary for an arbitrary small neighbourhood Ω' of the spectrum.

If the space is finite dimensional, then $\tilde{T}$ coincides with T, that is **dim**(H) = = **dim**($\mathcal{H}$). Indeed, since the operator T is diagonalizable, it sufficies to prove the equality of dimensions in the case **dim**(H) = 1. Let us assume then in addition that $\sigma(T) = \{0\}$. By the proof of Proposition 2.4 we obtain

$$\mathcal{H}_0(\Omega) \simeq \mathrm{Ran}(M)^{\perp} = \{u \in W^{-2}(\Omega) \,|\, \bar{z}u = 0\} = \mathbf{C}\cdot\delta,$$

and therefore **dim**$(H) = 1$.

The same duality argument shows that the operator $z - T : W_0^2(\Omega,H) \to W_0^2(\Omega,H)$ has not, in general, closed range.

Let us come back to an arbitrary hyponormal operator T. A simpler and a more canonical picture is obtained by dualizing the space H with respect to the sesquilinear form

$$\langle\cdot,\cdot\rangle : H_{co}^{-2}(\Omega,H) \times H_{loc}^2(\Omega,H) \to \mathbf{C},$$

defined in the preliminaries. Here $H_{co}^{-2}(\Omega,H)$ denotes the space of those distributions $u \in H^{-2}(\mathbf{C},H)$ with compact support in Ω. Let us define the space

$$W_T^{-2}(H) = H' = \{u \in H_{co}^{-2}(\Omega,H) | (\bar{z} - T^*)u = 0\}.$$

The space $W_T^{-2}(H)$ inherits a Hilbert space norm from $H_0^{-2}(\Omega,H)$, because the support of a distribution $u \in W_T^{-2}(H)$ is contained in $\sigma(T)$. Moreover, since the natural inclusion $H_0^{-2}(\Omega',H) \subset H_0^{-2}(\Omega,H)$ is isometric, whenever $\Omega' \subset \Omega$, this Hilbert space norm on $W_T^{-2}(H)$ doesn't depend on Ω.

On the other hand, the space $H^{-2}(\Omega,H)$ contains $W_T^{-2}(H)$, but not isometrically, so that the unitary operator $(1 - \Delta)^2 : H_0^2(\Omega,H) \to H^{-2}(\Omega,H)$ defines by pull back a universal norm on the space $[(1 - \Delta)^2]^{-1} W_T^{-2}(H)$. But a straightforward computation shows that

$$H \simeq H_0^2(\Omega,H) \ominus (z - T)H_0^2(\Omega,H) = [(1 - \Delta)^2]^{-1} W_T^{-2}(H).$$

In conclusion we state the following.

PROPOSITION 3.1. *Let* Ω *be a domain which contains the spectrum of a hyponormal operator* T. *The differential operator*

$$H \simeq H_0^2(\Omega,H) \ominus (z - T)H_0^2(\Omega,H) \xrightarrow{(1 - \Delta)^2} W_T^{-2}(H)$$

is invertible and the Hilbert space norm on H *which makes* $(1 - \Delta)^2$ *unitary doesn't depend on* Ω.

Let $W_T^2(H)$ denote the space H endowed with this canonical norm.

We point out that, although an element $f \in W_T^2(H)$, realized in virtue of the above proposition as a function in $H_0^2(\Omega,H)$, has a trivial extension F to $\mathbf{C}$, as well as every distribution $u \in W_T^{-2}(H)$, the relation $(1 - \Delta)^2 f = u$ holds only on Ω. When extending f and u to $\mathbf{C}$, the distribution $(1 - \Delta)^2 F - u$ is not necessarily identical zero, being supported by $\partial \Omega$.

PROPOSITION 3.2. *The natural map*

$$\rho : H^2(\Omega,H) \ominus (z - T)H^2(\Omega,H) \rightarrow W_T^2(H)$$

is an isometric isomorphism, whenever $\Omega \supset \sigma(T)$.

PROOF. Let Ω be a domain with smooth boundary, which contains the spectrum of the operator T. By the definition of the Sobolev spaces, the operator

$$H^2(\Omega,H) = H^2(\mathbf{C},H) \ominus H_0^2(\mathbf{C} \setminus \bar{\Omega},H) \xrightarrow{(1 - \Delta)^2} H_0^{-2}(\Omega,H)$$

is unitary. Let P_Ω denote the orthogonal projection of $H^2(\mathbf{C},H)$ onto $H^2(\Omega,H)$, and let $\chi \in D(\mathbf{C})$, $\chi \equiv 1$ on $\bar{\Omega}$. Then $(\chi z - T)(I - P_\Omega) = (I - P_\Omega)(\chi z - T)(I - P_\Omega)$.

For any functions $f \in H^2(\Omega,H) \ominus (z - T)H^2(\Omega,H)$ and $g \in H^2(\mathbf{C},H)$ we have

$$\langle(\bar{z} - T^*)(1 - \Delta)^2 f , g\rangle = \langle(1 - \Delta)^2 f , (\chi z - T)g\rangle = \langle f , (\chi z - T)g\rangle_{H^2} =$$

$$= \langle f , (z - T)P_\Omega g\rangle + \langle f , (I - P_\Omega)(\chi z - T)(I - P_\Omega)g\rangle = 0.$$

And conversely, the same computations show finally that

$$(1 - \Delta)^2[H^2(\Omega,H) \ominus (z - T)H^2(\Omega,H)] = W_T^{-2}(H).$$

Let Q denote the orthogonal projection of $H^2(\Omega,H)$ onto $H_0^2(\Omega,H)$. Then $Q[H^2(\Omega,H) \ominus (z - T)H^2(\Omega,H)] = H_0^2(\Omega,H) \ominus (z - T)H_0^2(\Omega,H)$, as easily follows from the equality $\langle Qf , (z - T)g\rangle = \langle f , (z - T)g\rangle$, where $f \in H^2(\Omega,H)$ and $g \in H_0^2(\Omega,H)$.

In conclusion, the map ρ in the statement, which coincides with Q when the space $W_T^2(H)$ is realized inside $H_0^2(\Omega,H)$, is unitary, **q.e.d.**

The second part of this section is devoted to a spectral property of the scalar extension $\tilde{T} \in L(W_T^2(H))$ of a hyponormal operator T.

THEOREM 3.3. *Let* T *be a hyponormal operator and let* $\tilde{T}$ *be its natural scalar extension. Then* $\sigma(\tilde{T}) = \sigma(T)$.

PROOF. Since the spectral distribution $\tilde{U}$ of the scalar operator $\tilde{T}$ is supported by $\sigma(T)$, the inclusion of local spectra $\sigma_{\tilde{T}}(Vh) \subset \sigma_T(h)$ holds true for every $h \in H$. We recall that the operator $V : H \to W^2_T(H)$ intertwines T and $\tilde{T}$.

Let $h \in H$ be fixed. In order to prove the converse inclusion, $\sigma_T(h) \subset \sigma_{\tilde{T}}(Vh)$, we identify $W^2_T(H)$ with the Hilbert space $H^2(\Omega,H) \ominus (z - T)H^2(\Omega,H)$, where Ω is a bounded domain which contains the spectrum of T.

Let $\lambda \notin \sigma_{\tilde{T}}(Vh)$. Then there exists an open neighbourhood ω of λ and an analytic function $\tilde{g} \in O(\omega, W^2_T(H))$, such that

$$(\zeta - T)\tilde{g}(\zeta) = Vh , \quad \zeta \in \omega.$$

Let $g \in O(\omega,H^2(\Omega,H))$ be a holomorphic lifting of $\tilde{g}$. Then for a fixed $\zeta \in \omega$,

$$h - (\zeta - z)g(\zeta,z) \in \overline{(z - T)H^2(\Omega,H)}.$$

But the dense range property of a Hilbert space operator is preserved by the topological tensor multiplication with a nuclear space. Therefore there is a sequence $f'_n \in O(\omega,H^2(\Omega,H))$, so that

$$\lim_n (h - (\zeta - z)g(\zeta,z) - (z - T)f'_n(\zeta,z)) = 0$$

in the Fréchet topology of the space $O(\omega,H^2(\Omega,H))$.

Let ω' be another open neighbourhood of the point λ, relatively compact in ω. Let m denote the unique continuous linear extension

$$m : O(\omega)\hat{\otimes}H^2(\Omega,H) \to H^2(\omega',H)$$

of the map $a\otimes b \to (a \cdot b)\,|\,\omega'$. Then $m(h - (\zeta - z)g(\zeta,z) - (z - T)f'_n(\zeta,z)) = h - (z - T)f_n(z)$, where $f_n(z) = f'_n(z,z)$ for $z \in \omega'$. Therefore $h = \lim(z - T)f_n$ in $H^2(\omega',H)$.

In view of the inequality (1) we obtain $\lim \|f_n - Pf_n\|_{2,\omega'} = 0$ and $\lim \|h - (z - T)Pf_n\|_{2,\omega'} = 0$, which in turn implies $h \in \overline{(z - T)O(\omega',H)}$. But the operator T satisfies Bishop's property (β), see [15] or for instance as a subscalar operator, so that the operator $(z - T)$ has closed range on $O(\omega',H)$.

Finally $h \in (z - T)O(\omega',H)$, or, in other terms, $\lambda \notin \sigma_T(h)$, **q.e.d.**

In fact we have proved more, namely:

COROLLARY 3.4. *The local spectra* $\sigma_T(h)$ *and* $\sigma_{\tilde{T}}(Vh)$ *coincide for every* $h \in H$.

The above spectral behaviour of a minimal scalar extension of a hyponormal operator is different from that of the normal extension of a subnormal operator. In particular, the natural scalar extension $\tilde{S}$ of a subnormal operator S doesn't coincide, in general, with the normal extension of S. On the essential resolvent set of S, the operator $\tilde{S}$ has not, in general, closed range.

§4. A DISTRIBUTION KERNEL

This section deals with the relationship between hyponormal operators and operator-valued distribution kernels on $\mathbf{C}^2$. The existence of a scalar extension of a hyponormal operator makes possible the analogy with the generalized Bergman kernels theory of Curto and Salinas [7]. Although the general framework developped in the sequel leads to rather tautological results, when applying it to particular hyponormal operators the fine invariants fit naturally into this scheme.

Let T be a hyponormal operator on the Hilbert space H. With the notations of the preceding section, the dual $V' : W_T^{-2}(H) \to H$ of the embedding V is onto. We recall that $W_T^{-2}(H)$ denotes the set of those distributions $u \in H^{-2}(\mathbf{C},H)$ which are annihilated by $\bar{z} - T^*$. The operator V' acts by the formula

$$V'(u) = (u\,,\,1)\,, \quad u \in W_T^{-2}(H),$$

where $(\cdot\,,\cdot)$ stands for the natural *bilinear* pairing

$$(\cdot\,,\cdot) : E'(\mathbf{C},H) \times E(\mathbf{C}) \to H.$$

We shall use the following continuity property of this bilinear form:

$$\text{(5)} \qquad \|(u\,,\,\phi)\| \leq \|u\|_{H^{-2}} \cdot \|\phi\|_{H^2}\,, \quad u \in E'(\mathbf{C},H),\ \phi \in E(\mathbf{C}),$$

which can be proved by a Fourier transform argument.

For the beginning we prove for its own interest the following.

PROPOSITION 4.1. *A cohyponormal operator* $T^* \in L(H)$ *has a generalized global resolvent, localized at an arbitrary vector* $h \in H$, i.e. *there exists a distribution* $v_h \in H^{-1}_{loc}(\mathbf{C},H)$ *so that* $(\bar{z} - T^*)v_h = h$ *on* $\mathbf{C}$.

The vectors of the range of the operator $[T^*\,,\,T]^{\frac{1}{2}}$ have even a global resolvent of a function type, [13].

PROOF. We choose for a fixed $h \in H$ a distribution $u \in W_T^{-2}(H)$ with the property $V'(u) = -\pi h$. Because the operator $\partial : H^{-1}_{loc}(\mathbf{C},H) \to H^{-2}_{loc}(\mathbf{C},H)$ is onto, there exists a distribution $u_1 \in H^{-1}_{loc}(\mathbf{C},H)$ such that $\partial u_1 = u$. In particular, the restriction of u_1 to the open set $\mathbf{C} \setminus \sigma(T)$ is an antiholomorphic function:

$$u_1(z) = \sum_{n=-\infty}^{\infty} a_n \bar{z}^n,$$

where the series converges for $|z|>||T||$ and $a_n \in H$. Then the expression

$$f(z) = \sum_{n=0}^{\infty} a_n \bar{z}^n$$

defines an antiholomorphic function on the whole complex plane.

The desired global resolvent is $v_h = u_1 - f$. Indeed, $\partial(\bar{z} - T^*)v_h = (\bar{z} - T^*)u = 0$, and the estimate

$$||(\bar{z} - T^*)(u_1(z) - f(z))|| = O(1) \quad \text{for } |z|\to\infty,$$

holds true. Then by Liouville's Theorem $(\bar{z} - T^*)v_h \equiv a_{-1}$.

It remains to compute the constant a_{-1}. Let $\chi \in D(\mathbf{C})$, $\chi \equiv 1$ on a neighbourhood of $\sigma(T)$. By Stokes Theorem we obtain

$$-\pi h = (u, 1) = (u, \chi) = (\partial v_h, \chi) = -(v_h, \partial\chi) =$$

$$= -\sum_{n=1}^{\infty} a_{-n} \int \bar{z}^{-n} \partial\chi(z)\, d\mu(z) = -\pi a_{-1},$$

and the proof is complete.

Throughout this section we denote for the sake of simplicity $L = W_T^{-2}(H)$ for a fixed hyponormal operator T.

DEFINITION 4.2. The *distribution kernel* $K_T \in D'(\mathbf{C}^2, L(L))$ associated to the hyponormal operator T is

(6) $$\langle K_T(\phi\otimes\psi)u, v\rangle_L = \langle(u, \psi), (v, \bar{\phi})\rangle,$$

where $\phi, \psi \in D$ and $u,v \in L$.

The distribution K_T is completely determined by the relation (6) because of the following estimate derived from (5):

(7) $$|\langle K_T(\phi\otimes\psi)u, v\rangle_L| \leq ||u||_L \cdot ||v||_L \cdot ||\phi||_{H^2} \cdot ||\psi||_{H^2},$$

with the same notations as above.

PROPOSITION 4.3. *The kernel* K_T *has the following properties:*

a) $K_T(\phi\otimes\psi)^* = K_T(\bar{\psi}\otimes\bar{\phi})$, $\phi, \psi \in D(\mathbf{C})$;

b) *For every finite sequence* $(\phi_i)_{i=1}^n$, $\phi_i \in D(\mathbf{C})$, $n \geq 1$, *the operator matrix* $(K_T(\phi_i \otimes \bar{\phi}_j))_{i,j}$ *is positive;*

c) $\mathbf{Supp}(K_T) \subset \sigma(T) \times \sigma(T)$;

d) $K_T \in H^{-2}(\mathbf{C}^2, L(H))$.

PROOF. a) Let $u, v \in L$. Directly by the definition (6),

$$\langle K_T(\phi \otimes \psi)^* u , v\rangle = \langle u , K_T(\phi \otimes \psi) v\rangle = \langle (u , \bar{\phi}) , (v , \psi)\rangle = \langle K_T(\bar{\phi} \otimes \bar{\psi}) u , v\rangle.$$

b) If $(\phi_i)_{i \in I}$, $\phi_i \in D(\mathbf{C})$ and $(u_i)_{i \in I}$, $u_i \in L$, are finite systems, then

$$\sum_{i,j} \langle K_T(\phi_i \otimes \bar{\phi}_j) u_j , u_i\rangle = \sum_{i,j} \langle (u_j , \bar{\phi}_j) , (u_i , \bar{\phi}_i)\rangle = \|\sum_i (u_i , \bar{\phi}_i)\|^2.$$

c) It suffices to recall that $\mathbf{supp}(u) \subset \sigma(T)$ whenever $u \in L$.

d) The distribution K_T belongs to $H^{-2}_{loc}(\mathbf{C}^2, L(L))$ by (7), and it has compact support, hence $K_T \in H^{-2}(\mathbf{C}^2, L(L))$.

The continuity and the positivity properties of the kernel K_T insure the existence of a scalar product on the space $D(\mathbf{C}, L)$, which extends continuously the following form

$$\langle K_T f , g\rangle = \int \langle K_T(z,w) f(w) , g(z)\rangle = \sum_{i,j} \langle K_T(\bar{\psi}_j \otimes \phi_i) u_i , v_j\rangle_L,$$

where $f = \sum_{i=1}^{n} \phi_i \otimes u_i \in D(\mathbf{C}, L)$, $g = \sum_{j=1}^{m} \psi_j \otimes v_j \in D(\mathbf{C}, L)$ and the integers n, m are finite.

Let us define the continuous evaluation map

$$\Psi : D(\mathbf{C}, L) \to H,$$

which acts on simple functions of the form $\phi \otimes u$ by the formula

$$\Psi(\phi \otimes u) = (u , \phi), \quad \phi \in D(\mathbf{C}),\ u \in L.$$

Then, for the above finite dimensional valued functions f and g, the relation

$$\langle K_T f , g\rangle = \langle \Psi(f) , \Psi(g)\rangle_H \tag{8}$$

holds true. The right side of (8) makes a good sense for arbitrary functions $f, g \in D(\mathbf{C}, L)$, and it can be taken as a second definition of the kernel K_T.

The separate completion of the space $D(\mathbf{C}, L)$ with respect to the seminorm derived from the scalar product (8) is isometrically isomorphic with the space H, through the linear extension of the operator Ψ. In this description, as a vector-valued function space, of the Hilbert space H, the operator T^* becomes the multiplication with $\bar{z}$. Moreover, the reproducing kernel K_T of this function space is a complete unitary invariant of the operator T, in the following sense.

LEMMA 4.4. *Two hyponormal operators* $T \in L(H)$ *and* $T' \in L(H')$ *are unitarily equivalent iff there exists a unitary operator* $U : W_T^{-2}(H) \to W_{T'}^{-2}(H')$ *such that* $UK_TU^* = K_{T'}$.

PROOF. Let K denote the separate completion of the space $D(\mathbf{C}, W_T^{-2}(H))$ in the norm $\langle K_T f , f\rangle$, so that the operator $\Psi : K \to H$ is unitary and has the property $\Psi^* T^* \Psi = \bar{z}$. Analogously, K' and Ψ' denote the corresponding objects associated to T'.

If U is a unitary operator as in the statement, then it induces a unitary transform $I \otimes U : K \to K'$ with the property $(I\otimes U)\bar{z} = \bar{z}(I\otimes U)$. Consequently $(I\otimes U)\Psi^* T^* \Psi(I\otimes U) = \Psi'^* T' \Psi'$, and the proof is complete.

There are examples which show that only a "thin" subspace of $W_T^{-2}(H)$ is necessary, in order to classify the hyponormal operator T up to unitary equivalence, as in Lemma 4.4. Thus we adopt the next.

DEFINITION 4.5. A closed subspace G of L (= $W_T^{-2}(H)$) is called *generating* for the operator T if $\Psi D(\mathbf{C},G)$ is a dense subspace of H.

In other words $G \subset L$ is generating if the linear space

$$\{(u , \phi) ; u \in G, \phi \in D(\mathbf{C})\}$$

is dense in H.

Let P_G denote the orthogonal projection of L onto a generating subspace G. Then the compression of the kernel K_T to G,

$$K_T^G := P_G K_T P_G \in D'(\mathbf{C}^2, L(G))$$

has all the properties listed in Proposition 4.3. Therefore the restriction Ψ_G of the operator Ψ to $D(\mathbf{C}, G)$ is related to the kernel K_T^G by the relation:

$$\langle K_T^G f , g\rangle = \langle \Psi_G(f) , \Psi_G(g)\rangle, \quad f,g \in D(\mathbf{C},G). \tag{9}$$

Now we may conclude with the following.

THEOREM 4.6. *Let* T *be a hyponormal operator on the Hilbert space* H *and let* G *be a generating space of eigendistributions of* T^*. *Let* $\mathbf{H}$ *denote the separate completion of the space* $D(\mathbf{C},G)$ *with respect to the seminorm* $\langle K_T f, f\rangle$ *and let* $U : \mathbf{H} \to H$ *be the continuous linear extension of the operator* Ψ_G. *Then* U *is unitary and* $U(\bar{z}f) = T^* U(f)$ *for every* $f \in \mathbf{H}$.

PROOF. The operator U is an isometry, which is onto because G was supposed a generating subspace of L. For a simple function of the form $\phi \otimes u \in D(\mathbf{C},G)$, we have

$$T^* \Psi_G(\phi \otimes u) = T^*(u , \phi) = (\bar{z}u , \phi) = \Psi_G(\bar{z}\phi \otimes u),$$

because $u \in W_T^{-2}(H)$. Then the proof is over by an approximation argument.

Lemma 4.4 and Theorem 4.6 lead to the next.

COROLLARY 4.7. Let $T \in L(H)$ *and* $T' \in L(H')$ *be hyponormal operators and let* G, G' *be generating subspaces of eigendistributions of* T^*, *respectively of* T'^*.

If there exists a unitary operator $U : G \to G'$, *so that* $UK_T^G U^* = K_{T'}^{G'}$, *then* T *is unitarily equivalent with* T'.

The necessary condition for unitary equivalence stated in Corollary 4.7 is also sufficient whenever the space G is canonically related to the operator T. In many cases **dim**(G) is finite, so that the classification of the related operators is a finite dimensional problem. We illustrate this statement with two examples.

EXAMPLE 1. Normal irreducible operators.

Let N be a normal operator on the Hilbert space H, with a cyclic vector ξ of norm one. We denote as usually by $\phi(N)$ the continuous functional calculus of N, $\phi \in C(\sigma(N))$.

Let Ω be a bounded domain which contains $\sigma(N)$. In virtue of the continuity of the functional calculus and of the Sobolev embedding theorem, the following estimate

$$\|\phi(T)\xi\| \leq \|\phi\|_{\infty,\Omega} \leq C\,\|\phi\|_{H^2(\Omega)}$$

holds true, with a positive constant C depending only on Ω. Thus the relation

$$(u , \phi) = \phi(T)\xi , \quad \phi \in D(\mathbf{C}),$$

defines a H-valued distribution $u \in H_0^{-2}(\Omega,H)$. Moreover, $u \in W_N^{-2}(H)$. Indeed, for every $\phi \in D(\mathbf{C})$,

$$((N^* - \bar{z})u , \phi) = N^*\phi(N)\xi - (\bar{z}\phi)(N)\xi = 0.$$

Since the linear space $\{\phi(N)\xi , \phi \in D(\mathbf{C})\}$ is dense in H, the one dimensional subspace $\mathbf{C} \cdot u$ of $W_N^{-2}(H)$ is generating for N. The compression k of the kernel K_N to this space is a scalar distribution which can be easily computed, as follows: let $\phi , \psi \in D(\mathbf{C})$,

$$k(\phi\otimes\psi) = \langle K_N(\phi\otimes\psi)u , u\rangle = \langle(u , \psi) , (u , \bar{\phi})\rangle = \langle\psi(N)\xi , \bar{\phi}(N)\xi\rangle =$$

$$= \langle\phi(N)\psi(N)\xi , \xi\rangle = \int \phi\psi \, d\mu_\xi,$$

where $d\mu_\xi = \langle dE\xi , \xi\rangle$ and E is the spectral measure of N. The norm of the space $\mathbf{C} \cdot u$ was chosen so that $\|u\| = 1$. Concluding, we have proved the formula

$$k(z,w) = \mu_\xi(z)\delta(z - w),$$

where $\delta(z - w)$ stands for the Dirac measure supported by the diagonal of the space $\mathbf{C}^2$.

The completion of the space $D(\mathbf{C})$ with respect to the seminorm given by the kernel k coincides with the space $L^2(\mu_\xi)$ and the unitary $U : L^2(\mu_\xi) \to H$ is the operator which diagonalizes N.

EXAMPLE 2. The class $A(\Omega)$ of Curto and Salinas [7].

The class $A(\Omega)$ of operators provides the localization of the class $B(\Omega)$ of Cowen and Douglas [6]. We restrict for the beginning to a particular element of $A(\Omega)$.

Let Ω be a bounded domain in the complex plane and let S be a hyponormal operator which has the properties:

(i) **Ran**$(z - S)$ is closed for all $z \in \Omega$;

(ii) **span**$\{$**Ker**$(\bar{z} - S^*) \mid z \in \Omega\}$ is dense in H; and

(iii) There exists a *bounded* coanalytic function $\Gamma : \Omega \to L(\mathbf{C}^n,H)$, such that Ran $\Gamma(z)$ = **Ker**$(\bar{z} - S^*)$ for every $z \in \Omega$.

For a relation between hyponormal operators and the class $B(\Omega)$ see [1].

Under the assumption (i) - (iii) the space $W_S^{-2}(H)$ contains the following privileged eigendistributions. Let e_i, $i = 1,\dots,n$, be the canonical basis in $\mathbf{C}^n$. Since the operatorial function Γ is uniformly bounded, the formula

$$(u_i , \phi) = \int_\Omega \phi(z)\Gamma(z)e_i \, d\mu(z) , \quad \phi \in D(\mathbf{C})$$

defines for every $i = 1,\dots,n$, a distribution $u_i \in H^{-2}(\mathbf{C},H)$. It is plain to check that $(\bar{z} - S^*)u_i = 0$, in the sense of distributions.

The linear span G of u_i, $i = 1,\dots,n$, is a generating subspace of $W_S^{-2}(H)$, because of the assumption (ii). Let us compute the corresponding distribution kernel.

We renorm the space G so that $(u_1,\dots,u_n)$ becomes an orthogonal basis. Let f, $g \in D(\mathbf{C},G)$, so that

$$f = \sum_{i=1}^{n} \phi_i \otimes u_i \quad \text{and } g = \sum_{i=1}^{n} \psi_j \otimes u_j,$$

where $\phi_k, \psi_k \in D(\mathbf{C})$, $k = 1,\dots,n$. Then,

$$\langle K_S^G f , g\rangle = \sum_{i,j} \langle K_S^G(\phi_i \otimes u_i) , \psi_j \otimes u_j\rangle = \sum_{i,j} \langle K_S(\bar{\psi}_j \otimes \phi_i)u_i , u_j\rangle =$$

$$= \sum \langle (u_i , \phi_i) , (u_j , \psi_j)\rangle = \sum \int_{\Omega\times\Omega} \langle \phi_i(z)\Gamma(z)e_i , \psi_j(w)\Gamma(w)e_j\rangle \, d\mu \times d\mu =$$

$$= \int_{\Omega\times\Omega} \langle \Gamma(w)^* \Gamma(z)f(z) , g(w)\rangle \, d\mu(z)\, d\mu(w).$$

In conclusion the distribution kernel K_S^G coincides on $\Omega\times\Omega$ with the generalized Bergman kernel of the function Γ, $K_\Gamma(z,w) = \Gamma(w)^* \Gamma(z)$.

In the general case, when the function Γ is not necessarily bounded, the Lebesgue measure must be multiplied by a weight, in order to annihilate the growth of $\|\Gamma(z)\|$ when $z \to \partial\Omega$.

The last subject of this section is a naturality formula of the kernel K_T, to analytic changes of coordinates.

Let $f : U \to V$ be a biholomorphic map between two domains of the complex plane. We recall for the beginning the operation of change of coordinates introduced by f at the level of distributions.

Let $\alpha \in D'(V)$ and let $\phi \in D(U)$. The distribution $\alpha\circ f \in D'(U)$ is defined by the formula

$$(\alpha\circ f , \phi) = (\alpha , \psi),$$

where

$$\psi(\zeta) = \phi(f^{-1}(\zeta)) \, |\partial f^{-1}(\zeta)|^2, \quad \zeta \in V.$$

The same definition applies to vector-valued distributions.

LEMMA 4.8. *Let* $T \in L(H)$ *be a hyponormal operator with* $\sigma(T) \subset U$. *If* $u \in W_T^{-2}(H)$, *then* $u\circ f^{-1} \in W_{f(T)}^{-2}(H)$.

PROOF. The spectrum of the operator $f(T)$ is, by the spectral mapping theorem, contained in V. Let us assume $u \in W_T^{-2}(H)$. For every function $\phi \in D(V,H)$ there exists by standard arguments a function $\psi \in D(U,H)$, so that

$$(z - T)\psi(z) = (f(z) - f(T))\phi(f(z))|\bar\partial f(z)|^2, \quad z \in U.$$

Then

$$\langle(\bar\zeta - f(T)^*)(u \circ f^{-1}), \phi\rangle = \langle u \circ f^{-1}, (\zeta - f(T))\phi\rangle =$$

$$= \langle u, (f - f(T))(\phi \circ f)|\partial f|^2\rangle = \langle u, (z - T)\psi\rangle = 0,$$

and therefore $u \circ f^{-1} \in W_{f(T)}^{-2}(H)$, **q.e.d.**

Since the support of a distribution $u \in W_T^{-2}(H)$ is contained in the compact subset $\sigma(T)$ of U, the linear operator

$$C_f : W_{f(T)}^{-2}(H) \to W_T^{-2}(H), \quad C_f(v) = v \circ f,$$

is bounded. Notice that $C_{f^{-1}}$ is a two-sided inverse of C_f.

THEOREM 4.9. *Let* $T \in L(H)$ *be a hyponormal operator and let* $f : U \to V$ *be a biholomorphic map, defined on a domain which contains the spectrum of* T. *Then*

$$K_{f(T)} \circ (f \times f) = C_f^* K_T C_f.$$

PROOF. Let $\phi, \psi \in D(U,H)$ and let $u, v \in W_{f(T)}^{-2}(H)$. By Definition 4.2 we have

$$\langle K_{f(T)} \circ (f \times f)(\phi \otimes \psi)u, v\rangle = \langle K_{f(T)}((\phi \circ f^{-1})|\partial f^{-1}|^2 \otimes (\psi \circ f^{-1})|\partial f^{-1}|^2)u, v\rangle =$$

$$= \langle(u, (\psi \circ f^{-1})|\partial f^{-1}|^2), (v, (\bar\phi \circ f^{-1})|\partial f^{-1}|^2)\rangle = \langle(u \circ f, \psi), (v \circ f, \bar\phi)\rangle =$$

$$= \langle(C_f u, \psi), (C_f v, \bar\phi)\rangle = \langle K_T(\phi \otimes \psi)C_f u, C_f v\rangle = \langle C_f^* K_T(\phi \otimes \psi)C_f u, v\rangle,$$

and the proof is complete.

The compression of the kernel K_T to a generating subspace G of $W_T^{-2}(H)$ has a similar formula, if one takes the compression of the kernel $K_{f(T)}$ to the generating subspace $C_{f^{-1}}(G) \subset W_{f(T)}^{-2}(H)$.

§ 5. OPERATORS WITH ONE-DIMENSIONAL SELF-COMMUTATOR

The compression of the distribution kernel of a hyponormal operator with one-dimensional self-commutator to a space generated by a single eigendistribution of its adjoint coincides with a scalar kernel already existing in the literature, cf.[3], [11], [18]. We use this kernel in order to obtain a concrete functional description of the operator in terms of its principal function.

Let us recall for the beginning some facts and notations concerning the operators with rank one self-commutator. The reader is refered to [4], [8] and [10] for details.

Let T be a bounded linear operator on the Hilbert space H, so that

$$[T^* , T] = \xi \otimes \xi,$$

where $\xi \otimes \xi$ is a rank one, positive operator: $(\xi \otimes \xi)h = \langle h , \xi \rangle \xi$, $h \in H$.

The complete invariant to unitary equivalence of an irreducible operator with one-dimensional self-commutator was discovered by Pincus [10] as a scalar function on the spectrum, called the principal function. The *principal function* g_T of the operator T is a compactly supported, real valued measurable function on $\mathbf{C}$, characterized by Helton and Howe formula [8]:

$$\mathrm{tr}[p(T,T^*) , q(T,T^*)] = (1/\pi) \int_{\mathbf{C}} (\bar{\partial} p \, \partial q - \partial p \, \bar{\partial} q) g_T \, d\mu,$$

where p, q are polynomials in two variables.

The order of the factors T and T^* in the monomials of p and q doesn't affect the trace of the commutator $[p(T,T^*) , q(T,T^*)]$, which exists whenever $[T^* , T]$ is trace-class. In our case, when T is hyponormal and $\mathbf{rk}[T^* , T] = 1$, the principal function satisfies the additional condition $0 \leq g_T \leq 1$.

Putnam proved in [13] that for an arbitrary complex number $z \in \mathbf{C}$, the equation

$$(\bar{z} - T^*)x = \xi$$

has a unique solution $x \in \mathbf{Ker}(\bar{z} - T^*)^{\perp}$. Moreover, this solution depends weakly continuous on $z \in \mathbf{C}$, and it will be denoted in the sequel by $(\bar{z} - T^*)^{-1}\xi$.

Recently, Clancey [3] proved the identity

$$(10) \qquad 1 - \langle (\bar{w} - T^*)^{-1}\xi , (\bar{z} - T^*)^{-1}\xi \rangle = \exp\{ - (1/\pi) \int [g_T(\zeta)/(\zeta - z)(\bar{\zeta} - \bar{w})] d\mu(\zeta) \},$$

which holds on the whole $\mathbf{C}^2$, see also [11]. The integral of the right side of (10) has removable singularities off the diagonal of $\mathbf{C}^2$. In case $z = w$ and $\int |\zeta - z|^2 g_T(\zeta) d\mu(\zeta) = \infty$, the right side of (10) is taken to be zero. The kernel

$$T_*(z,w) = 1 - \langle \bar{w} - T^*)^{-1}\xi , (\bar{z} - T^*)^{-1}\xi \rangle$$

was used in [4] in connection with a distributional model for T. We use the next.

THEOREM 5.1. (Clancey [4]) *An operator* T *with* $[T^*, T] = \xi\otimes\xi$ *is irreducible iff the values of the function* $(\bar z - T^*)^{-1}\xi$ *span* H.

The theorem has, with the terminology of the preceding section, the following.

COROLLARY 5.2. *Let* T *be an irreducible hyponormal operator with* $[T^*, T] = \xi\otimes\xi$. *The distribution* $\partial(\bar z - T^*)^{-1}\xi$ *belongs to* $W_T^{-2}(H)$ *and it spans a one-dimensional generating space for* T.

PROOF. The function $(\bar z - T^*)^{-1}\xi$ belongs to $L^2_{loc}(\mathbf{C},H)$ and

$$(\bar z - T^*)\partial(\bar z - T^*)^{-1}\xi = \partial[(\bar z - T^*)(\bar z - T^*)^{-1}\xi] = \partial\xi = 0,$$

therefore $\partial(\bar z - T^*)^{-1}\xi \in W_T^{-2}(H)$.

Furtheron we prove that the space

$$M = \{(\partial(\bar z - T^*)^{-1}\xi, \phi) \mid \phi \in D(\mathbf{C})\}$$

is dense in H.

The Cauchy (anti)transform of a test function $\phi \in D(\mathbf{C})$ is

$$f(z) = (1/\pi)\int \phi(\zeta)/(\bar\zeta - \bar z)\, d\mu(\zeta).$$

It is a antianalytic function off **supp**(ϕ), vanishing at ∞, and $\partial f = -\phi$. Let Ω be a bounded complex domain with smooth boundary, which contains **supp**(ϕ) and $\sigma(T)$. Then by Stokes Theorem

$$\int_{\mathbf{C}} \phi(z)(\bar z - T^*)^{-1}\xi\, d\mu(z) = -\int_{\Omega} \partial f(z)(\bar z - T^*)^{-1}\xi\, d\mu(z) =$$

$$= (1/2i)\int_{\partial\Omega} f(z)(\bar z - T^*)^{-1}\xi\, d\bar z + (\partial(\bar z - T^*)^{-1}\xi, f).$$

The contour integral is zero since $f(z)(\bar z - T^*)^{-1}$ is a antianalytic function on $\mathbf{C}\setminus\Omega$, vanishing of second order at ∞. Therefore $\int\phi(\bar z - T^*)^{-1}\xi\, d\mu \in M$.

By Theorem 5.1 the vectors $\int\phi(\bar z - T^*)^{-1}\xi\, d\mu$, $\phi \in D(\mathbf{C})$, span the Hilbert space H, hence the space M is also dense in H, and the proof is over.

The above corollary and Theorem 4.6 imply that the compression k of the distribution kernel K_T to the space generated in $W_T^{-2}(H)$ by the distribution $u = \partial(\bar z - T^*)^{-1}\xi$ is a scalar kernel which reproduces H and diagonalizes T^*. Let us compute this kernel. Let consider the test functions $\phi, \psi \in D(\mathbf{C})$. Then,

$$k(\phi\otimes\psi) = \langle K_T(\phi\otimes\psi)u\ ,\ u\rangle = \langle (u\ ,\ \psi)\ ,\ (u\ ,\ \overline{\phi})\rangle =$$

$$= \int \langle (\bar{w} - T^*)^{-1}\xi\, \partial\psi(w)\ ,\ (\bar{z} - T^*)^{-1}\xi\, \partial\overline{\phi}(z)\rangle\, d\mu(z)\, d\mu(w) =$$

$$= \int \langle (\bar{w} - T^*)^{-1}\xi\ ,\ (\bar{z} - T^*)^{-1}\xi\rangle\, \partial\psi(w)\, \overline{\partial}\phi(z)\, d\mu(z)\, d\mu(w) =$$

$$= -\int \overline{\partial}_z \partial_w T_*(z,w)\psi(w)\phi(z)\, d\mu(z)\, d\mu(w).$$

In conclusion, we have proved the equality

$$k(z,w) = -\, \overline{\partial}_z \partial_w T_*(z,w).$$

THEOREM 5.3. *Let* $T \in L(H)$ *be an irreducible hyponormal operator with* $[T^*\ ,T] = \xi\otimes\xi$, *and let* g *be the principal function of* T.

The separate completion $\mathbf{H}$ *of the space* $D(\mathbf{C})$ *with respect to the seminorm*

$$\|\phi\|_g^2 = -\int \exp\{(-1/\pi)\int g(\zeta)/(\zeta - z)(\overline{\zeta} - \bar{w})\, d\mu(\zeta)\}\, \partial\phi(w)\, \overline{\partial}\,\overline{\phi}(z)\, d\mu(z)\, d\mu(w)$$

is unitarily equivalent with the space H, *via the operator* $U : \mathbf{H} \to H$,

$$U(\phi) = \int \partial\phi(z)(\bar{z} - T^*)^{-1}\xi\, d\mu(z).$$

The operators T *and* T^* *become in that identification:*

(11) $$U^*TU(\phi)(z) = z\phi(z) - (1/\pi)\int \phi(\zeta)g(\zeta)/(\overline{\zeta} - \bar{z})\, d\mu(\zeta),$$

and

$$U^*T^*U(\phi)(z) = \bar{z}\phi(z).$$

The right side of (11) should be taken as the class of the respective function in **H**.

PROOF. Theorem 4.5 and the preceding computation of the kernel k prove the assertions of the statement, with the exception of the relation (11).

In order to prove (11), we recall from Propsition 7 of [4] the identity

(12) $$T\partial(\bar{z} - T^*)^{-1}\xi = z\partial(\bar{z} - T^*)^{-1}\xi - g(z)(\bar{z} - T^*)^{-1}\xi,$$

which holds at the level of distributions.

Let $\chi \in D(\mathbf{C})$, so that $\chi \equiv 1$ on a neighbourhood of $\sigma(T)$. Because the kernel k is supported by $\sigma(T)\times\sigma(T)$, it is sufficient to prove that

$$U(z\phi - (\chi(z)/\pi)\int \phi(\zeta)g(\zeta)/(\overline{\zeta} - \bar{z})\, d\mu(\zeta)) = TU(\phi),$$

for every function $\phi \in \mathcal{D}(\mathbf{C})$. But the explicit form of the operator U gives

$$U(z\phi - (\chi(z)/\pi) \int \phi(\zeta)g(\zeta)/(\bar{\zeta} - \bar{z})\, d\mu(\zeta)) =$$

$$= \int \partial[z\phi - (\chi/\pi) \int \phi(\zeta)g(\zeta)/(\bar{\zeta} - \bar{z})\, d\mu(\zeta)](\bar{z} - T^*)^{-1}\xi\, d\mu(z) =$$

$$= - (\partial(\bar{z} - T^*)^{-1}\xi , z\phi) + ((\bar{z} - T^*)^{-1}\xi , g\phi) +$$

$$+ \int \partial[(1 - \chi)/\pi \int \phi(\zeta)g(\zeta)/(\bar{\zeta} - \bar{z})\, d\mu(\zeta)](\bar{z} - T^*)^{-1}\xi\, d\mu(z) =$$

$$= -(T\partial(\bar{z} - T^*)^{-1}\xi , \phi) + I = TU(\phi) + I,$$

where we have used (12) and we have denoted the last singular integral by I.

Stokes Theorem and the observation that the antianalytic functions $\int \phi(\zeta)g(\zeta)/(\bar{\zeta} - \bar{z})\, d\mu(\zeta)$ and $(\bar{z} - T^*)^{-1}\xi$ vanish at infinity, imply $I = 0$, and the proof is complete.

FINAL REMARKS.

a) The formulae of T and T^* obtained in Theorem 5.3 are dual to those given by Pincus-Xia-Xia [11] on their analytic model.

b) Because the vector ξ has a privileged place in the Hilbert space H, relative to the operator T, Theorem 4.6 shows that the scalar kernel k is a complete unitary invariant of the operator T.

c) Conversely, a compactly supported, measurable function g on $\mathbf{C}$, with $0 \leq g \leq 1$, produces by the formulae of Theorem 5.3 an irreducible hyponormal operator with rank one self-commutator and with principal function $g_T = g$.

REFERENCES

1. **Clancey, K. :** Completeness of eigenfunctions of seminormal operators, *Acta Sci. Math. (Szeged)* **39**(1977), 31-37.

2. **Clancey, K. :** *Seminormal operators*, Lecture Notes in Math. **742**, Springer, Berlin–Heidelberg–New York, 1979.

3. **Clancey, K. :** A kernel for operators with one-dimensional self-commutator, *Integral Equations Operator Theory* **7**(1984), 441-458.

4. **Clancey, K. :** Hilbert space operators with one-dimensional self-commutator *J. Operator Theory* **13**(1985), 265-289.

5. **Colojoară, I. ; Foiaş, C. :** *Theory of generalized spectral operators*, Gordon and Breach, New York, 1968.

6. **Cowen, M.J. ; Douglas, R.G. :** Complex geometry and operator theory, *Acta. Math.* **141**(1978), 187-261.

7. **Curto, R.E. ; Salinas, N.** : Generalized Bergman kernels and the Cowen-Douglas theory, *Amer. J. Math.* **106**(1984), 447-488.

8. **Helton, J.W. ; Howe, R.** : Integral operators, commutator traces, index and homology, in *Lecture Notes in Math.* **345**, Springer, Berlin–Heidelberg–New York, 1973.

9. **Hörmander, L.** : *Linear partial differential operators*, Springer, Berlin–Heidelberg–New York, 1963.

10. **Pincus, J.D.** : Commutators and systems of singular integral equations. I, *Acta Math.* **121**(1968), 219-249.

11. **Pincus, J.D. ; Xia, D. ; Xia, J.** : The analytic model of a hyponormal operator with rank one self-commutator, *Integral Equations Operator Theory* **7**(1984), 516-535.

12. **Putinar, M.** : Hyponormal operators are subscalar, *J. Operator Theory* **12**(1984), 385-395.

13. **Putnam, C.R.** : Resolvent vectors, invariant subspaces and sets of zero capacity, *Math. Ann.* **205**(1973), 165-171.

14. **Simon, B.** : *The* $P(\phi)_2$ *Euclidean (quantum) field theory*, Princeton University Press, New Jersey, 1974.

15. **Stampfli, J.G.** : A local spectral theory for operators. V: Spectral subspaces for hyponormal operators, *Trans. Amer. Math. Soc.* **217**(1976), 285-296.

16. **Xia, D.** : On non-normal operators, *Chinese Math.* **3**(1963), 232-246.

17. **Xia, D.** : On the analytic model of a class of hyponormal operators, *Integral Equations Operator Theory* **6**(1983), 134-157.

18. **Xia, D.** : On the kernels associated with a class of hyponormal operators, *Integral Equations Operator Theory* **6**(1983), 444-452.

Mihai Putinar

Department of Mathematics, INCREST
Bdul. Păcii 220, 79622 Bucharest,
Romania.

Operator Theory:
Advances and Applications, Vol.17

THE INVARIANT SUBSPACE PROBLEM: A DESCRIPTION WITH FURTHER APPLICATIONS OF A COMBINATORIAL PROOF

C.J.Read

0. INTRODUCTION

The paper falls conveniently into three sections: first, the motivation for the definitions made in the original proof [1] (together with a clear outline of how the proof proceeds); second, a description of how we obtain an operator on the space ℓ_1; and third, an account of the spectrum of our operator on ℓ_1, of how it is related to a weighted shift operator, and of how it can be chosen in such a way that, though T has no invariant subspaces, T^2 does.

So we begin with a discussion of the original proof in [1].

1. THE MOTIVATION FOR THE DEFINITIONS MADE IN THE ORIGINAL PROOF

1.1. Naturally, we are looking for an example of an operator on an infinite dimensional Banach space X without closed invariant subspaces other than the trivial ones $\{0\}$ and X. In our first attempt at the problem, we did not at all care what Banach space X happened to be.

Throughout our search, it made no difference at all whether the underlying field was real or complex: we shall denote it throughout by the symbol Ω.

Furthermore, it is easily shown that a continuous operator $T : X \to X$ is without invariant subspaces if and only if every nonzero vector $x \in X$ is cyclic (that is, the closed linear span of the vectors x, Tx, T^2x, . . . is equal to X). It is therefore natural to define ourselves a cyclic vector $e_0 \in X$ by writing $e_i = T^i e_0$ $(i = 1, 2, 3, \ldots)$, and determining that X shall be the closed linear span of the vectors e_i, under a norm that we shall specify. The operator T then becomes a "right shift" operator $T : e_i \to e_{i+1}$. Hence in Section 1 of [1] we write F for the collection of finite linear combinations of some symbols e_i $(i = 0, 1, 2, \ldots)$ ([1], Section 1.6); T then is defined as the right shift operator on F ([1], Section 1.7); and we commit ourselves to find a norm $\|\cdot\|$ on F such that we may take X to be the completion of $(F, \|\cdot\|)$. Our invariant subspace free operator will then be the continuous extension $T^\sim$ of T to the completion $F^\sim$ ([1], Note 1.8).

The choices thus left to us are the choice of the norm only. We have to choose our norm, and then show that T does indeed extend continuously to the completion $F^{\sim}$, without acquiring any closed invariant subspaces other than $\{0\}$ and $F^{\sim}$.

1.2. The history of the norm

Throughout the early days of researching into this problem, the author felt that it was advisable to start from the "smallest" norm that one could reasonably impose on the space F, and add on conditions in the form of further operators which must be continuous, in order to obtain the correct norm $\|\cdot\|$ on F which will achieve the result.

As a brief sketch of how this programme developed, let us give the following account.

1.2.1. Let us take the bare conditions that

(1) T must be bounded, say $\|T\| \leq 2$.

(2) A certain chosen functional x^* on F must be continuous, let us say

$$\| x^* : (F, \|\cdot\|) \rightarrow \Omega \| \leq 1.$$

It is not difficult to see that the smallest norm on F satisfying these conditions is the norm

$$\|x\| = \sup_{n \geq 0} 2^{-n} \cdot |x^*(T^n x)|$$

(provided that this expression is finite for each $x \in X$).

In fact, by choosing x^* correctly, one can take exactly this norm, and one finds that every nonzero vector in F is cyclic, so F has no nontrivial closed T invariant subspace. But this property does not extend to the completion X. However, in the process of investigating this simple norm, the author found an effective method of defining suitable linear functionals on F, which will be described more fully below.

1.2.2. In order to take our very simple norm in (1.2.1), and modify it so that we know something about the behaviour of vectors in the completion X, it is good that we define a uniformly bounded sequence of finite rank projections $(Q_n)_{n=1}^{\infty}$ on F. Provided that the images $Q_n(F)$ of our projections Q_n satisfy

(1) $\mathrm{Im}\, Q_n \subset \mathrm{Im}\, Q_{n+1}$ for each n, and

(2) $\bigcup_{n=1}^{\infty} \mathrm{Im}\, Q_n = F$,

it follows immediately that $Q_n(x) \rightarrow x$ for every $x \in X$; so we can approximate any vector x in X by one of its images $Q_n(x)$ in F.

Within the vector space F of countable dimension, it is not difficult to find a polynomial p such that

$$\| p(T)y - e_0 \| < \varepsilon$$

where $y = Q_n(x)$, and ε may be chosen arbitrarily small. Since the vector e_0 is certainly cyclic, it is sufficient to find a reasonable bound on the norm of the vector $p(T)(I - Q_n)(x)$. We hoped to achieve this by obtaining a reasonable bound on the norm of the operator $p(T)(I - Q_n)$ (see [1], Section 5.6.2). The required inequality for this to be possible seemed to be the following:

(3) For each $n > 0$,

$$\| T^{a_n+b_n} \circ (I - Q_n) \| \leq 2$$

where a_n and b_n are some constants which we use in the definition of the functional x^*, and also in the definition of the projections Q_n. However, there is a problem with this condition, which is that it is incompatible with the other conditions on our norm, namely that x^* should be bounded, T should be bounded and the projections Q_n should be uniformly bounded. A modification was thus required.

1.2.3. An appropriate modification turns out to be that one replaces the condition

$$\| T^{a_n+b_n} \circ (I - Q_n) \| \leq 2$$

with the condition

$$\| T^{a_n+b_n} \circ (I - Q_n^0) \| \leq 2,$$

where the new collection $(Q_n^0)_{n=1}^\infty$ is another collection of finite rank projections, quite closely related to $(Q_n)_{n=1}^\infty$, but not a uniformyly bounded collection of projections. So our definition of the norm become something like the following:

Let $\| \cdot \|$ be the smallest norm on F such that

(1) $\| x^* : (F, \| \cdot \|) \to \Omega \| \leq 1$,

(2) $\| T \| \leq 2$,

(3) $\| Q_n \| \leq 2$ for all $n \geq 1$,

(4) $\| T^{a_n+b_n} \circ (I - Q_n^0) \| \leq 2$ for all $n \geq 1$.

1.2.4. The last condition which is required is one which enables us to get a handle on the relationship between Q_n and Q_n^0. It turns out that $Q_n^0 = Q_n \circ (I - \bar{R}_n)$ where

$\bar{R}_n$ is another projection, which in turn may conveniently be written as an infinite sum of "smaller" projections,

$$\bar{R}_n = \sum_{m=n+1}^{\infty} R_{n,m} .$$

Writing $\bar{R}_{n,m}$ for the partial sum $\sum_{j=n+1}^{m} R_{n,j}$ $(m > n \geq 1)$, our final condition on our norm becomes

$$\|\bar{R}_{n,m}\| \leq a_n , \quad m > n \geq 1.$$

Thus in Section 3.0 of [1], we make the correct definition: Let $\|\cdot\|$ be the smallest norm on F such that

(1) $\|x^* : (F, \|\cdot\|) \to \Omega\| \leq 1$,

(2) $\|T\| \leq 2$,

(3) $\|Q_n\| \leq 2, \quad n > 1$,

(4) $\|T^{a_n+b_n} \circ (I - Q_n^0)\| \leq 2, \quad n \geq 1$,

(5) $\|\bar{R}_{n,m}\| \leq a_n, \quad m > n \geq 1$.

Before going on to describe the consequences of t ese definitions, let us first give precise definitions of x^*, Q_n, Q_n^0, and the other projections $R_{n,m}$.

1.3. Rapidly increasing sequences and their properties

A key idea in the whole construction is that our functional x^*, and, later on, our various operators Q_n, $R_{n,m}$, and so on, are defined in terms of a rapidly increasing sequence $\mathbf{d} = (d_i)_{i=1}^{\infty}$. Out of this sequence we take odd index and even index subsequences $(a_i)_{i=1}^{\infty}$, $a_i = d_{2i-1}$, and $(b_i)_{i=1}^{\infty}$, $b_i = d_{2i}$.

In order that, throughout the proof, we may be able to write statements like "for all i, $a_i > 2^{b_{i-1}}$ (provided **d** increases sufficiently rapidly)", we define first what we mean by "increasing sufficiently rapidly".

We say ([1], 1.1 -3) that a proposition P(**d**) is true "for all **d** increasing sufficiently rapidly", if there is a constant $c \in \mathbf{N}$, and functions $(f_i)_{i=1}^{\infty}$, $f_i : \mathbf{N}^i \to \mathbf{N}$, such that whenever the strictly increasing sequence **d** satisfies $d_1 > c$ and, for all $r > 1$,

$$d_r > f_{r-1}(d_1, d_2, \ldots, d_{r-1}),$$

the proposition P(**d**) is true.

It is then obvious that if P(**d**) is true "for all **d** increasing sufficiently rapidly", then there is a sequence **d** such that P(**d**) is true; moreover if $P_1(\mathbf{d}), P_2(\mathbf{d}), \ldots, P_k(\mathbf{d})$ are all propositions which are true for all **d** increasing sufficiently rapidly, then the proposition $\bigwedge_{i=1}^{k} P_i(\mathbf{d})$ is true for all **d** increasing sufficiently rapidly.

1.4. The definitions of x^*, Λ and $M_{\boldsymbol{\delta}}$

The careful choice of the linear functional x^* is of great importance; here is an account of how the choice came to be made.

Let us recall that our ultimate goal is to ensure that for all $x \in X$ and all $\varepsilon > 0$, there is a polynomial p such that

$$\| p(T)x - e_0 \| < \varepsilon.$$

Now if $x \in F$, say $x \in \mathrm{lin}(e_k, e_{k+1}, \ldots, e_m)$, then certainly $p(T)x \in \mathrm{lin}\{e_j : j \geq k\}$. It is therefore clear that there are linear combinations $\sum_{i=j}^{N} \lambda_i e_i$, with large indices j, which must be good approximations to the first vector e_0. For the sake of simplicity we have in fact arranged that the sequence $e_{a_1}, e_{a_2}, e_{a_3}, \ldots, e_{a_n}, \ldots$ tends to e_0. This dictates that the operator T^{a_n} acts quite nearly like the identity on a finite dimensional subspace of F. Similarly it is quite useful for our construction that the operator $T^{a_n+b_n}$ should act like b_n times the identity map, on a finite dimensional subspace. Ideas like this are built into the definition of x^* in the following way.

If **d** increases sufficiently rapidly, then the finitely nonzero linear combinations $\sum_{i=1}^{\infty} \delta_i d_i$ ($\delta_i \in \mathbb{Z}$, $\delta_i \geq 0$), such that (let as say) each $\delta_i \leq i$, will all be distinct. If we rewrite $\sum_{i=1}^{\infty} \delta_i d_i$ as $\sum_{i=1}^{\infty} (\alpha_i a_i + \beta_i b_i)$, then we can write

$$\boldsymbol{\delta} = (\delta_i)_{i=1}^{\infty}, \quad \boldsymbol{\delta} = \{\boldsymbol{\alpha}, \boldsymbol{\beta}\},$$

where

$$\boldsymbol{\alpha} = (\alpha_i)_{i=1}^{\infty} = (\delta_{2i-1})_{i=1}^{\infty}$$

$$\boldsymbol{\beta} = (\beta_i)_{i=1}^{\infty} = (\delta_{2i})_{i=1}^{\infty}.$$

We can then write $\sigma_{\boldsymbol{\delta}}$ for the sum $\sum_{i=1}^{\infty} \delta_i d_i$ and $M_{\boldsymbol{\delta}}$ for the product $\prod_{i=1}^{\infty} b_i^{\beta_i}$.

Our basic idea is that the value $x^*(e_i)$ should mostly be zero, except on the (fairly sparse) set of values i which are among a collection $\{\sigma_{\boldsymbol{\delta}} : \boldsymbol{\delta} \in \Lambda\}$. If i is a member of this set then we define $x^*(e_i) = M_{\boldsymbol{\delta}}$, where $i = \sigma_{\boldsymbol{\delta}}$.

The reason for this definition is that if Λ is chosen suitably, the idea that the operator T^{a_n} should be like the identity map, and the operator $T^{a_n+b_n}$ should be quite like b_n times the identity map, is fulfilled, at least in the values taken by x^*. So in [1], Definition 1.25, we define the linear functional x^* on F;

$$x^* : F \to \Omega$$
$$: \sum_{i=0}^{\infty} \lambda_i e_i \to \sum_{\delta \in \Lambda} M_{\delta} \lambda_{\sigma_{\delta}} .$$

Equivalently, we have

$$x^*(e_i) = \begin{cases} M_{\delta}, & \text{if } i = \sigma_{\delta} \text{ for some } \delta \in \Lambda. \\ 0, & \text{otherwise.} \end{cases}$$

We must define the set Λ ([1], Definition 1.16).

DEFINITION. Let Λ be the collection of all finitely nonzero sequence $\boldsymbol{\delta} = [\boldsymbol{\alpha}, \boldsymbol{\beta}]$, such that the δ_i's are integers, and

(1) $0 \le \beta_i \le \alpha_i \le i$ for each i,

(2) If $\alpha_i \neq 0$, $\alpha_j \neq 0$, $i > j$, then $i - \alpha_i \geq j$.

Condition (1) ensures that, provided **d** increases sufficiently rapidly, all the σ_{δ}'s, $\delta \in \Lambda$, are distinct , and come in a certain specific order ([1], Note 1.17). Condition (1) also ensures that the only combinations of just two elements a_n and b_n of the sequence **d**, which are in Λ, are

$$\{ra_n + sb_n : 0 \le s \le r \le n\}.$$

Condition (2) is rather strange, but its usefulness will become apparent when we discuss the effect of the operator $T^{a_n+b_n}$.

In [1], § 1.22, we define $S = \{\sigma_{\delta} : \delta \in \Lambda\}$. Provided **d** increases sufficiently rapidly, the first few elements of the set S are

$$0 < a_1 < a_1 + b_1 < a_2 < a_2 + a_1 < a_2 + b_1 + a_1 < 2a_2 <$$
$$< b_2 + a_2 < b_2 + a_2 + a_1 < b_2 + a_2 + b_1 + a_1 < b_2 + 2a_2 <$$
$$< 2b_2 + 2a_2 < a_3 < a_3 + a_1 < a_3 + a_1 + b_1 < a_3 + a_2 <$$
$$< a_3 + a_2 + a_1 < a_3 + a_2 + b_1 + a_1 < a_3 + 2a_2 < a_3 + b_2 + a_2 <$$
$$< a_3 + b_2 + a_2 + a_1 < a_3 + b_2 + a_2 + b_1 + a_1 < a_3 + b_2 + 2a_2 <$$

$$< a_3 + 2b_2 + 2a_2 < 2a_3 < 2a_3 + a_1 < 2a_3 + b_1 + a_1 <$$
$$< 3a_3 < b_3 + a_3 < \dots$$

([1], § 1.24).

Thus we define the collections Λ and S, and the functions σ_δ and M_δ.

The next step is to define our projections Q_n.

1.5. The description of the projections Q_n

Each Q_n is a projection whose image is the linear span of the first $na_n + 1$ vectors e_j; $\mathrm{Im}\,Q = \mathrm{lin}\{e_j : 0 \le j \le na_n\}$. A useful way to describe Q_n is to examine first its effect on the slightly larger subspace $\mathrm{lin}\{e_j : 0 \le j \le n(a_n + b_n)\}$.

DEFINITION 1.5.1. The map $\hat{Q}_n : \mathrm{lin}\{e_j : 0 \le j \le n(a_n + b_n)\} \xrightarrow{\text{onto}} \mathrm{lin}\{e_j : 0 \le j \le na_n\}$ shall be a projection. Furthermore, its action on the subspace $\mathrm{lin}\{e_j : na_n < j \le n(a_n + b_n)\}$ shall be as follows.

(1) If for some $i \le r \le n$, we have $j \in [r(a_n + b_n), na_n + rb_n]$, then $\hat{Q}_n(e_j) = b_n^r e_{j-rb_n}$.

(2) Otherwise, $\hat{Q}_n(e_j) = 0$.

NOTE 1.5.2. These requirements are easily seen to define $\hat{Q}_n$ uniquely. We shall require that Q_n restricted to the domain of $\hat{Q}_n$ shall be equal to $\hat{Q}_n$. Let us extend our definition to a slightly larger subspace.

DEFINITION 1.5.3. The map

$$\check{Q}_n : \mathrm{lin}\{e_j : 0 \le j \le (n+1)a_{n+1}\} \xrightarrow{\text{onto}} \mathrm{lin}\{e_j : 0 \le j \le na_n\}$$

shall be a projection. It will act as $\hat{Q}_n$ on the subspace $\mathrm{lin}\{e_j : 0 \le j \le n(a_n + b_n)\}$. Furthermore, its action on the subspace

$$\mathrm{lin}\{e_j : n(a_n + b_n) < j < (n+1)a_{n+1}\}$$

shall be as follows. Let $a_0 = b_0 = 0$.

(1) If for some $1 \le r \le n + 1$ we have $j \in [ra_{n+1}, ra_{n+1} + (n + 1 - r)(a_{n+1-r} + b_{n+1-r})]$, then

$$\check{Q}_n(e_j) = e_{j-ra_{n+1}+(r-1)a_n};$$

(2) Otherwise, $\check{Q}_n(e_j) = 0$.

NOTE 1.5.4. These requirements define $\check{Q}_n$ uniquely. We can go straight from here to a definition of Q_n.

DEFINITION 1.5.5. The map $Q_n : F \to \text{lin}\{e_j : 0 \leq j \leq na_n\}$ shall act on each subspace $\text{lin}\{e_j : 0 \leq j \leq ma_m\}$ $(m > n)$ as the map

$$\check{Q}_n \circ \check{Q}_{n+1} \circ \check{Q}_{n+2} \circ \dots \circ \check{Q}_{m-1}.$$

NOTE 1.5.6. This requirement defines Q_n uniquely, and it is easy to see that the collection $(Q_n)_{n=1}^{\infty}$ is a commuting family of "nested" projections, such that $\bigcup_{n=1}^{\infty} \text{Im}\, Q_n = F$.

In [1], we adopted a slightly different approach when defining the projections Q_n, seeking for a direct definition rather than an "inductive" one. In this paper we have put in the inductive definition because we feel that it makes much clearer what is actually going on. Let us recall briefly how our other definition went.

We partially orderded Λ ([1]), § 1.18) by the lexicographic ordering $\boldsymbol{\delta}' > \boldsymbol{\delta}$ if $\delta'_J > \delta_J$ where $J = \max\{j : \delta'_j \neq \delta_j\}$. Then we remarked ([1]), § 1.23(a)), that provided **d** increases sufficiently rapidly, for each $\boldsymbol{\delta}, \boldsymbol{\delta}' \in \Lambda$, we have $\boldsymbol{\delta} > \boldsymbol{\delta}'$ if and only if $\sigma_{\boldsymbol{\delta}} > \sigma_{\boldsymbol{\delta}'}$.

Thus our partial ordering on Λ reflects the order of appearance of the $\sigma_{\boldsymbol{\delta}}$'s, as long as our underlying sequence increases moderately rapidly. After a little discussion about successors in this ordering ([1], § 2.3) it was natural ([1], § 2.8) to define functions $\boldsymbol{\delta}(i) : \mathbf{Z}^+ \to \Lambda$, $\boldsymbol{\delta}'(i) : \mathbf{Z}^+ \to \Lambda$, by

$$\boldsymbol{\delta}(i) = \max\{\boldsymbol{\delta} \in \Lambda : \sigma_{\boldsymbol{\delta}} \leq i\}$$

$$\boldsymbol{\delta}'(i) = \min\{\boldsymbol{\delta} \in \Lambda : \sigma_{\boldsymbol{\delta}} \geq i\}.$$

So each integer i sits in an appropriate interval $[\sigma_{\boldsymbol{\delta}(i)}, \sigma_{\boldsymbol{\delta}'(i)}]$.

The reader will observe that, under our projections Q_n, each vector e_j either goes to zero or a multiple of another vector e_k. In order to give a direct definition of the Q_n's, one must define a function $J(i)$ ([1], § 2.10),

$$J(i) = \max\{j : \delta_j \neq \delta'_j, \text{ where } \boldsymbol{\delta}(i) = \boldsymbol{\delta},\ \boldsymbol{\delta}'(i) = \boldsymbol{\delta}'\},$$

and an appropiate map q_n ([1], § 2.12),

$$q_n : \Lambda \to \{\boldsymbol{\delta} \in \Lambda : \sigma_{\boldsymbol{\delta}} \leq na_n\}.$$

Then the definition becomes:

$$Q_n(e_j) = \begin{cases} 0, & \text{if } J(j) > 2n-1 \\ (M_{\boldsymbol{\delta}} / M_{q_n(\boldsymbol{\delta})}) e_{j - \sigma_{\boldsymbol{\delta}} + \sigma_{q_n(\boldsymbol{\delta})}}, & \text{if } J(j) \leq 2n-1, \end{cases}$$

where $\boldsymbol{\delta} = \boldsymbol{\delta}(j)$.

The reader may quite easily check that these two definitions come to the same thing; for of course, there is only one commuting family of projections Q_n which act in a certain way on subspaces $\mathrm{lin}\{e_j : na_n < j \leq (n+1)a_{n+1}\}$.

1.6. The projections Q_n^0

As we have mentioned, it would be nice for our purposes if we could have a norm on F so that $\|T^{a_n + b_n} \circ (I - Q_n)\| \leq 2$, $\|Q_n\| \leq 2$, and various vectors e_j are close to e_0.

We have mentioned that these conditions are not consistent, and that in order for our conditions to work well, we must replace the first of them by the condition

$$\|T^{a_n + b_n} \circ (I - Q_n^0)\| \leq 2,$$

where Q_n^0 is a projection closely related to Q_n, but without such a nice bound on its norm. Q_n^0 may be defined in a similar style to Q_n as follows.

DEFINITION 1.6.1. (Compare 1.5.). The map

$$\check{Q}_{n,m} : \mathrm{lin}\{e_j : 0 \leq j \leq (m+1)a_{m+1}\} \to \mathrm{lin}\{e_j : 0 \leq j \leq ma_m\}$$

shall be a projection onto the latter subspace. It will act as $\check{Q}_m$ on the subspace $\mathrm{lin}\{e_j :$ $: 0 \leq j \; m(a_m + b_m)\}$. Furthermore, its action on the subspace $\mathrm{lin}\{e_j : m(a_m + b_m) < j \leq$ $\leq (m+1)a_{m+1}\}$ shall be as follows.

(1) If for some $1 \leq r \leq m - n + 1$ we have $j \in [ra_{m+1}, ra_{m+1} + (m+1-r)(a_{m+1-r} +$ $+ b_{m+1-r})]$, then

$$\check{Q}_{n,m}(e_j) = e_{j - ra_{m+1} + (r-1)a_m}.$$

(2) Otherwise, $\check{Q}_{n,m}(e_j) = 0$.

DEFINITION 1.6.2. (Compare 1.5.) The map $Q_n^0 : F \to \mathrm{lin}\{e_j : 0 \leq j \leq na_n\}$ shall act on each subspace $\mathrm{lin}\{e_j : 0 \leq j \leq ma_m\}$ $(m > n)$ as the map

$$\check{Q}_{n,n} \circ \check{Q}_{n,n+1} \circ \check{Q}_{n,n+2} \circ \cdots \circ \check{Q}_{n,m-1}.$$

NOTE 1.6.3. As before, this definition gives Q_n^0 uniquely. However, the Q_n^0's do not commute with each other. The reader will observe that the definitions of Q_n and Q_n^0 are identical except that in Definition 1.6.1., more vectors e_j are sent to zero because the condition on the parameter r in 1.6.1.(1) is the more restrictive $1 \leq r \leq m - n + 1$, rather than the condition $1 \leq r \leq m + 1$ in 1.5.3(1). In fact, let R_n be the projection which sends each e_j to either e_j or zero, according to whether the following holds:

(*) For some $m > n$, $Q_m(e_j)$ is a nonzero multiple of e_k, $k \in [(m - n + 1)a_m, ma_m]$.

It is not difficult to see that $Q_n^0 = Q_n \circ (I - \bar{R}_n)$.

The condition (*) is equivalent to the following:

(**) For some $m > n$, $\delta(j) = [\alpha, \beta]$ satisfies $\alpha_m > m - n$.

Inspecting Definition 1.16, we see that there can never be more than one value of m satisfying $\alpha_m > m - n$. Hence we can split up the projection $\bar{R}_n$ into a countable sum of smaller projection $R_{n,m}$ where

$$R_{n,m}(e_j) = \begin{cases} e_j, \text{ if } \delta(j) = [\alpha, \beta] \text{ satisfies } \alpha_m > m - n \\ 0 \quad \text{otherwise.} \end{cases}$$

So $\sum_{m=n+1}^{\infty} R_{n,m} = \bar{R}_n$, in the sense that at each point $x \in F$, the sum $\sum_{m=n+1}^{\infty} R_{n,m}$ is finitely nonzero, and is eventually equal to $\bar{R}_n(x)$. This is a very useful way to look at $\bar{R}_n$ later on.

In order that the relationship

$$\sum_{m=n+1}^{\infty} R_{n,m}(x) = \bar{R}_n(x)$$

should hold for every x when we pass to the completion X, it is necessary that the partial sums

$$\bar{R}_{n,m} = \sum_{j=n+1}^{m} R_{n,j}$$

should be uniformly bounded as $m \to \infty$. So it is not surprising that among our conditions in [1], 3.0, is the condition

$$\| \bar{R}_{n,m} \| \leq a_n.$$

It is now time to discuss the consistency of the conditions on our norm.

1.7. The definition of the norm

As in [1], 3.0, we define our norm to be the unique smallest norm on F such that

(1) $\| x^* : F \to \Omega \| \leq 1$,

(2) $\| T \| \leq 2$,

(3) $\| Q_n \| \leq 2$ for each $n \geq 1$,

(4) $\| T^{a_n+b_n} \circ (I - Q_n^0) \| \leq 2$ for each $n \geq 1$,

(5) $\quad ||\bar{R}_{n,m}|| \leq a_n$ for each $1 \leq n < m$.

The details of how this norm always exists are given in [1], Section 4; here we give just a concise sketch of the method.

As far as we know, the only reasonable way of establishing that this norm exists, is to take a guess at what the unit ball might look like, and prove one's guess correct.

Specifically, let us consider the operators

$$\tfrac{1}{2}T,\ \tfrac{1}{2}Q_n,\ \tfrac{1}{2}\cdot T^{a_n+b_n} \circ (I - Q_n^0),\ (1/a_n)\bar{R}_{n,m}$$

whose norms must be less than or equal to 1 in $(F, ||\cdot||)$.

If we can find a set $U_0 \subset F$ such that the absolutely convex hull $\Delta(U_0)$ is invariant under each of them, then the seminorm

$$p(x) = \inf\{M > 0 : (1/M)x \in \Delta(U_0)\}$$

will satisfy (2), (3), (4), and (5). If, further, $|x^*(x)| \leq 1$ for all $x \in \Delta(U_0)$ then (1) will also be satisfied.

Next to choosing the right set of conditions (1) to (5), the choice of the set U_0 is the other major choice in the proof. U_0 contains elements which are either multiples $\mu(i)e_i$ of one vector e_i, or else differences $\alpha e_j - \beta e_k$ of two vectors e_j and e_k. The precise definitions are as follows.

DEFINITION 1.7.1. (4.1 in [1]) Given $i \in \mathbf{N}$, $i \notin S$, let $J(i) = J$, $\boldsymbol{\delta}(i) = \boldsymbol{\delta}$, $\boldsymbol{\delta}'(i) = \boldsymbol{\delta}'$ $J(i) = j$, $K(i) = k = [J/2]$. We define

(a) $\quad \gamma(i) = (\sigma_{\boldsymbol{\delta}'} - i)/(2b_{k-1})$, where we write $b_0 = 1$ if $k = 1$.

(b) $\quad \mu(i) = (2^{\gamma(i)}/M_{\boldsymbol{\delta}})d_J^{\delta_J - 3J}$.

DEFINITION 1.7.2. If $i \in S$, $i = \sigma_{\boldsymbol{\delta}}$, then define $\mu(i) = 1/M_{\boldsymbol{\delta}}$.

The idea behind these definitions is that $1/\mu(i)$ is a reasonable estimate of the norm of e_i. As such, it usually rises gradually with increasing i, but sometimes falls sharply. We define $U_1 = \{\mu(i)e_i : i \geq 0\}$.

U_1 is one of two sets which go to make up U_0. The other one, U_2 has the following rather complicated definition in [1], §4.7:

$$U_2 = \{b_n(e_{t+r(a_n+b_n)} - (1/M_{\boldsymbol{\delta}})e_{t+r(a_n+b_n)+\sigma_{\boldsymbol{\delta}}}) : \boldsymbol{\delta}[\boldsymbol{\alpha}, \boldsymbol{\beta}] \in \Lambda_0,\ r \geq 0,\ n \geq 1,$$

$$\beta_n a_n \leq t \leq (n - r)a_n,\ \boldsymbol{\delta} - \beta_n B_n \in \Lambda,\ \delta_j = 0 \text{ for all } j < 2n,\ \alpha_k \leq k - n \text{ for all } k > n\}.$$

Here is a slightly simpler equivalent definition.

DEFINITION 1.7.3.

$$U_2 = \{y = b_n(e_j - \alpha e_k) : j \leq k,\ j \in \bigcup_{r=0}^{n} [r(a_n + b_n), na_n + rb_n],\ Q_n^0(y) = 0\}.$$

We hope that this alternative definition gives an indication of why we chose U_2 in this way. Each element of $U_0 = U_1 \cup U_2$ is going to have norm less than or equal to 1; the inclusion of these differences $b_n(e_j - \alpha e_k)$ embodies the idea that certain vectors e_k are very close to being multiples of earlier vectors e_j; it turns out that if we want to have $\|e_j - \alpha e_k\| \leq 1/b_n$, with j being the earlier index, $j \in \bigcup_{r=0}^{n} [r(a_n + b_n), na_n + rb_n]$ then the values α, k which will do are those such that Q_n^0 gives zero when applied to the vector $e_j - \alpha e_k$.

So the proof in [1] now proceeds along the following lines:

Let $U_0 = U_1 \cup U_2$ ([1], Definition 4.7.). Provided **d** increases sufficiently rapidly, the following will be true.

(1) $|x^*(u)| \leq 1$ for all $u \in U_0$.

(2) $h(u) \in \Delta(U_0)$ for all $u \in U_0$, $h \in \{\frac{1}{2}T, \frac{1}{2}Q_n, \frac{1}{2}T^{a_n+b_n} \circ (I - Q_n^0), (1/a_n)\bar{R}_{n,m} : 1 \leq n < m\}$.

(3) $\mathrm{lin}(U_0) = F$.

Hence, there is a unique smallest norm $\|\cdot\| = \|\cdot\|^{(\mathbf{d})}$ on F such that $\|x^* : (F, \|\cdot\|) \to \Omega\| \leq 1$, $\|T\| \leq 2$, $\|Q_n\| \leq 2$, $\|T^{a_n+b_n} \circ (I - Q_n^0)\| \leq 2$, $\|\bar{R}_{n,m}\| \leq a_n$.

Moreover, $\|u\| \leq 1$ for all $u \in U_0$ (Lemma 4.14 in [1]).

The proof of (3) is very elementary given (1) and (2); the proofs of (1) and especially (2) are very long, but there is nothing essentially difficult about them. We omit these proofs from our present discussion, and continue by describing how one proves that, if $T^\sim$ is the continuous extension of T to the completion $F^\sim$ of F, then $T^\sim$ has no nontrivial closed invariant subspace.

1.8. The rest of the proof

Recall that it is sufficient to prove that for every $x^\sim \in F^\sim$ and every $\varepsilon > 0$, there is a polynomial p such that

$$(*) \qquad \|p(T^\sim)x^\sim - e_0\| < \varepsilon .$$

This assertion (∗) is proved in the following way. Let us write $P_{n,m}$ for the operator $(I - R_{n+1,m}) \circ Q_m$. We finish the proof in the following two stages:

LEMMA 1.8.1. (Lemma 5.6 of (1)) *Provided* **d** *increases sufficiently rapidly, the space* $(F^\sim, \|\cdot\|)$ *has the following property. For all* $x^\sim \in F^\sim$ *with* $\|x^\sim\| = 1$ *such that for some* $m > n + 1 > 2$, $\|P^\sim_{n,m} \circ Q^{0\sim}_m(x^\sim)\| > 1/a_m$ *there is a polynomial* p *such that*

$$\| p(T^\sim)x^\sim - e_0 \| < 2/b_{n-1}$$

(where $P^\sim_{n,m}$, $Q^{0\sim}_m$ *are the continuous extensions of* $P_{n,m}$, Q^0_m *to* $F^\sim$*).*

LEMMA 1.8.2. (Lemma 5.13 of [1]) *If* **d** *increases sufficiently rapidly, the following is true. For every* $x^\sim \in S(F^\sim)$, $n > 1$, *there is an integer* $m > n + 1$ *such that*

$$\| P^\sim_{n,m} \circ Q^{0\sim}_m(x^\sim) \| > 1/a_m.$$

It is clear that these two lemmas together give us what we want. Let us then decribe how Lemma 1.8.1 is proved in [1].

Now $P_{n,m}$ acts as the projection $Q_m : F \to \text{lin}\{e_j : j \le ma_m\}$ followed by the "truncation" Ψ,

$$\Psi\Big(\sum_{j=0}^{ma_m} \lambda_j e_j\Big) = \sum_{j=0}^{(m-n)a_{m-1}} \lambda_j e_j.$$

So given an $x^\sim \in S(F^\sim)$, such that $\|P^\sim_{n,m} \circ Q^{0\sim}_m(x^\sim)\| > 1/a_m$, we first need to work within the finite dimensional subspace $\text{lin}\{e_j : 0 \le j \le ma_m\}$ and show that if $P^\sim_{n,m} \circ Q^{0\sim}_m(x^\sim)$ is large, (that is, $Q^{0\sim}_m(x^\sim)$ has "weight" greater than or equal to $1/a_m$ in the first $(m - n)a_m$ coordinates), then we can take a linear combination $p(T^\sim) \circ Q^{0\sim}_m(x^\sim)$ of translates $(T^\sim)^k$ of the vector $y = Q^{0\sim}_m(x^\sim)$, and get close to $e_{(m-n+1)a_m}$. For the sake of convenience later on it is desirable that p should have degree less that or equal to ma_m, and should also be divisible by T^{a_m}.

The actual statement of the result is in [1], Lemma 5.10, which we restate here:

LEMMA. *There is a function* $C: \mathbf{N}^3 \to \mathbf{N}$ *with the following property. For all* **d** *increasing sufficiently rapidly, the following is true. For all* $x^\sim \in S(F^\sim, \|\cdot\|)$ *such that*

$$\| P^\sim_{n,m} \circ Q^{0\sim}_m(x^\sim) \| \ge 1/a_m,$$

$m > n + 1 > 2$, *there is a polynomial* $p(t) = \sum_{j=a_m}^{ma_m} \rho_j t^i$, *with* $|p| = \sum |\rho_j| \le C(n,m,a_m)$ *such that*

$$\|\tau \circ p(T^{\sim}) \circ Q_m^{0\sim}(x^{\sim}) - e_{(m-n+1)a_m}\| \leq 1/a_m;$$

where τ is the truncation operator,

$$\tau : e_j \to \begin{cases} e_j & \text{if } j \leq ma_m \\ 0 & \text{otherwise.} \end{cases}$$

The proof of this statement involves, once again, some not too dificult combinatorics, which we omit. The truncation operator is introduced, not because it has any great significance in itself, but rather because it has the effect of "truncating" the proof of Lemma 1.8.1 into four pieces of short length. The next piece is the following.

LEMMA. (5.11 in [1]) *If the sequence* **d** *increases sufficiently rapidly, then for all* m,n *for which* $m > n + 1 > 2$, *the following is true. For all polynomials* $p(t) = \sum_{j=a_m}^{ma_m} \rho_j t^j$, *such that* $|p| \leq C(n,m,a_m)$, *and all* $z \in \text{lin}\{e_0, e_1, \ldots, e_{ma_m}\}$ *for which* $\|z\| \leq 2(1 + a_m)$, *we have*

$$\|(T^{b_n}/b_n) \circ \tau \circ p(T)z - \tau \circ p(T)z\| \leq 1/a_m.$$

Note that we shall apply this lemma with $z = Q_m^{0\sim}(x^{\sim})$; it is known that $\|x^{\sim}\| = 1$ and $\|Q_m^0\| \leq 2(1 + a_m)$ so $\|z\| \leq 2(1 + a_m)$.

The next, fairly similar lemma is the following:

LEMMA. (5.12 in [1]) *If the sequence* **d** *increases sufficiently rapidly then the following also is true. For all* n,m,p,z *satisfying the conditions of the previous lemma, we have*

$$\|(T^{b_m}/b_m) \circ (I - \tau) \circ p(T)z\| \leq 1/a_m.$$

Adding together our three lemmas we obtain:

$$\|(T^{b_m}/b_m) \circ p(T)z - e_{(m-n+1)a_m}\| \leq 3/a_m.$$

It then remains to estimate first

$$(*) \qquad \|(T^{b_m}/b_m) \circ p(T)(z - x^{\sim})\| = \|(T^{b_m}/b_m)p(T) \circ (I - Q_n^0)x^{\sim}\|,$$

and then to estimate

$$\|e_{(m-n+1)a_m} - e_0\|.$$

The first estimate is easy because the operator

$$T^{a_m+b_m} \circ (I - Q_m^0)$$

has norm less than or equal to 2, and T^{a_m} divides $p(T)$, hence $T^{a_m+b_m}$ divides $(T^{b_m}/b_m)p(T)$.

Expression (*) is less than or equal to $1/a_m$, provided **d** increases sufficiently rapidly.

The second estimate is based on the fact that

$$b_{n-1}\left(e_0 - e_{(m-n+1)a_m}\right) \varepsilon\, U_2,$$

hence

$$\| e_0 - e_{(m-n+1)a_m} \| \leq 1/b_{n-1}.$$

Adding up our various estimates we obtain, writing

$$q(T) = (T^{b_m}/b_m)p(T),$$

that

$$\| q(\tilde{T})\tilde{x} - e_0 \| \leq 4/a_m + 1/b_{n-1} < 2/b_{n-1}$$

provided **d** increases sufficiently rapidly. This concludes the proof of Lemma 1.8.1.

To conclude Section 1 it now remains to indicate how Lemma 1.8.2. is proved. First we observe (5.14 in [1]), that since the collection

$$(\tilde{P}_{n,m})_{m=n+2}^{\infty}$$

is uniformly bounded, and since for each $x \varepsilon F$, $P_{n,m}(x) = x$ for all but finitely many values of m, it follows that

$$\tilde{P}_{n,m}(\tilde{x}) \xrightarrow{n\to\infty} \tilde{x}$$

for every $\tilde{x} \varepsilon \tilde{F}$.

Given $\tilde{x} \varepsilon S(\tilde{F})$, and $n \varepsilon \mathbf{N}$, let us begin our proof of Lemma 1.8.2 by choosing an $r > n+1$ such that $\| \tilde{P}_{n,r}(\tilde{x}) \| > 1/2$. If

$$\| \tilde{P}_{n,r} \circ Q_r^{\alpha}(\tilde{x}) \| \geq 1/4,$$

then we are home. If not then

$$\|(P^{\sim}_{n,r} - P^{\sim}_{n,r} \circ Q^{\mathcal{Q}}_r)x^{\sim}\| = \|P^{\sim}_{n,r} \circ (Q^{\sim}_r - Q^{\mathcal{Q}}_r)x^{\sim}\| > 1/4.$$

But

$$Q_r - Q^0_r = Q_r \circ \bar{R}_r = Q_r \circ \left(\sum_{k=r+1}^{\infty} R_{r,k} \right).$$

Therefore

$$\sum_{k=r+1}^{\infty} \|P^{\sim}_{n,r} \circ R^{\sim}_{r,k}(x^{\sim})\| > 1/4$$

(Lemma 5.16 of [1]).

However, it is very beautifully true that for all $n + 1 < r < k$,

$$P_{n,r} \circ R_{r,k} = Q_r \circ R_{r,k} \circ P_{n,k} \circ Q^0_k$$

(Lemma 5.17 of [1]). Hence, since

$$\|Q_r \circ R_{r,k}\| \leq 4a_r,$$

we have

$$\sum_{k=r+1}^{\infty} 4a_r \|P^{\sim}_{n,k} \circ Q^{\mathcal{Q}}_k(x^{\sim})\| > 1/4.$$

However, if (1.8.2) were to fail for these values of x, n then for each $k > n + 1$, $\|P^{\sim}_{n,k} \circ Q^{\mathcal{Q}}_k(x^{\sim})\|$ may be assumed less than $1/a_k$. So we must have

$$\sum_{k=r+1}^{\infty} 4a_r / a_k > 1/4,$$

which is a contradiction, if **d** increases sufficiently rapidly.

This proves Lemma 5.13, and draws our proof that $T^{\sim}$ has no invariant subspace to a close.

It remains to discuss our solution on the space ℓ_1 in Section 2, and then, in Section 3, to describe the spectrum of our operator, and its relationship to a weighted shift. We shall also show that there is an operator T such that T has no invariant subspaces, but T^2 does have invariant subspaces.

2. THE SOLUTION ON THE SPACE ℓ_1

Our solution to the invariant subspace problem on the space ℓ_1 occurs as a variant of the original construction in the following way.

In the latter sections of [1], we proved in detail how any norm on F which satisfied

(1) $\| x^* : (F, \|\cdot\|) \to \Omega \| \leq 1,$

(2) $\| T \| \leq 2,$

(3) $\| Q_n \| \leq 2$ for all $n \geq 1,$

(4) $\| T^{a_n+b_n} \circ (I - Q_n^0) \| \leq 2$ for all $n \geq 1,$

(5) $\| \bar{R}_{n,m} \| \leq a_n$ for all $m > n \geq 1,$

and also satisfied

$$\| u \| \leq 1 \quad \text{for all } u \in U_0,$$

had the following property. When the completion $(F^{\sim}, \|\cdot\|)$ was taken, and T continuously extended to a map $T^{\sim}$ on $F^{\sim}$, we found that $T^{\sim}$ had no nontrivial closed invariant subspace.

It was a suggestion of my supervisor, colleague and close friend Bela Bollobas that I should examine the norm obtained by taking $\Delta(U_0)$ as unit ball. More precisely, the function

$$p_{U_0}(x) = \inf\{\varepsilon > 0 : (1/\varepsilon)x \in \Delta(U_0)\}$$

is a norm on F satisfying all the above conditions; but what space is the completion $(F^{\sim}, p_{U_0})$?

The answer to this question was surprisingly simple; the completion of $(F^{\sim}, p_{U_0})$ is isomorphic to the sequence space ℓ_1. Thus there is an operator on ℓ_1 with no nontrivial closed invariant subspaces; in the closing sections of this paper we shall find out quite a lot about this operator.

First we sketch the proof that the completion of $(F^{\sim}, p_{U_0})$ is isomorphic to ℓ_1. The proof is given in detail in [2].

We observe that if B is any basis of F (which has countable dimension), then the associated norm

$$p_B(x) = \inf\{\varepsilon > 0 : (1/\varepsilon)x \in \Delta(B)\}$$

satisfies the condition that the completion of (F, p_B) is isomorphic to ℓ_1, (indeed, the isomorphism is an isometry with the canonical basis of ℓ_1 corresponding to the elements of B).

In order to prove that (F, p_{U_0}) also has this property, it is sufficient to find a subset $B = U_3 \subset U_0$ which is a basis of F, such that

$$\Delta(U_0) \supset \Delta(B) \supset \varepsilon \Delta(U_0)$$

for some $\varepsilon > 0$. (F, p_{U_0}) is then $1/\varepsilon$ isomorphic to (F, p_B), hence its completion also is isomorphic to ℓ_1.

We proceed to give the definition of the set U_3; the details of the proof that condition (*) holds (indeed, it holds for any fixed value of $\varepsilon < 1$ provided **d** increases sufficiently rapidly) are given in [2].

DEFINITION 2.1. Given $i \in \mathbf{N}$, let

$$E_i = \{(t, r, n, \delta) : b_n(e_{t+r(a_n+b_n)} - (1/M_\delta)e_{t+r(a_n+b_n)+\sigma_\delta}) \in U_2,\ \delta \neq \mathbf{0},$$
$$t + r(a_n + b_n) + \sigma_\delta = i\}.$$

E_i may be empty for a given value of i. Partially order E_i in the following way:

$$(t, r, n, \delta) > (t', r', n', \delta')$$

if

$$n > n', \text{ or } n = n' \text{ and } r > r'.$$

It is not difficult to see that E_i is finite for each i, and if nonempty E_i has a unique maximal element which we shall denote by max E_i.

DEFINITION 2.2. For each $i \geq 0$, let

$$f_i = \begin{cases} \mu(i)e_i & \text{if } E_i = \emptyset, \\ b_n(e_{t+r(a_n+b_n)} - (1/M_\delta)e_{t+r(a_n+b_n)+\sigma_\delta}) & \text{if } E_i \neq \emptyset,\ (t, r, n, \delta) = \max E_i. \end{cases}$$

DEFINITION 2.3. Let $U_3 = \{f_i : i \geq 0\}$.

Then U_3 is a countable basis for F, and in [2], Lemma 5.13 we establish the following.

THEOREM. *For every* $C > 1$ *the following is true. Provided* **d** *increases sufficiently rapidly,*

$$\Delta(U_3) \subset \Delta(U_0) \subset C\,\Delta(U_3).$$

Hence, $(F^\sim, p_{U_0})$ is C isomorphic to ℓ_1, where as in [1], $F^\sim$ denotes the completion of F under the appropriate norm. So the operator T on F gives, in a constructive way, an operator on which has no nontrivial closed invariant subspaces. We shall now examine our operator on ℓ_1 quite closely, with a view to finding its spectrum and so on. Since this part of our research is not duplicated elsewhere, we shall give somewhat more detailed proofs of our assertions from here on the close of the paper.

3. THE SPECTRUM OF THE OPERATOR ON ℓ_1, AND OTHER TOPICS.

3.1. The spectrum of T

LEMMA. *Provided* **d** *increases sufficiently rapidly, we have*

$$\| T^{a_n+b_n} : (F, p_{U_0}) \to (F, p_{U_0}) \| \leq G(n, a_n) b_n$$

where $G : \mathbf{N}^2 \to \mathbf{N}$ *is a fixed function independent of the choice of* a_n *and* b_n.

PROOF.

$$\| T^{a_n+b_n} \| \leq \| T^{a_n+b_n} \circ (I - Q_n^0) \| + \| T^{a_n+b_n} \circ Q_n^0 \| \leq 2 + \| T^{a_n+b_n} \circ Q_n^0 \| \leq$$

$$\leq 2 + \| Q_n^0 \| \, \| T^{a_n+b_n} | \mathbf{Im}\, Q_n^0 \| \leq 2 + 2(1 + a_n) \| T^{a_n+b_n} | \mathbf{Im}\, Q_n^0 \| ,$$

([1], 3.0 and Lemma 5.3).

Now $\mathbf{Im}\, Q_n^0 = \mathrm{lin}\{e_i : i \leq na_n\} = \mathrm{lin}\{f_i : i \leq na_n\}$. It is therefore sufficient to establish that there is a function G such that for all $0 \leq i \leq na_n$,

$$\| T^{a_n+b_n}(f_i) \| \leq G_1(n, a_n) b_n .$$

It is not difficult to check that this is so, using the fact that $|||x|||$ is $N(n, a_n)$ isomorphic to $\|x\|_{c_0}$ for $x \in \mathrm{lin}(e_0, e_1, \ldots, e_{na_n})$ ([1], Lemma 5.8), and checking that for all $x \in \mathrm{lin}(e_{a_n+b_n}, e_{a_n+b_n+1}, \ldots, e_{(n+1)a_n+b_n})$,

$$|||x||| \leq b_n N_1(n, a_n) \|x\|_{c_0} .$$

We omit the details of this proof.

COROLLARY. *Provided* **d** *increases sufficiently rapidly, the spectral radius of* T *is less than or equal to 1.*

PROOF.

$$\rho(T) \leq \liminf \| T^{a_n+b_n} \|^{1/(a_n+b_n)} \leq \liminf (b_n G(n, a_n))^{1/(a_n+b_n)} \leq \liminf (b_n^2)^{1/(a_n+b_n)} = 1$$

(provided **d** increases sufficiently rapidly).

So the spectrum of T is contained in the unit disk $\mathbf{D} = \{z \in \mathbf{C} : |z| \leq 1\}$. We can identify the spectrum of T as being the whole of **D** in the following way.

LEMMA. *Every* $z \in \mathbf{D}$ *is in the approximate point spectrum of* T, *provided* **d** *increases sufficiently rapidly.*

PROOF. Given $z \in \operatorname{int} \mathbf{D}$, let

$$u_n = \sum_{j=1}^{n} z^{j-1} e_{a_n - j}$$

(each $n > 1$).

Provided **d** increases sufficiently rapidly, $\delta(a_n - j)$ is always $(n-1)(A_{n-1} + B_{n-1})$ and $E_{a_n - j}$ is empty; hence

$$f_{a_n - j} = \mu(a_n - j)\, e_{a_n - j} = (1/b_{n-1}^{n-1}) 2^{j/2b_{n-1}} a_n^{-6n+1} e_{a_n - j}$$

([1], Definition 4.1).

Hence

$$|||\, a_n \,||| = \sum_{j=1}^{n} b_{n-1}^{n-1} a_n^{6n-1} 2^{-j/2b_{n-1}} |z|^{j-1} \geq a_n^{6n-1} b_{n-1}^{n-1} 2^{-1/2b_{n-1}} \geq \tfrac{1}{2} a_n^{6n-1} b_{n-1}^{n-1} \tfrac{1}{2}$$

(considering the term $j = 1$ only of the sum).

However,

$$(T - zI) u_n = e_{a_n} - z^n e_{a_n - n}$$

and since $\mu(a_n) = 1$, $|||\, e_{a_n} \,||| \leq 1$. So

$$|||\, (T - zI) u_n \,||| \leq 1 + |z|^n \| e_{a_n - n} \| \leq 1 + |z|^n a_n^{6n-1} b_{n-1}^{n-1}.$$

As $n \to \infty$, $|||\, (T - zI) u_n \,||| / |||\, u_n \,||| \to 0$, therefore every $z \in \operatorname{int} \mathbf{D}$ is in the approximate point spectrum of T.

Therefore, by standard arguments, every $z \in \mathbf{D}$ is in the approximate point spectrum.

So $\sigma(T) = \mathbf{D}$, and every $\lambda \in \mathbf{D}$ is an approximate eigenvalue.

This conclude our investigation of the spectrum of T. We proceed to establish another interesting fact, that there is an operator T such that T has no invariant subspace but T^2 does.

3.2. T^2 may have invariant subspaces

LEMMA. *If* **d** *is chosen suitably, then the operator* $T^{\sim}$ *on* ℓ_1 *will have no invariant subspaces, but* $(T^{\sim})^2$ *will have invariant subspaces.*

PROOF. Choose **d** so that

(1) **d** increases sufficiently rapidly that the proofs in [1] and [2] hold true.

(2) every d_i is even.

Then in view of (1), $T^{\sim}$ has no nontrivial closed invariant subspaces. However, the subspace

$$E = \overline{\text{lin}}\{e_i : i \text{ is odd}\}$$

is plainly $(T^{\sim})^2$ invariant, it is not equal to $\{0\}$ and furthermore it is not the whole space. For examining the definition of x^* we see that if every d_i is even then $x^*(e_j)$ is zero for every odd j, so $x^*(y)$ is zero for all $y \in E$, hence, E is not the whole space.

To conclude our paper, we establish a relationship between our operator and a weighted shift on ℓ_1.

3.3. $T = W + N$, **where** W **is a weighted shift and** N **is a nuclear operator**

LEMMA. *The set* U_3 *defined in* §2.2 *may alternately be characterised as follows.* $U_3 = U_4 \cup U_5 \cup U_6$, *where*

$$U_4 = \{b_n(e_j - e_{j+ma_{n+m}}) : j \in \bigcup_{r=0}^{n} [r(a_n + b_n), rb_n + na_n], m > 0\}$$

$$U_5 = \{b_n \cdot (e_j - (1/b_n)e_{j+b_n}) : j \in \bigcup_{r=0}^{n-1} [r(a_n + b_n), rb_n + (n - 1)a_n]\}$$

$$U_6 = \{\mu(k)e_k : k \notin (\bigcup_{n,m=1}^{\infty} \bigcup_{r=0}^{n} [r(a_n + b_n) + ma_{n+m}, rb_n + na_n + ma_{n+m}]) \cup \\ \cup (\bigcup_{n=1}^{\infty} \bigcup_{r=1}^{n} [r(a_n + b_n), rb_n + na_n])\}.$$

PROOF. It is not difficult to see that every element in the sets above is in U_0. Referring to Definition 2.2, we see that it is necessary to check the following.

(1) Each element $u = b_n(e_j - (1/M_\delta)e_k)$ in the sets U_4, U_5 above is maximal in the sense that an appropriate (t, r, n, δ) is maximal in E_k.

(2) For every j with $\mu(j)\, e_j \in U_6$ we have $E_j = \emptyset$ (the empty set).

PROOF of (1).

<u>Case 1.</u> $u \in U_4$, let us say

$$u = b_n(e_{t+r(a_n+b_n)} - e_{t+r(a_n+b_n)+ma_{n+m}}),$$

where $t \in [0, (n-r)a_n]$.

We wish to check that (t, r, n, mA_m) is maximal in E_k, $k = t + r(a_n + b_n) + ma_{n+m}$. If n is not maximal then by [1] Lemma 4.10(j) we have $Q^0_m(y) = 0$ for all $y = b_m \cdot (e_{t+r(a_m+b_m)} - (1/M_{\boldsymbol{\delta}})e_{t+r(a_m+b_m)+\sigma_{\boldsymbol{\delta}}}) \in U_2$; yet by inspection $Q^0_m(e_{t+r(a_m+b_m)}) = b^r_m e_{t+ra_m}$. So $Q^0_m(e_k)$ cannot be either zero or e_k; it must be $M_{\boldsymbol{\delta}} b^r_m e_{t+ra_m}$.

However it is easy to check from Definition 2.18 of [1], that for $j \in [r(a_n + b_n) + ma_{n+m}, ra_n + nb_n + ma_{n+m}]$, we have

$$Q^0_{\ell}(e_j) = \begin{cases} 0, & n < \ell < n+m \\ e_j, & \ell \geq n+m \end{cases}.$$

This is a contradiction, so in fact n is maximal. Given n, r is also maximal since for all j in the above interval, we have

$$r(A_n + B_n) + mA_{n+m} \leq \boldsymbol{\delta}(j) \leq \boldsymbol{\delta}'(j)' \leq rB_n + nA_n + mA_{n+m};$$

by [2], 4.10 (b), if there were a $(t', r', n', \boldsymbol{\delta}') \in E_k$, $r' > r$, then

$$(r' + \beta'_n)A_n + r'B_n + \boldsymbol{\delta}' \leq \boldsymbol{\delta}(j) \leq \boldsymbol{\delta}'(j) \leq nA_n + r'B_n + \boldsymbol{\delta}'.$$

Comparing these two inequalities and bearing in mind that each coordinate δ'_n of $\boldsymbol{\delta}'$ is nonnegative, we have $r' \leq r$; so r is in fact maximal.

<u>Case 2</u>. $u \in U_5$. It is now obvious that n is maximal since if $u = b_n(e_j - (1/M_{\boldsymbol{\delta}}) \cdot e_{j+b_n})$, then $k = j + b_n < b_{n+1}$. r is also maximal since if $(t', r', n, \boldsymbol{\delta}') \in E_k$ we cannot have $\boldsymbol{\delta}' = 0$, yet $k > \sigma_{\boldsymbol{\delta}'}$, so $\delta'_s = 0$ for all $s > 2n$. But by [1], Definition 4.7, we have $\delta'_s = 0$ for all $s < 2n$ therefore $\delta_{2n} = \beta_n \geq 1$. So $b_n + t' + r'(a_n + b_n) \leq k = b_n + t + r(a_n + b_n)$; hence $r'a_n + r'b_n \leq rb_n + na_n$.

Hence $r' \leq r$ because, as usual, a_n is much smaller than b_n.

PROOF of (2).

<u>Case (a).</u> For some n, m, r we have $j \in (rb_n + na_n + ma_{n+m}, (r+1)(a_n + b_n) + ma_{n+m})$. Then $J(j) = 2n$ so if $(t', r', n', \boldsymbol{\delta}') \in E_j$ then by 4.10(b), $(r' + \beta_{n'})A_{n'} + r'B_{n'} + \boldsymbol{\delta}' \leq \boldsymbol{\delta}(j) \leq \boldsymbol{\delta}'(j) \leq n'A_{n'} + r'B_{n'} + \boldsymbol{\delta}'$, and it follows that $n' > n$. But inspecting Definition 2.18 of [1] we see that $Q_{n'}(e_j)$ is zero for $n < n' < n+m$ and e_j for $n + m \leq n'$. This is a contradiction, as in Case 1 of the proof of assertion (1); so $E_j = \boldsymbol{\phi}$.

Case (b). For some n, $r \geq 0$ we have $j \in (rb_n + na_n, (r+1)(a_n + b_n))$. Then $\delta(j)=2n$ so as in Case (a), we must have $n' > n$. However $Q^0_{n'}(e_j) = e_j$ for all $n' > n$.

Case (c). For some n we have $j = na_n$. Then

$$Q^0_r(e_j) = \begin{cases} 0, & r < n \\ e_j, & r \geq n. \end{cases}$$

Therefore as before, E_j is empty because we need an r such that $Q^0_r(e_j)$ is not zero or e_j.

Case (d). For some n we have $j \in (n(a_n + b_n), a_{n+1})$. Then $J(j) = 2n + 1$ so $n' > n$. But $Q'_n(e_j) = e_j$ for all such n'. Note that these four cases cover all the possibilities. Thus Lemma 3.3.1 is proved.

LEMMA 3.3.2. *If* $k \in S$, *where*

$$S = \bigcup_{n=1}^{\infty} \bigcup_{r=0}^{n} \bigcup_{m>0} [r(a_n + b_n) + ma_{n+m}, rb_n + na_n + ma_{m+n}) \cup$$

$$\cup \bigcup_{n=1}^{\infty} \bigcup_{r=1}^{n} [r(a_n + b_n), rb_n + na_n) \cup$$

$$\cup \bigcup_{n,m=1}^{\infty} \bigcup_{r=0}^{n-1} (rb_n + na_n + ma_{n+m}, (r+1)(a_n + b_n) + ma_{n+m} - 1) \cup$$

$$\cup \bigcup_{n=1}^{\infty} \bigcup_{r=0}^{n-1} (rb_n + na_n, (r+1)(a_n + b_n) - 1) \cup$$

$$\cup \bigcup_{n=1}^{\infty} \{na_n\} \bigcup_{n=1}^{\infty} (n(a_n + b_n), a_{n+1} - 1);$$

then $Tf_k = w_k f_{k+1}$, $|w_k| \leq 2$.

PROOF. Case 1. $k \in \bigcup_{n=1}^{\infty} \bigcup_{r=0}^{n} \bigcup_{m>0} [r(a_n + b_n) + ma_{n+m}, rb_n + na_n + ma_m)$.

Then for suitable j, n, m, $f_k = b_n(e_j - e_{j+ma_{n+m}})$, and $f_{k+1} = b_n(e_{j+1} - e_{j+1+ma_{n+m}}) = Tf_k$.

We argue similarly if

$$j \in \bigcup_{n=1}^{\infty} \bigcup_{r=1}^{n} [r(a_n + b_n), rb_n + na_n).$$

In all other cases, $f_j = \mu(j)e_j$ and

$$f_{j+1} = \mu(j+1)e_{j+1} = (\mu(j+1)/\mu(j))Tf_j.$$

So

$$Tf_j = (\mu(j)/\mu(j+1))f_{j+1} = w_j\, f_{j+1}$$

where $|w_j| \leq 2$ by [1], Lemma 4.9(c). This proves Lemma 3.3.2.

LEMMA 3.3.3. *For all* m, n $\in$ **N**, $0 \leq r \leq n$, *the action of* T *on* f_j, $j = rb_n + na_n + ma_{n+m}$ *is to send it to* $w_j f_{j+1} + b_n e_{rb_n+na_n+1}$ *where* $|w_j| < 2$.

PROOF.

$$f_{j+1} = \mu(j+1)e_{j+1};\ Tf_j = b_n(e_{j+1} - e_{rb_n+na_{n+1}}).$$

Using Lemma 4.9 of [1] it is easily shown that

$$|w_j| = b_n/\mu(j+1) < 2.$$

(note that w_j is negative).

LEMMA 3.3.4. *The action of* T *on* f_j, $j = rb_n + na_n$ *is to send it to* $w_j f_{j+1} + b_n e_{(r-1)b_n+na_n+1}$ *(each* $1 \leq r \leq n$*), where* $|w_j| < 2$.

The proof is very similar to that of Lemma 3.3.3.

LEMMA 3.3.5. *The action of* T *on* f_j, $j = (r+1)(a_n + b_n) + ma_{n+m} - 1$, $(0 \leq r \leq n-1)$ *is to send it to* $w_j e_{j+1} + \mu(j)e_{(r+1)(a_n+b_n)}$, *where* $|w_j| \leq 2$.

PROOF. $f_j = \mu(j)e_j$, and $f_{j+1} = b_n(e_{(r+1)(a_n+b_n)} - e_{j+1})$. By [1], Lemma 4.9, $\mu(j) \leq 2^{-\sqrt{b_n}}$, so provided **d** increases sufficiently rapidly, $|w_j| < 2$.

LEMMA 3.3.6. *The action of* T *on* f_j, $j = (r+1)(a_n + b_n) - 1$ *is to send it to* $w_j f_{j+1} + \mu(j)e_{(r+1)a_n+rb_n}$, *where* $|w_j| < 2$.

The proof is very similar to the proof of Lemma 3.3.5.

LEMMA 3.3.7. *The action of* T *on* f_j, $j = a_{n+1} - 1$ *is to send it to* $w_j e_{j+1} + b_n\mu(j)e_0$, *where* $|w_j| < 2$.

Once again, the proof is very similar to that of Lemma 3.3.5.

LEMMA 3.3.8. *Provided* **d** *increases sufficiently rapidly,* T *is the sum of the weighted shift*

$$W : f_j \rightarrow w_j f_{j+1}$$

and the nuclear operator

$$N = \sum_{n=1}^{\infty} \sum_{r=0}^{n} \Big(\Big(\sum_{m=1}^{\infty} f^*_{rb_n+na_n+ma_{n+m}}\Big) \otimes b_n e_{rb_n+na_n+1}\Big) +$$

$$+ \sum_{n=1}^{\infty} \sum_{r=1}^{n} f^*_{rb_n+na_n} \otimes b_n e_{(r-1)b_n+na_n+1} +$$

$$+ \sum_{n=1}^{\infty} \sum_{r=0}^{n-1} \Big(\sum_{m=1}^{\infty} \mu((r+1)(a_n+b_n)+ma_{n+m}-1) f^*_{(r+1)(a_n+b_n)+ma_{n+m}-1}\Big) \otimes e_{(r+1)(a_n+b_n)} +$$

$$+ \sum_{n=1}^{\infty} \sum_{r=0}^{n-1} \mu((r+1)(a_n+b_n)-1) f^*_{(r+1)(a_n+b_n)-1} \otimes e_{(r+1)a_n+rb_n} + \sum_{n=1}^{\infty} b_n \mu(a_{n+1}-1) f^*_{a_{n+1}-1} \otimes e_0.$$

PROOF. In view of Lemmas 3.3.3 to 3.3.7, we need only check that, provided **d** increases sufficiently rapidly, the nuclear norm of N is indeed finite.

Using the fact that $\| \mu(j) e_j \| \leq 1$, and using estimates from [1], Lemma 4.9, we obtain the following estimates.

Provided **d** increases sufficiently rapidly, we have

(a) $$\| b_n e_{rb_n+na_n+1} \| \leq b_n 2^{-\sqrt{b_n}},$$

(b) $$\| e_{(r+1)(a_n+b_n)} \| \leq b_n^{r+1}, \quad \| e_{(r+1)a_n+rb_n} \| \leq b_n^{r},$$

(c) $$\mu((r+1)(a_n+b_n)+ma_{m+n}-1) < 2^{-\sqrt{b_n}} \quad (m \geq 0),$$

(d) $$b_n \mu(a_{n+1}-1) < b_n 2^{-\sqrt{b_n}}.$$

Hence the nuclear norm of N is less than or equal to

$$2 \sum_{n=1}^{\infty} \sum_{r=0}^{n} b_n 2^{-\sqrt{b_n}} + \sum_{n=1}^{\infty} \sum_{r=0}^{n} b_n^{r+1} 2^{-\sqrt{b_n}} + \sum_{n=1}^{\infty} \sum_{r=0}^{n} b_n^{r} 2^{-\sqrt{b_n}} + \sup_n b_n 2^{-\sqrt{b_n}}.$$

This quantity is arbitrarily small provided **d** increases sufficiently rapidly. So we have even the slightly stronger:

THEOREM. T = W + N *where* W *is a weighted shift and* N *is nuclear; moreover* T *may be chosen so that the nuclear norm of* N *is arbitrarily small compared to the spectral radius (* = 1*) or the norm (* $\leq$ 2*) of* T.

4. NOTES

It is clear that there must be a simpler proof of the invariant subspace problem counterexample, starting with Lemma 5.1 as the definition of an isomorphism between F and a dense subset of ℓ_1, and then working entirely in the context of an operator on ℓ_1.

This will be the subject of a future paper, and it seems not unreasonable to hope that it will be a simpler paper than our previous papers on the same subject.

The fact that T is a perturbation of a weighted shift by a nuclear operator shows that the infinite dimensional "matrix" of T with respect to the basis (f_i) will be very sparse indeed, except on the diagonal which contains the weights.

So we now have T constructively defined on ℓ_1, and we know quite a lot about the structure of T.

REFERENCES

1. **Read, C.J. :** A solution to the invariant subspace problem, *Bull. London Math. Soc.* **16**(1984), 337-401.

2. **Read, C.J. :** A solution to the invariant subspace problem on the space ℓ_1, *Bull. London Math. Soc.*, to appear.

C.J.Read
Department of Pure Mathematics and Mathematical Statistics
Cambridge University
England.

Operator Theory:
Advances and Applications, Vol.17

SOME RESULTS ON COHOMOLOGY WITH BOREL COCHAINS, WITH APPLICATIONS TO GROUP ACTIONS ON OPERATOR ALGEBRAS

Jonathan Rosenberg *)

0. INTRODUCTION

This paper is an outgrowth of joint work with Richard Herman [5] and Iain Raeburn [13]. In both of these projects, questions concerning group actions on operator algebras naturally led to a study of obstruction classes in $H^2(G,U(A))$, where U(A) is the (suitably topologized) unitary group of an abelian operator algebra A. The appropriate cohomology theory here is the "Borel cochain" theory of C.C.Moore, as developed and systematized in [8]. In case A is a von Neumann algebra, U(A) is essentially what Moore calls $U(X,\mathbf{T})$ (X here is some standard measure space), and machinery for computing the relevant cohomology groups is developed and applied in [8] and [9].

We were interested, however, in problems concerning separable C^*-algebras. In this case, U(A) becomes $C(X,\mathbf{T})$, the continuous functions into the circle group $\mathbf{T}$ on a second-countable topological space X. The study of $H^n(G,C(X,\mathbf{T}))$ now becomes more a matter of topology than of measure theory, and techniques different from those of [8] are called for. The purpose of this paper is to compute the Moore cohomology groups in certain cases relevant to operator algebraists, and then to translate some of these calculations back into statements about operator algebras.

I wish to thank my coworkers Richard Herman and Iain Raeburn for their help in getting me started on the work described here. It will be obvious that my results depend heavily on the machines developed in [8] and [17]. Finally, I wish to thank the Mathematical Sciences Research Institute for its congenial and stimulating environment.

1. NOTATION AND REVIEW OF KNOWN FACTS

In this section we shall establish notation and review some known facts about "cohomology with Borel cochains" and its relation to problems concerning group actions on C^*-algebras. If G is a second-countable locally compact group and A a Polish G-

*) Research supported in part by NSF Grants DMS-8400900 and 8120790.

-module (that is, a metrizable topological G-module complete in its two-sided uniformity – see [8], §2), then $H^n(G,A)$ $(n \geq 0)$ will always denote the cohomology groups defined by C.C.Moore in [8]. As noted there, these are the cohomology groups of either of the complexes $\{C^n_{Borel}(G,A), \delta\}$ or $\{\underline{C}^n(G,A), \delta\}$, where

$$C^n_{Borel}(G,A) = \{\text{Borel functions } G^n \to A\},$$

δ is the usual coboundary operator, and $\underline{C}^n(G,A)$ denotes the quotient of $C^n_{Borel}(G,A)$ by the equivalence relation $\sim$, where $f_1 \sim f_2$ if $f_1 = f_2$ almost everywhere (with respect to Haar measure on G^n).

We shall sometimes also refer to the "continuous cohomology" groups $H^n_{cont}(G,A)$, in other words, the cohomology groups of the complex $\{C^n_{cont}(G,A), \delta\}$ of continuous cochains. Continuous cohomology does not in general have good functorial properties, because of the fact that the functor $A \mapsto C^n_{cont}(G,A)$ is only left exact. Thus a short exact sequence of Polish G-modules need not give a long exact sequence in continuous cohomology, unless one considers only short exact sequences which are topologically split (see [4] for a systematic development). Nevertheless, continuous cohomology is sometimes more computable than Borel cochain cohomology. This makes it appropriate to study the relationship between the two theories. One always has $H^n(G,A) = H^n_{cont}(G,A)$ for $n \leq 1$ (this is essentially [8], Theorem 3 and corollaries); $H^2(G,A)$ classifies topological group extensions of G by A ([8], Theorem 10), whereas $H^2_{cont}(G,A)$ classifies extensions which split topologically.

If X is a paracompact topological space and G a topological group, $H^n(X,\underline{G})$ will denote the sheaf cohomology, which may be computed by the Čech process, of X with coefficients in the sheaf $\underline{G}$ of germs of continuous functions with values in G. This makes sense for all n if G is abelian, and only for n = 0 or 1 if G is non-commutative. Recall that via the correspondence between bundles and systems of transition functions, $H^1(X,\underline{G})$ classifies equivalence classes of locally trivial principal G-bundles over X. If G is discrete, $\underline{G}$ is the constant sheaf G and we have usual Čech cohomology $H^n(X,G)$.

Now suppose A is a separable C^*-algebra, not necessarily unital. We denote by M(A) the multiplier algebra of A (see [11], § 3.12), which is separable and metrizable in the *strict* topology defined by the semi-norms

$$x \mapsto \|xa\| + \|ax\|, \quad a \in A.$$

Aut(A) will denote the group of $*$-automorphisms of A; this is a Polish group in the topology of pointwise convergence. Inn(A) denotes the *inner* automorphisms, i.e., those of the form

$$a \mapsto (\mathrm{Ad}\, u)(a) = uau^*,$$

where $u \in U(M(A))$, the unitary group of $M(A)$. Note that the map

$$\mathrm{Ad} : U(M(A)) \to \mathrm{Aut}(A)$$

is a continuous homomorphism for the appropriate Polish topologies, with image Inn(A) and kernel

$$Z(U(M(A))) = U(Z(M(A))) \simeq C(\mathrm{Prim}\, A, \mathbf{T})$$

(by the Dauns-Hofmann Theorem, [11], Corollary 4.4.8). Here $\mathbf{T}$ is the circle group and $C(\mathrm{Prim}\, A, \mathbf{T})$ is the group of continuous functions from the primitive ideal space of A (or what is the same, its maximal Hausdorff quotient X) into $\mathbf{T}$. The strict topology on U(M(A)) restricts to the compact-open topology on $C(X,\mathbf{T})$. In general, Inn(A) is not closed in Aut(A). However, the following results from [13] will be useful.

THEOREM 1.1. ([13], §0). a) *If* A *is any separable* C^**-algebra,* Inn(A) *is a Borel subset of* Aut(A), *and any continuous homomorphism or crossed homomorphism*

$$\phi : G \to \mathrm{Aut}(A)$$

from a Polish group G *which takes its values in* Inn(A) *is automatically continuous for the Polish topology on* Inn(A) *coming from its identification with the quotient* $U(M(A))/C(\mathrm{Prim}\, A,\mathbf{T})$.

b) *If* X *is a second-countable locally compact space with* $H^2(X,\mathbf{Z})$ *countable (in particular, if* X *is compact or has a compact deformation retract) and if* A *is a separable continuous-trace algebra with spectrum* X, *then* Inn(A) *is closed in* Aut(A). *Furthermore,* Inn(A) *is open in the subgroup* $\mathrm{Aut}_{C_0(X)}A$ *of automorphisms of* A *that leave* X *pointwise fixed.* ■

Now let G be a second-countable locally compact group, A a separable C^*-algebra. An *action* of G on A (sometimes called a C^**-dynamical system* or *locally compact automorphism group*) means a continuous homomorphism $\alpha: G \to \mathrm{Aut}(A)$, where Aut(A) as usual has the topology of pointwise convergence. From such an action one can construct a crossed product $A \rtimes_\alpha G$, and when G is abelian, a dual action of $\hat{G}$. The action α is said to be *inner* or *unitary* if there is a (continuous) homomorphism $u : G \to U(M(A))$ such that $\alpha_g = \mathrm{Ad}\, u_g$ for all $g \in G$. In this case $A \rtimes_\alpha G \simeq A \otimes C^*(G)$ (the crossed product for the trivial action of G), and the isomorphism is equivariant for the dual action of $\hat{G}$ in case G is abelian. More generally, two actions α and β of G are said to be *exterior equivalent* if $\alpha_g\beta_g^{-1} = \mathrm{Ad}\, u_g$ for some 1-cocycle $u : G \to U(M(A))$ (with respect to the action of G on A given by β), or equivalently if α and β may be realized as opposite "corners" of an action of G on $M_2(A)$ ([11], Lemma 8.11.2). Thus α is unitary if and only if α is exterior equivalent to the trivial action. A weaker equivalence relation, which still implies isomorphism of $A \rtimes_\alpha G$ with $A \rtimes_\beta G$ (equivariant for $\hat{G}$ when G is abelian), is exterior equivalence composed with conjugacy, i.e., exterior equivalence of α with $g \mapsto \gamma\beta_g\gamma^{-1}$, for some $\gamma \in \mathrm{Aut}\, A$. Further refinements of this will be studied in § 4 below.

REMARK 1.2. Given an action α of G on A, there are two obstructions to its being unitary. First of all, α(G) must be contained in Inn(A), and in particular must act trivially on $\hat{A}$. (Note that if A has continuous trace and $H^2(\hat{A}, \mathbf{Z})$ is countable, then this is automatic by Theorem 1.1 (b) if G is connected, once $\alpha(G) \subseteq \mathrm{Aut}_{C_0(\hat{A})}A$.) Then by Theorem 1.1 (a), α may be viewed as a homomorphism into $U(M(A))/C(\mathrm{Prim}\, A, \mathbf{T})$, and one must be able to lift this homomorphism to some

$$u : G \to U(M(A)) .$$

By the cohomology exact sequence for

$$1 \to C(\mathrm{Prim}\, A, \mathbf{T}) \to U(M(A)) \to U(M(A))/C(\mathrm{Prim}\, A, \mathbf{T}) \to 1,$$

which is valid as far as $H^2(G, C(\mathrm{Prim}\, A, \mathbf{T}))$ despite non-commutativity of $U(M(A))$, the obstruction to this is precisely a class in $H^2(G, C(\mathrm{Prim}\, A, \mathbf{T}))$ (where the G-module has trivial G-action). Similarly, given two actions α and β of G on A, there are two obstructions to their being exterior equivalent. First, we must have $\alpha_g\beta_g^{-1} \in \mathrm{Inn}(A)$ for all $g \in G$, and secondly, once this is the case, there is an additional obstruction in $H^2(G, C(\mathrm{Prim}\, A, \mathbf{T}))$ (this time the G-action on the module is non-trivial and comes from the action of either of α or β on Prim A). In the special case $A = \mathcal{K}$, the algebra of compact operators on a separable infinite-dimensional Hilbert space, every automorphism of A is inner, and an action α of G on A just amounts to a projective unitary representation of G. In this case the obstruction class of α in $H^2(G, \mathbf{T})$ is the usual Mackey obstruction to lifting a projective representation to an ordinary representation, and the obstruction to exterior equivalence of α and β is the difference between their Mackey obstructions. ■

Remark 1.2 was implicitly used in [5], where we noted that if α and β are two actions of a connected group G on a separable C*-algebra A, all of whose derivations are inner (this applies both to simple C*-algebras and to continuous-trace algebras), such that $\|\alpha_g - \beta_g\| < 2$ for all g in a neighborhood of the identity element e in G, then

$\alpha_g \beta_g^{-1} \in \mathrm{Inn}(A)$ for all g and the obstruction to exterior equivalence of α and β lies in $H^2(G,C(\mathrm{Prim}\, A,\mathbf{T}))$. This is somewhat relevant to certain problems of stability in quantum field theory and quantum statistical mechanics; for a discussion of physical implications, see the references quoted in [5].

A final notion we shall need is that of a *locally unitary* group action. Given a type I C^*-algebra A, an action α of G on A is said to be *pointwise unitary* if for each $\pi \in \hat{A}$ there exists a strongly continuous unitary representation u of G on H_π such that

$$\pi(\alpha_g(a)) = u_g \pi(a) u_g^* \qquad \text{for all } g \in G,\ a \in A .$$

Equivalently, α fixes $\hat{A}$ pointwise, and all Mackey obstructions in $H^2(G,\mathbf{T})$ vanish. Then α is said to be *locally unitary* if there is a covering $\{U_i\}$ of $\hat{A}$ by open sets (locally closed sets will also do if their interiors cover $\hat{A}$) and if the restrictions of α to the corresponding ideals (or subquotients) of A are unitary. Locally unitary actions were studied in great detail in [12] (using a slightly stronger definition: the implementing map u_i : $: G \to U(M(A_i))$, where A_i is the subquotient of A with spectrum U_i, was supposed to come from a map $G \to M(A)$. This was never used in an essential way, and anyway we shall only be interested in the case where A is separable and $\hat{A}$ Hausdorff, in which case the two definitions are equivalent. The reason is that then we may refine our original covering to a covering by compact, hence *closed* sets, so that the A_i's are all *quotients* of A. Then the natural maps $M(A) \to M(A_i)$ are surjective ([11], Proposition 3.12.10) continuous maps of Fréchet spaces, and so have continuous sections by a selection theorem of E.Michael ([6], Corollary 7.3)).

2. COHOMOLOGY WITH COEFFICIENTS IN A TRIVIAL MODULE AND OBSTRUCTIONS TO AN ACTION BEING UNITARY

The elegant description of locally unitary group actions given in [12] makes it natural to ask when a pointwise unitary action is locally unitary. We shall give a substantial improvement of [12], Proposition 1.1, showing that this is automatic under very mild conditions. The tool needed is a generalization of Theorem 2.6 of [5], which in view of Example 1.2 of [12] is best possible if G is abelian.

THEOREM 2.1. *Let* X *be a second-countable locally compact space and let* G *be a second-countable locally compact group with* $H^2(G,\mathbf{T})$ *Hausdorff and with the abelianization* $G_{ab} = G/[G,G]^-$ *compactly generated. (It suffices, but is not necessary, for* G *to satisfy one of the following:* a) G *is abelian and compactly generated, or* b) $G = [G,G]$, *or* c) G_{ab} *compactly generated and* $H^2(G,\mathbf{T})$ *countable, or* d) G *a connect-*

-ed, simply connected Lie group.) Let $C(X,\mathbf{T})$ *be given the compact-open topology and the trivial action of* G, *and let* $[\alpha] \in H^2(G,C(X,\mathbf{T}))$ *be pointwise trivial. (In other words, for each* $x \in X$, *we assume that* $(e_x)_*[\alpha] = 0$ *in* $H^2(G,\mathbf{T})$, *where* $e_x : C(X,\mathbf{T}) \to \mathbf{T}$ *is evaluation at* x.*) Then* $[\alpha]$ *is locally trivial,* i.e., *there is an open covering* $\{V_i\}$ *of* X *such that the image of* $[\alpha]$ *is zero in each* $H^2(G,C(V_i,\mathbf{T}))$.

PROOF. Let $\alpha : G \times G \to C(X,\mathbf{T})$ be a Borel 2-cocycle representing the class $[\alpha]$. Proceeding as in the proof of [5], Theorem 2.6, we may view α as a continuous map

$$f_\alpha : X \to \underline{B}^2(G,\mathbf{T}) \simeq \underline{C}^1(G,\mathbf{T})/\mathrm{Hom}(G,\mathbf{T}).$$

Now *a priori*, f_α is continuous (using [8], Proposition 6) only for the relative topology on $\underline{B}^2(G,\mathbf{T})$ as a subset of $\underline{C}^2(G,\mathbf{T})$, but this coincides with the quotient topology if $H^2(G,\mathbf{T})$ is Hausdorff. So to prove the theorem, it is enough to show that the quotient map

$$q : \underline{C}^1(G,\mathbf{T}) \to \underline{C}^1(G,\mathbf{T})/\mathrm{Hom}(G,\mathbf{T})$$

is locally trivial (topologically), for given U open in X and sufficiently small, it will follow that $f_\alpha|U$ can be lifted to a continuous map $U \to \underline{C}^1(G,\mathbf{T})$ which will provide (for reasons we will discuss below) an element of $\underline{C}^1(G,C(U,\mathbf{T}))$ of which $\alpha|U$ is the coboundary. By the Palais local cross-section theorem ([10], §4.1), local triviality of q is automatic provided $\mathrm{Hom}(G,\mathbf{T})$ is a Lie group. But $\mathrm{Hom}(G,\mathbf{T}) \simeq \mathrm{Hom}(G_{ab},\mathbf{T})$ is just the Pontryagin dual of G_{ab}, which will be a Lie group if and only if G_{ab} is compactly generated.

There is a point still to be settled, which is to see why a continuous map $\bar{\psi} : U \to \underline{C}^1(G,\mathbf{T})$ with coboundary $f_\alpha|U \in C(U,\underline{C}^2(G,\mathbf{T}))$ gives an element of $\underline{C}^1(G,C(U,\mathbf{T}))$ with coboundary $\alpha|U \in \underline{C}^2(G,C(U,\mathbf{T}))$. (There is no obvious reason why $\underline{C}^n(G,C(U,\mathbf{T}))$ and $C(U,\underline{C}^n(G,\mathbf{T}))$ should be isomorphic, though the first includes in the second.) The point is that by [8], $\{C_{\mathrm{Borel}}(G,\cdot)\}$ and $\{\underline{C}(G,\cdot)\}$ give the same cohomology groups, so we may represent $\bar{\psi}$ by a Borel function $\psi : G \times U \to \mathbf{T}$ with $\delta(\psi) = \alpha|U$ (as a Borel function on $G \times G \times U$) everywhere, not just almost everywhere. The problem is then to show that if $x_n \to x$ in U, then $\psi(g,x_n) \to (g,x)$ for all g. By definition of the topology of $C(U,\underline{C}^1(G,\mathbf{T}))$, we know that $\psi(\cdot,x_n) \to \psi(\cdot,x)$ in measure (as functions on G). Furthermore, since $\delta(\psi) = \alpha|U$ is a Borel function from $G \times G$ to $C(U,\mathbf{T})$, we have for $g_1,g_2 \in G$,

$$\delta(\psi)(g_1,g_2,x_n) \to \delta(\psi)(g_1,g_2,x),$$

i.e.,

$$(*) \qquad \psi(g_1,x_n)\psi(g_2,x_n)\psi(g_1g_2,x_n)^{-1} \to \psi(g_1,x)\psi(g_2,x)\psi(g_1g_2,x)^{-1}.$$

Now if there is a $g \in G$ for which $\psi(g,x_n) \not\to \psi(g,x)$, we may pass to a subsequence and assume $|\psi(g,x_n) - \psi(g,x)| \geq \varepsilon > 0$ for all n. By [8], Proposition 6, however, after passage to a further subsequence there is a null set N in G such that $\psi(g_1,x_n) \to \psi(g_1,x)$ for $g_1 \notin N$. Since $\psi(g,x_n) \not\to \psi(g,x)$, it follows from (*) that also $\psi(gg_1,x_n) \not\to \psi(gg_1,x)$ for $g_1 \notin N$. Since $(G\setminus N) \cap g(G\setminus N) \neq \emptyset$, this is a contradiction, and so ψ has the necessary continuity.

Finally, we comment on sufficiency of conditions (a) – (d). If G is abelian, $H^2(G,\mathbf{T})$ is Hausdorff by [9], Theorem 7. If $G = [G,G]$, $H^2(G,\mathbf{T})$ is Hausdorff by [9], Theorem 13. If $H^2(G,\mathbf{T})$ is countable, it is automatically Hausdorff by [9], Proposition 6. And if G is a connected, simply connected Lie group, $H^2(G,\mathbf{T})$ is a (Hausdorff) vector group by [7], Theorem A and subsequent remarks. ■

COROLLARY 2.2. *Let* G *be as in Theorem* 2.1. *Then any pointwise unitary action* α *of* G *on a separable continuous-trace algebra* A *is automatically locally unitary.*

PROOF. Since $\hat{A}$ is locally compact and the problem is local, we may assume without loss of generality that $\hat{A}$ is compact. Then $H^2(\hat{A},\mathbf{Z})$ is countable and Inn(A) is open in $\mathrm{Aut}_{C_0(\hat{A})}A$ by Theorem 1.1; in fact

$$\mathrm{Aut}_{C_0(\hat{A})}A/\mathrm{Inn}\,A \hookrightarrow H^2(\hat{A},\mathbf{Z})$$

topologically by [12], Corollary 3.12 together with [13], §0. Since α is pointwise unitary, in particular $\alpha(G) \subseteq \mathrm{Aut}_{C_0(\hat{A})}A$, and by passage to the quotient we get a continuous map

$$\bar{\alpha}: G \to H^2(\hat{A},\mathbf{Z}).$$

This map must factor through G_{ab}, which by assumption is compactly generated, so since $H^2(\hat{A},\mathbf{Z})$ is countable and discrete, the image of $\bar{\alpha}$ is *finitely* generated. Choose a finite set of generators for $\bar{\alpha}(G)$ and an open covering $\{U_i\}$ of $\hat{A}$ which simultaneously trivializes all these cocycles. (Just choose a covering for each generator and take appropriate intersections.) Then $\bar{\alpha}$ is trivial over each U_i, i.e., α takes G into $\mathrm{Inn}(A|U_i)$ for each i. By Remark 1.2, the only obstruction to α being unitary over each U_i is a class $[\alpha|U_i]$ in $H^2(G,C(U_i,\mathbf{T}))$, and this class is pointwise trivial since α was assumed pointwise unitary. The conclusion now follows from Theorem 2.1. ■

The most interesting case of the above occurs when G is abelian and compactly generated, in which case the above proof suggests a link between the Moore cohomology group $H^2(G,C(X,\mathbf{T}))$ and the Phillips-Raeburn classification of locally unitary G-actions on continuous-trace algebras with spectrum X in terms of locally trivial $\hat{G}$-bundles. The key to a more precise analysis is the appearance in the proof of the quotient map q. The significance of this may be seen in the following results, due essentially to D.Wigner, which will also be used in § 3 below.

LEMMA 2.3. *Let* (Y,μ) *be a standard measure space without atoms and let* A *be a Polish group. Then* $U(Y,A)$ *as defined on* p.5 *of* [8], *that is, the set of equivalence classes (modulo agreement* μ-a.e.*) of* A*-valued Borel functions on* Y*, with the topology of convergence in measure, is contractible.*

PROOF. We may assume Y is the unit interval [0,1] and μ is Lebesgue measure. Then as pointed out on pp.86-87 of [17],

$$h_t(f)(x) = \begin{cases} f(x) & \text{if } x \geq t, \\ e, \quad \text{the identity of } A, & \text{if } x < t, \end{cases}$$

gives an explicit homotopy from the identity map h_0 on $U([0,1],A)$ to the map h_1 collapsing $U([0,1],A)$ to a point. ■

PROPOSITION 2.4. a) *If* G *is a second-countable locally compact abelian group which is compactly generated and non-discrete, then*

$$q : \underline{C}^1(G,\mathbf{T}) \to \underline{C}^1(G,\mathbf{T})/\mathrm{Hom}(G,\mathbf{T}) \simeq \underline{B}^2(G,\mathbf{T})$$

is a universal $\hat{G}$*-bundle, and* $\underline{B}^2(G,\mathbf{T})$ *is a classifying space for* $\hat{G}$.

b) *Let* G *be any Lie group with countably many components, or more generally, any second-countable Banach Lie group (e.g., the unitary group of a separable unital* C^**-algebra). Then*

$$U([0,1],G) \to U([0,1],G)/G$$

is a universal G*-bundle and* $U([0,1],G)/G$ *is a classifying space for* G.

c) *If* A *is a locally arcwise connected Polish abelian group, then so is* $BA = U([0,1],A)/A$, *and* BA *is a weak classifying space for* A*. (This means* $H^1(X,A) \simeq [X,BA]$ *for* X *a* CW*-complex, where* $[X,BA]$ *denotes homotopy classes of maps from* X *into* BA*. Equivalently,* BA *has the weak homotopy type of a classifying space for* A*.)*

PROOF. a) Since G is non-discrete, G is non-atomic with respect to Haar measure. But as a space $\underline{C}^1(G,\mathbf{T})$ is the same as $U(G,\mathbf{T})$, which is contractible by Lemma 2.3. The map q is a locally trivial principal $\hat{G}$-bundle by [10], § 4.1, and of course $\underline{B}^2(G,\mathbf{T})$ is metrizable, hence paracompact. Thus all conditions for a universal $\hat{G}$-bundle are satisfied ([3], Theorem 7.5).

b) works exactly the same way, except that if G is not a Lie group, local triviality of the bundle can still be proved by [6], Corollary 7.3. (In case G is the unitary group of a unital separable C^*-algebra, G is locally isomorphic to the real Banach space of skew-adjoint elements, via the exponential map.)

c) is similar except that the hypothesis is not strong enough to guarantee that the quotient map to BA is locally trivial. It is, however, a Serre fibration by [17], Proposition 3. As is well known, this is sufficient to make BA a weak classifying space, but for completeness we give the argument here. Let $EA = U([0,1],A)$. Then

$$0 \to A \to EA \to BA \to 0$$

is a short exact sequence of Polish abelian groups. If X is a CW-complex (it's enough to consider the case of a finite complex), we obtain an exact sequence

$$0 \to \underline{A} \to \underline{EA} \to \underline{BA}$$

of sheaves over X. The map $\underline{EA} \to \underline{BA}$ is surjective as a map of sheaves since $EA \to BA$ is a Serre fibration so that a continuous map $X \to BA$ can be lifted (by the HLP, homotopy lifting property) in a neighborhood of any point in X. Thus we have a long exact sequence in sheaf cohomology

$$H^0(X,\underline{EA}) \to H^0(X,\underline{BA}) \to H^1(X,\underline{A}) \to H^1(X,\underline{EA}).$$

Here $H^1(X,\underline{EA}) = 0$ since EA is contractible (either by the theory of [3] or else by Lemma 4 of [2]), so $H^1(X,\underline{A})$ is a quotient of

$$H^0(X,\underline{BA}) = C(X,BA).$$

But by contractibility of EA, a continuous map $X \to BA$ with a lifting $X \to EA$ must be null-homotopic, and conversely, any null-homotopic map $X \to BA$ has a lifting by the HLP. Thus

$$H^1(X,A) \cong C(X,BA)/\{\text{null-homotopic maps}\} \cong [X,BA]. \qquad \blacksquare$$

Parts (b) and (c) of the above proposition were only included here for completeness, since they logically belong with Lemma 2.3. However, part (a) can be used immediately to prove the following.

THEOREM 2.5. *Let G be a connected, second-countable, locally compact abelian group, and let X be a second-countable locally compact space with $H^2(X, \mathbf{Z})$ countable (for instance, a compact metric space). Let A be any separable continuous-trace alge-*

bra with spectrum X *and* $\alpha : G \to \mathrm{Aut}_{C_0(X)}A$ *any action of* G *on* A *inducing the trivial action on* X. *Then:*

a) $\alpha(G)$ *consists of inner automorphisms, and* α *is locally unitary if and only if it is pointwise unitary (which is automatic if* G *is compact),*

b) *the pointwise trivial part of* $H^2(G,C(X,\mathbf{T}))$ *is naturally isomorphic to* $H^1(X,\underline{\hat{G}})$, *and*

c) *when* α *is pointwise unitary, the obstruction in* $H^2(G,C(X,\mathbf{T}))$ *to* α *being unitary may be identified under the isomorphism of* (b) *to the Phillips-Raeburn obstruction* $\zeta(\alpha) \in H^1(X,\underline{\hat{G}})$.

PROOF. (a) Since $\alpha(G)$ is a connected subgroup of $\mathrm{Aut}_{C_0(X)}A$, it must lie in Inn(A) by Theorem 1.1 (b). Thus an obstruction $[\alpha] \in H^2(G,C(X,\mathbf{T}))$ is defined by Remark 1.2, and this is the only obstruction to α being unitary. We know α will be pointwise unitary if and only if the Mackey obstructions $(e_x)_*[\alpha] \in H^2(G,\mathbf{T})$, $x \in X$, all vanish. Thus it is useful to note that by [7], Proposition 2.1 and Theorem 2.1, $H^2(G,\mathbf{T}) = 0$ if G is compact, since then G is an inverse limit of tori. If α is pointwise unitary, it is then locally unitary by Corollary 2.2.

(b) The proof of Theorem 2.1 shows that if α is pointwise unitary, the obstruction $[\alpha]$ may be identified with the obstruction to finding a lifting in the diagram

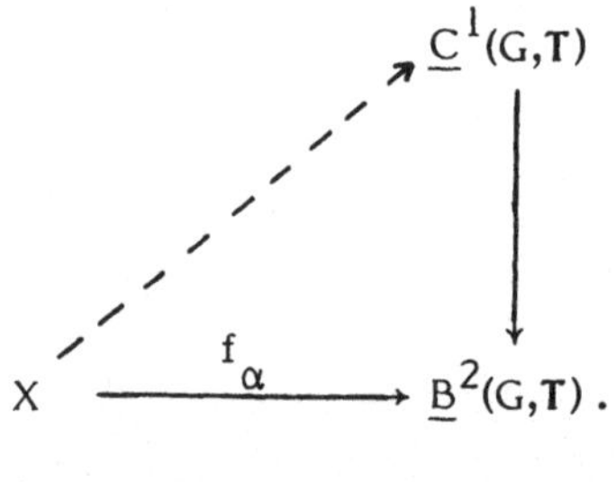

By Proposition 2.4 (a), this is the same as determining the homotopy class of f_α in $[X,B\hat{G}] \simeq H^1(X,\underline{\hat{G}})$. In this way one obtains an injection from the pointwise trivial part of $H^2(G,C(X,\mathbf{T}))$ into $H^1(X,\underline{\hat{G}})$, and the map is obviously surjective since any continuous $f : X \to \underline{B}^2(G,\mathbf{T})$ gives rise to a pointwise trivial element of $\underline{Z}^2(G,C(X,\mathbf{T}))$. (This part of the argument doesn't use connectedness of G except to imply G is non-discrete.)

(c) Note by the structure theory of locally compact abelian groups that if K is the maximal compact subgroup of G, then $G \simeq K \times \mathbf{R}^n$ for some n, so that actually $H^1(X,\underline{\hat{G}}) \simeq H^1(X,\hat{K})$. Thus it is only the compact part of G that really matters, although we shall not use this.

Let's compare $[\alpha] \in H^2(G,C(X,\mathbf{T}))$ with $\zeta(\alpha) \in H^1(X,\underline{\hat{G}})$. The latter is computed by covering X by sets X_j over which α is unitary, so that one has maps

$$v^j : G \to U(M(A\,|\,X_j))$$

implementing α locally. On the overlap sets $X_{jk} = X_j \cap X_k$, v^j and v^k must implement the same automorphism group, hence they differ by some

$$v^{jk} \in \mathrm{Hom}(G,C(X_{jk},\mathbf{T})) \simeq C(X_{jk},\hat{G}).$$

Then $\{v^{jk}\}$ is a Čech cocycle in $Z^1(X,\underline{\hat{G}})$ whose cohomology class is the obstruction $\zeta(\alpha)$ to patching the v^j's to a global homomorphism $G \to U(M(A))$. But the X_j's also trivialize $[\alpha] \in H^2(G,C(X,\mathbf{T}))$, which under the isomorphism of (b) goes to the "patching data" for local liftings $X_j \to \underline{C}^1(G,\mathbf{T})$ of f_α. This is clearly the same cohomology class obtained from the $\{v^{jk}\}$. ∎

3. CONTINUOUS VS. BOREL COHOMOLOGY AND INDUCTION

In this section we first show that $H^n(G,A)$ often coincides with $H^n_{cont}(G,A)$ if G is a vector group. Then we discuss "continuous induction" and the expected form of a "Shapiro's Lemma" for continuously induced modules. This is applied to the computation of the cohomology groups of certain **R**-modules that sometimes arise in applications. We also mention some results on cohomology of modules for compact Lie groups, complementing some results in [5] and [13], and explain how to generalize to general Lie groups.

Our first result is an interesting curiosity that is perhaps known to topologists, but which we haven't seen in the literature. The hypothesis on A could be weakened provided BA is an infinite loop space.

PROPOSITION 3.1. *Let* X *be a* CW *complex,* A *a locally arcwise connected Polish abelian group. Then for* $q \geq 1$, *there is a natural isomorphism*

$$H^q(X,\underline{A}) \simeq [X,B^qA],$$

where the iterated (weak) classifying spaces BA, $B^2A = B(BA), \ldots, B^qA = B(B^{q-1}A)$ *may be constructed to satisfy the same conditions as* A *by the procedure of Proposition* 2.4 (c). *In particular,* $H^q(X,\underline{A}) = 0$ *for all* $q \geq 1$ *provided either* X *or* A *is contractible.*

PROOF. Let us show, using Proposition 2.4 (c), that for all $q > 1$,

$$H^q(X,\underline{A}) \simeq H^{q-1}(X,\underline{BA}).$$

Since we already know that $H^1(X,\underline{A}) = [X,BA]$ and that BA can be constructed to have all the same properties as A, the result then follows by iteration. But recall from the proof of Proposition 2.4 (c) that we have a short exact sequence of sheaves over X

$$0 \to \underline{A} \to \underline{EA} \to \underline{BA} \to 0.$$

Furthermore, since EA is contractible, the sheaf $\underline{EA}$ is acyclic by [2], Lemma 4. So we get the result immediately from the long exact sequence

$$0 = H^{q-1}(X,\underline{EA}) \to H^{q-1}(X,\underline{BA}) \to H^q(X,\underline{A}) \to H^q(X,\underline{EA}) = 0.$$ ∎

PROPOSITION 3.2. *Suppose* G *is a vector group (i.e.,* $\mathbf{R}^n$ *for some* n*) and* A *is a Polish* G*-module with "property F" of* [17]. *(It suffices for* A *to be locally arcwise connected.) Then the natural map* $C^*_{cont}(G,A) \to C^*_{Borel}(G,A)$ *induces isomorphisms*

$$H^n_{cont}(G,A) \to H^n(G,A)$$

for all n. *In particular, every extension of* G *by* A *splits topologically.*

PROOF. By [17], Theorem 2, there is a natural spectral sequence converging to $H^*(G,A)$ with

$$E_1^{p,q} \simeq H^q_{sheaf}(G^p,\underline{A})$$

and

$$d_1 : E_1^{p,0} = C^p_{cont}(G,A) \to E_1^{p+1,0} = C^{p+1}_{cont}(G,A)$$

the usual coboundary operator δ. (Caution to the reader: there is a misprint on p.91, 1.6 of [17]; p and q are reversed there.) Since G, hence G^p is contractible, we have $E_1^{p,q} = 0$ for $q > 0$ by Proposition 3.1. Thus the spectral sequence degenerates at E_2 and gives the desired result. Of course, the statement about topological splitting of extensions, which may be read off from the equality $H^2_{cont}(G,A) \simeq H^2(G,A)$, also follows directly from the HLP built into the definition of property F, except for the problem that the definition deals only with abelian extensions. ∎

Now if G is a locally compact group, among the most commonly encountered G--modules are modules induced from a closed subgroup. When the induction process is the "Borel induction" of [8], p.14, the cohomology of the induced module is given by Moore's version of "Shapiro's Lemma", [8], Theorem 6. Often, however, one is interested in *continuous* induction, so we outline here the analogous theory.

Suppose H is a closed subgroup of a second-countable locally compact group G, and A is a Polish G-module. We define the *continuously induced* G-module to be

$$\mathrm{Ind}_{H\uparrow G}A = \{f \in C(G,A) \mid f(gh) = h^{-1}\cdot f(g) \text{ for all } h \in H\},$$

equipped with the topology of uniform convergence on compacta and with the action of G by left translation. When G/H is discrete, $\mathrm{Ind}_{H\uparrow G}A$ may be identified with Moore's $I_H^G(A)$ (which is defined similarly using measurable functions in place of continuous functions). Usually, however, the two will differ. For instance, if $A = \mathbf{T}$ with trivial H-action, $I_H^G(\mathbf{T})$ is the unitary group (with the weak topology) of the von Neumann algebra $L^\infty(G/H)$, whereas $\mathrm{Ind}_{H\uparrow G}(\mathbf{T}) = C(G/H,\mathbf{T}) = U(M(C_0(G/H)))$, equipped with the strict topology. Thus continuous induction is related to problems in C^*-algebra theory in the same way Borel induction is related to problems in von Neumann algebra theory.

We would like to prove a variant of Shapiro's Lemma for continuous induction; however, it is *not* true in general that $H^n(G,\mathrm{Ind}_{H\uparrow G}A) \simeq H^n(H,A)$. The problem is that the functor $A \mapsto \mathrm{Ind}_{H\uparrow G}A$ (from Polish H-modules to Polish G-modules) is left exact but not usually right exact. For in the simplest case where

$$0 \to A \to B \to C \to 0$$

is a short exact sequence of Polish abelian groups with trivial H-action, exactness of $\mathrm{Ind}_{H\uparrow G}$ would mean that every continuous map $G/H \to C$ can be lifted to a continuous map $G/H \to B$. This will rarely be the case, even if $B \to C$ is a fibration, since G/H may not be contractible.

The correct formalism is suggested by homological algebra. If $\mathcal{A}$, $\mathcal{B}$, and $\mathcal{C}$ are abelian categories with enough injectives and $T : \mathcal{A} \to \mathcal{B}$, $S : \mathcal{B} \to \mathcal{C}$ are left-exact functors, recall that one can define right-derived functors $R^nT : \mathcal{A} \to \mathcal{B}$, $R^nS : \mathcal{B} \to \mathcal{C}$, and $R^n(S\circ T) : \mathcal{A} \to \mathcal{C}$. Under a mild technical condition, one has a "composition of functors" spectral sequence

$$R^pS(R^qT(A)) \Longrightarrow R^{p+q}(S\circ T)(A),$$

and in fact most of the familiar spectral sequences of homological algebra are of this type. Here is a standard example: the Hochschild-Serre spectral sequence in group cohomology. Suppose $\Delta \triangleleft \Gamma$ are abstract groups, and take

$\mathcal{A}$ = category of Γ-modules,

$\mathcal{B}$ = category of Γ/Δ-modules,

$\mathcal{C}$ = category of abelian groups,

$T : \mathcal{A} \to \mathcal{B}$ the functor $A \mapsto A^{\Delta}$, "fixed points under Δ",

$S : \mathcal{B} \to \mathcal{C}$ the functor $B \mapsto B^{\Gamma/\Delta}$, "fixed points under Γ/Δ".

Then $S \circ T : A \mapsto A^{\Gamma}$, $R^nT = H^n(\Delta,\_\_)$, $R^nS = H^n(\Gamma/\Delta,\_\_)$, and the spectral sequence is the usual one

$$H^p(\Gamma/\Delta, H^q(\Delta,A)) \Longrightarrow H^{p+q}(\Gamma,A).$$

In our situation we do not have abelian categories with enough injectives, but for G a Lie group, $\mathcal{A}$ the quasi-abelian S-category of Polish H-modules with property F, $\mathcal{B}$ the category of Polish G-modules with property F, $\mathcal{C}$ the category of abelian groups, $T = \mathrm{Ind}_{H\uparrow G}$, and $S : A \mapsto A^G$, the same formalism suggests a spectral sequence of the sort

$$H^p(G, \mathcal{R}^q\mathrm{Ind}_{H\uparrow G}A) \Longrightarrow H^{p+q}(H,A),$$

which should enable one to compute $H^*(G,\mathrm{Ind}_{H\uparrow G}A)$ in terms of $H^*(H,A)$, provided one can obtain enough information about the "derived functors" of $\mathrm{Ind}_{H\uparrow G}$. (Compare the situation for Zuckerman's "cohomological induction functor" in [15], Theorem 6.2.14.)

For lack of compelling applications, we refrain from trying to work out the theory in this generality, and content ourselves with a few special cases. First we use Proposition 3.2 to prove "Shapiro's Lemma" in the one case one would expect it in the usual form.

PROPOSITION 3.3. *Let* A *be a locally arcwise connected Polish abelian group,* G *a vector group. Then* $H^n(G,\mathrm{Ind}_{1\uparrow G}A) = 0$ *for* $n > 0$. *(Since, clearly,* $H^0(G,\mathrm{Ind}_{1\uparrow G}A) \simeq$ $\simeq \{\text{constant functions } G \to A\} \simeq A$, *this is the same as saying* $H^n(G,\mathrm{Ind}_{1\uparrow G}A) \simeq H^n(1,A)$ *for all* n.*)*

PROOF. Observe that if A is locally arcwise connected and A_0 is the path component of 0 in A, then A_0 is open in A and $C(G,A_0) = \mathrm{Ind}_{1\uparrow G}A_0$ is open in $\mathrm{Ind}_{1\uparrow G}A$. Since G is contractible, every continuous function $G \to A_0$ is homotopic to a constant function, and so $C(G,A_0)$ is path connected. Thus $\mathrm{Ind}_{1\uparrow G}A$ is locally arcwise connected and so has property F by [17], Proposition 3. Applying Proposition 3.2, we obtain $H^n(G,\mathrm{Ind}_{1\uparrow G}A) \simeq H^n_{\mathrm{cont}}(G,\mathrm{Ind}_{1\uparrow G}A)$. The vanishing is now implicit in [17] and much easier than Theorem 4 in [8]: given $f : G^n \to \mathrm{Ind}_{1\uparrow G}A$ a continuous n-cocycle, we define a continuous (n - 1)-cochain with values in Ind A by

$$h(g_1,\ldots,g_{n-1})(x) = f(g_1,\ldots,g_{n-1},g_{n-1}^{-1}\cdots g_j^{-1}\cdots g_1^{-1}x)(x),$$

and a direct calculation with the cocycle identity shows

$$(\delta h)(g_1,\dots,g_n) = \pm f(g_1,\dots,g_n). \qquad \blacksquare$$

We specialize now to the case $H = \mathbf{Z}$, $G = \mathbf{R}$, A a Polish abelian group with trivial H-action, so that $\mathrm{Ind}_{\mathbf{Z}\uparrow\mathbf{R}}A$ is just $C(\mathbf{R}/\mathbf{Z},A)$. Now if A has property F and we have a short exct sequence

$$0 \to A \to B \to C \to 0$$

of Polish groups, this is a Serre fibration and we obtain the long exact homotopy sequence

$$\dots \to \pi_1(B) \to \pi_1(C) \to \pi_0(A) \to \pi_0(B) \to \pi_0(C) \to 0.$$

This means that we should regard $\pi_0(A) = A/A_0$ as $R^1\mathrm{Ind}_{\mathbf{Z}\uparrow\mathbf{R}}A$, since this is what we must add on the right to continue the exact sequence

$$0 \to \mathrm{Ind}_{\mathbf{Z}\uparrow\mathbf{R}}A \to \mathrm{Ind}_{\mathbf{Z}\uparrow\mathbf{R}}B \to \mathrm{Ind}_{\mathbf{Z}\uparrow\mathbf{R}}C.$$

In other words, we should expect a degenerate spectral sequence

$$H^p(\mathbf{R},R^q\mathrm{Ind}_{\mathbf{Z}\uparrow\mathbf{R}}A) \Longrightarrow H^*(\mathbf{Z},A),$$

with $E_2^{p,0} = H^p(\mathbf{R},\mathrm{Ind}_{\mathbf{Z}\uparrow\mathbf{R}}A)$, $E_2^{p,1} = H^p(\mathbf{R},A/A_0)$, and $E_2^{p,q} = 0$ for $q > 1$. We proceed to establish this by mimicing the usual process of resolving A, applying $\mathrm{Ind}_{\mathbf{Z}\uparrow\mathbf{R}}$, and resolving again to get a double complex. Actually carrying out the process and indentifying explicitly the modules obtained leads to the following lemma. We make the technical hypothesis that the path component of 0 in A should be closed, since it is not clear if this is always the case in a Polish group.

LEMMA 3.4. *Let* A *be a Polish abelian group, viewed as a* $\mathbf{Z}$*-module with trivial action. Assume the path component* A_0 *of* 0 *in* A *is closed, hence Polish in the relative topology. Then the following sequence of Polish* $\mathbf{R}$*-modules is exact (topologically)*

$$0 \to \mathrm{Ind}_{\mathbf{Z}\uparrow\mathbf{R}}A \to \mathrm{Ind}_{1\uparrow\mathbf{R}}A \xrightarrow{\beta} \mathrm{Ind}_{1\uparrow\mathbf{R}}A_0 \to 0,$$

where β *is given by* $(\beta f)(s) = f(s+1) - f(s)$, *for* $f \in C(\mathbf{R},A)$.

PROOF. Identifying periodic functions $\mathbf{R} \to A$ with a subset of the continuous functions gives an obvious embedding of $\mathrm{Ind}_{\mathbf{Z}\uparrow\mathbf{R}}A$ as a closed submodule of $\mathrm{Ind}_{1\uparrow\mathbf{R}}A$. Furthermore, the indicated formula for β gives an $\mathbf{R}$-module map from $\mathrm{Ind}_{1\uparrow\mathbf{R}}A$ to itself whose kernel consists precisely of functions $f : \mathbf{R} \to A$ satisfying $f(s+1) = f(s)$ for all s, i.e., $f \in C(\mathbf{R}/\mathbf{Z}, A) = \mathrm{Ind}_{\mathbf{Z}\uparrow\mathbf{R}}A$. Thus it's enough to show that the image of β is precisely $\mathrm{Ind}_{1\uparrow\mathbf{R}}A_0$.

First of all note that if $f : \mathbf{R} \to A$ is continuous, its image must lie in a single path-component of A; hence the difference of two values of f lies in A_0. Thus β maps *into* $\mathrm{Ind}_{1\uparrow\mathbf{R}}A_0$. To show β maps *onto* this module, choose any continuous $\phi : \mathbf{R} \to A_0$. We shall construct a continuous $f : \mathbf{R} \to A$ with $\phi = \beta(f)$. To do this, we begin by setting $f(0) = 0$, $f(1) = \phi(0)$. Since $\phi(0) \in A_0$, there exists a continuous path from $f(0)$ to $f(1)$; choose one and let it define $f(s)$ for $0 \le s \le 1$. It's then easy to see that the functional equation

$$f(s+1) - f(s) = \phi(s)$$

has a unique continuous solution defined for all real s and extending f as already defined on $[0,1]$. This shows β maps onto $\mathrm{Ind}_{1\uparrow\mathbf{R}}A_0$, and of course β is continuous. Since all groups involved are Polish, the open mapping theorem says β is open, i.e., the exact sequence is topological. ■

THEOREM 3.5. *Let* A *be a Polish abelian group viewed as a* $\mathbf{Z}$*-module with trivial action, and let* A_0 *be the path-component of the identity in* A. *There are natural isomorphisms*

$$H^n(\mathbf{R}, \mathrm{Ind}_{\mathbf{Z}\uparrow\mathbf{R}}A) \simeq \begin{cases} A, & \text{if } n = 0, \\ A_0, & \text{if } n = 1 \text{ and } A_0 \text{ is closed in } A, \\ 0, & \text{if } n > 1 \text{ and } A_0 \text{ is open in } A . \end{cases}$$

PROOF. Consider the long exact cohomology sequence coming from the short exact sequence of Lemma 3.4. Recall that $H^1(\mathbf{R}, \mathrm{Ind}_{1\uparrow\mathbf{R}}A) = H^1_{\mathrm{cont}}(\mathbf{R}, \mathrm{Ind}_{1\uparrow\mathbf{R}}A) = 0$ (by Proposition 3.3). Thus we have the exact sequence

$$0 \to H^0(\mathbf{R}, \mathrm{Ind}_{\mathbf{Z}\uparrow\mathbf{R}}A) \to A \xrightarrow{\beta_*} A_0 \to H^1(\mathbf{R}, \mathrm{Ind}_{\mathbf{Z}\uparrow\mathbf{R}}A) \to 0.$$

It is clear that $H^0(\mathbf{R}, \mathrm{Ind}_{\mathbf{Z}\uparrow\mathbf{R}}A) \to A$ is an isomorphism and that $\beta_* = 0$, which gives us the result $H^1(\mathbf{R}, \mathrm{Ind}_{\mathbf{Z}\uparrow\mathbf{R}}A) \simeq A_0$ (when A_0 is closed in A).

When also A_0 is open in A, then A and A_0 are locally path-connected and we may apply Proposition 3.3 to conclude that all higher cohomology groups of $\mathrm{Ind}_{1\uparrow\mathbf{R}}A$ and of $\mathrm{Ind}_{1\uparrow\mathbf{R}}A_0$ vanish. Plugging this back into the exact cohomology sequence coming from Lemma 3.4, we conclude that $H^n(\mathbf{R}, \mathrm{Ind}_{\mathbf{Z}\uparrow\mathbf{R}}A) = 0$ for $n \ge 2$. ■

COROLLARY 3.6. *Let* A *be a locally path-connected Polish abelian group, with* A_0 *the path-component of the identity in* A. *Then there are natural isomorphisms*

$$H^n(\mathbf{T},\mathrm{Ind}_{1\uparrow\mathbf{T}}A) \simeq \begin{cases} A & \text{if } n = 0, \\ A/A_0 & \text{if } n = 2,4,6,\ldots, \\ 0 & \text{if } n \text{ odd.} \end{cases}$$

PROOF. If we view $\mathrm{Ind}_{1\uparrow\mathbf{T}}A$ as the $\mathbf{R}$-module $\mathrm{Ind}_{\mathbf{Z}\uparrow\mathbf{R}}A$, then $\mathbf{Z}$ acts trivially and so

$$H^q(\mathbf{Z},\mathrm{Ind}_{\mathbf{Z}\uparrow\mathbf{R}}A) \simeq \begin{cases} \mathrm{Ind}_{1\uparrow\mathbf{T}}A & \text{if } q = 0 \text{ or } 1, \\ 0 & \text{if } q > 1. \end{cases}$$

Since these groups are always Hausdorff, we may apply the spectral sequence of [8], Theorem 9, p.29. We obtain a spectral sequence with E_2-terms

$$E_2^{p,q} = \begin{cases} H^p(\mathbf{T},\mathrm{Ind}_{1\uparrow\mathbf{T}}A) & \text{if } q = 0 \text{ or } 1, \\ 0 & \text{if } q > 1 \end{cases}$$

converging to $H^{p+q}(\mathbf{R},\mathrm{Ind}_{\mathbf{Z}\uparrow\mathbf{R}}A)$. The only differential which is possibly nonzero is

$$d_2 : E_2^{p,1} \to E_2^{p+2,0}.$$

But by Theorem 3.5, we must have $E_\infty^{0,0} = A$, $E_\infty^{p,q} = 0$ for $p > 1$ or $p = 1$, $q = 1$. Finally, we must have an exact sequence

$$0 \to E_\infty^{1,0} \to A_0 \to E_\infty^{0,1} \to 0.$$

But

$$H^1(\mathbf{T},\mathrm{Ind}_{1\uparrow\mathbf{T}}A) \simeq H^1_{\mathrm{cont}}(\mathbf{T},\mathrm{Ind}_{1\uparrow\mathbf{T}}A) \simeq 0,$$

so $E_\infty^{1,0} = 0$ and $E_\infty^{0,1} \simeq A_0$. This gives the exact sequence

$$0 \to A_0 \to E_2^{0,1} \simeq A \xrightarrow{d_2} E_2^{2,0} \to 0,$$

so $E_2^{2,0} = H^2(\mathbf{T},\mathrm{Ind}_{1\uparrow\mathbf{T}}A) \simeq A/A_0$. Since

$$d_2 : E_2^{2,1} \to E_2^{4,0}$$

must be an isomorphism to give $E_\infty^{2,1} = 0$ and $E_\infty^{4,0} = 0$, we have

$$E_2^{4,0} = H^4(\mathbf{T},\mathrm{Ind}_{1\uparrow\mathbf{T}}A) \simeq E_2^{2,1} \simeq A/A_0.$$

Continuing this way by induction we obtain the result. ∎

REMARK 3.7. Recall that if

$$0 \to A \to B \to C \to 0$$

is a Serre fibration, then the sequence

$$\pi_1(A) \to \pi_1(B) \to \pi_1(C) \to \pi_0(A) \to \pi_0(B) \to \pi_0(C) \to 0$$

is exact. This suggests that for A locally path connected, one should have

$$\begin{cases} R^1 \mathrm{Ind}_{1 \uparrow \mathbf{T}} A = A/A_0 \ (= \pi_0(A)), \\ R^q \mathrm{Ind}_{1 \uparrow \mathbf{T}} A = 0 \ \text{ for } q > 1. \end{cases}$$

Then one should have the "composition of functors" spectral sequence

$$H^p(\mathbf{T}, R^q \mathrm{Ind}_{1 \uparrow \mathbf{T}} A) \Longrightarrow H^{p+q}(1, A),$$

and once again one could obtain the same result as before for $E_2^{p,0} = H^p(\mathbf{T}, \mathrm{Ind}_{1 \uparrow \mathbf{T}} A)$, knowing that it must cancel against

$$E_2^{p,1} = H^p(\mathbf{T}, A/A_0).$$

However, A/A_0 is discrete and carries trivial **T**-action, so by [17], Theorem 4,

$$H^p(\mathbf{T}, A/A_0) \cong H^p_{\text{top}}(B\mathbf{T}, A/A_0),$$

which gives A/A_0 for $p = 2, 4, \ldots$ by the universal coefficient theorem. ($H^*_{\text{top}}(B\mathbf{T}, \mathbf{Z})$ is a polynomial ring on a single generator in degree 2.) ■

REMARK 3.8. One can also compute $H^1(\mathbf{R}, \mathrm{Ind}_{\mathbf{Z} \uparrow \mathbf{R}} A)$ directly from the definition, using the fact ([8], Theorem 3) that any Borel 1-cocycle is automatically continuous. An element of $Z^1(\mathbf{R}, \mathrm{Ind}_{\mathbf{Z} \uparrow \mathbf{R}} A)$ may be viewed as a function $\phi : \mathbf{R} \times \mathbf{R}/\mathbf{Z} \to A$ which is (jointly) continuous and satisfies the cocycle identity $\phi(s + t, \dot{r}) = \phi(s, \dot{r}) + \phi(t, \dot{r} - \dot{s})$. From this it follows that $\phi(0, \dot{r}) = 0$ for all $\dot{r} \in \mathbf{R}/\mathbf{Z}$ and that $\phi(n, \dot{r})$ is independent of $\dot{r} \in \mathbf{R}/\mathbf{Z}$ for fixed $n \in \mathbf{Z}$. The isomorphism $H^1(\mathbf{R}, \mathrm{Ind}_{\mathbf{Z} \uparrow \mathbf{R}} A) \to A_0$ of Theorem 3.5 is then given by

$$[\phi] \mapsto \phi(1, \dot{r}).$$

This is independent of $\dot{r}$ and lies in A_0 since it is path-connected to $\phi(0, \dot{r}) = 0$. ■

Using Proposition 3.3 and Theorem 3.5 in the case where A is of the form $C(Y, \mathbf{T})$ (where $H^1(Y, \mathbf{Z})$ should be countable so that A is locally path-connected) it is easy to deduce the vanishing of $H^n(\mathbf{R},C(X,\mathbf{T}))$, $n \geq 2$, for suitable $\mathbf{R}$-spaces X on which the action of $\mathbf{R}$ is proper. In [13], Theorem 4.1, we gave a stronger vanishing theorem that allows arbitrary $\mathbf{R}$-action on X. We conclude the present section with an analogous vanishing theorem for actions of compact semisimple Lie groups (e.g., SU(N)).

THEOREM 3.9. *Let* G *be a connected, simply connected compact Lie group, and let* X *be any second-countable locally compact* G*-space with* $H^0(X,\mathbf{Z})$ *and* $H^1(X,\mathbf{Z})$ *countable (it suffices for* X *to be compact). Give* $C(X,\mathbf{T})$ *the topology of uniform convergence on compacta and the action of* G *coming from the* G*-action on* X*. Then the cohomology groups* $H^n(G,C(X,\mathbf{T}))$ *are countable for* $n \geq 1$ *and vanish for* $n = 1,2$.

PROOF. As in the proof of [13], Theorem 4.1, we use the short exact sequences of G-modules

$$1 \to C(X,\mathbf{T})_0 \to C(X,\mathbf{T}) \to H^1(X,\mathbf{Z}) \to 1$$

and

$$1 \to H^0(X,\mathbf{Z}) \to C(X,\mathbf{R}) \to C(X,\mathbf{T})_0 \to 1.$$

These are topological short exact sequences of Polish groups by the countability assumptions on $H^0(X,\mathbf{Z})$ and $H^1(X,\mathbf{Z})$, together with Proposition 6 of [9]. Now since $C(X,\mathbf{R})$ is a topological vector space and G is compact, $H^n(G,C(X,\mathbf{R})) = 0$ for $n \geq 1$ by a generalization of the "averaging argument" of [7], p.60, or else by reduction to continuous cochains using [17], Theorem 3, followed by [4], Corollaire III.2.1. On the other hand, since $H^0(X,\mathbf{Z})$ and $H^1(X,\mathbf{Z})$ are discrete and G is connected, the G-action on them is trivial. But for A a countable group with trivial G-action, Theorem 4 of [17] gives $H^n(G,A) \cong H^n(BG,A)$, which is countable for all n. By the long exact cohomology sequences, we now get an exact sequence

$$H^{n+1}(BG,H^0(X,\mathbf{Z})) \to H^n(G,C(X,\mathbf{T})) \to H^n(BG,H^1(X,\mathbf{Z}))$$

for any $n \geq 1$. This proves the countability. Furthermore, the assumptions on G guarantee that G is 2-connected ([16], p.198), hence BG is 3-connected and we deduce vanishing of $H^n(G,C(X,\mathbf{T}))$ for $n = 1$ or 2. ■

COROLLARY 3.10. *If* G *and* X *are as in Theorem* 3.9 *and* α *and* β *are actions of* G *on a separable continuous-trace algebra* A*, such that* $\hat{A} = X$ *and* α *and* β *induce the same action of* G *on* X *, then* α *and* β *are exterior equivalent.*

PROOF. This follows from Theorem 3.9 together with Remark 1.2. ■

COROLLARY 3.11. *If* G *and* X *are as in Theorem* 3.9 *and* A *is a separable* C^*-*-algebra with* $\hat{A} = X$ *and with all derivations inner, then any two norm-close actions of* G *on* A *are exterior equivalent.*

PROOF. As we noted in § 1, this follows from Theorem 3.9 together with the analysis in [5]. ∎

REMARK 3.12. In fact, compactness of G was only used at one point in the proof of Theorem 3.9. Since any connected, simply connected Lie group is 2-connected even if non-compact, the same argument together with Theorem 3 of [17] and Corollaire III.7.5 of [4] (a form of Van Est's Theorem) shows that for any such G (and X as above)

$$H^2(G,C(X,\mathbf{T})) \simeq H^2(G,C(X,\mathbf{T})_0) \simeq H^2(\underline{g},\underline{k};C(X,\mathbf{R})_\infty),$$

the relative Lie algebra cohomology for the Lie algebra of G relative to the Lie algebra of a maximal compact subgroup, with coefficients in the Fréchet space of continuous real-valued functions on X which are smooth along orbits of G. There are some cases in which this will be computable. For instance, if X = G/H is a homogeneous space of G, we get

$$H^2(G,C(G/H,\mathbf{T})) \simeq H^2(\underline{g},\underline{k};C^\infty(G/H)),$$

which may be computed by the methods of [1]. For instance, if H is a lattice subgroup and G is semisimple of real rank ≥ 3, the cohomology must vanish by [1], Theorem V.3.3. This has implications such as those of Corollaries 3.10 and 3.11 for G-actions on C^*-algebras with spectrum G/H (as a G-space). Of course, vanishing of the cohomology when G acts trivially on X was already proved in [5], Theorem 2.6. ∎

4. BUNDLES AND PULL-BACKS

This section is based on ideas of [13] and a suggestion of Iain Raeburn, for which I am grateful. It concerns the situation of a principal G-bundle $p : \Omega \to X$, where G is a suitable group (the most interesting case being $G = \mathbf{T}$), and a comparative analysis of $H^*(G,C(\Omega,\mathbf{T}))$, where G acts via the free G-action on Ω, and of $H^*(G,C(X,\mathbf{T}))$, with G acting trivially on $C(X,\mathbf{T})$. Then we apply this to study certain group actions on C^*-algebras. Though it would probably be possible to work with a somewhat larger class of groups (using some of the techniques discussed elsewhere in this paper), we limit the discussion to tori and "solenoids". The finite-dimensionality hypothesis on G is only needed so that we can apply Theorem 4 of [17], and is probably unnecessary.

THEOREM 4.1. *Let* G *be a compact finite-dimensional connected metrizable abelian group, and let* X *be a locally compact second-countable space with* $H^0(X,\mathbf{Z})$ *and* $H^1(X,\mathbf{Z})$ *countable (for instance, a compact metrizable space). Let* $p:\Omega \to X$ *be a principal* G*-bundle, and let* G *act trivially on* $C(X,\mathbf{T})$ *and via the action on* Ω *on* $C(\Omega,\mathbf{T})$*. Then* $H^0(G,C(X,\mathbf{T})) \cong H^0(G,C(\Omega,\mathbf{T})) \cong C(X,\mathbf{T})$*, and for all* $n>0$*, the groups* $H^n(G,C(X,\mathbf{T}))$ *and* $H^n(G,C(\Omega,\mathbf{T}))$ *are countable. In particular,* $H^1(G,C(X,\mathbf{T})) \cong H^0(X,\hat{G})$ *maps via* p^* *onto*

$$H^1(G,C(\Omega,\mathbf{T})) \cong H^0(X,\hat{G})/\text{image of } H^1(\Omega,\mathbf{Z}),$$

and

$$H^1(X,\hat{G}) \cong H^2(G,C(X,\mathbf{T})) \hookrightarrow H^2(G,C(\Omega,\mathbf{T})) \text{ via } p^* .$$

PROOF. Note that the assumption on G means $\hat{G}$ is a countable torsion-free abelian group of finite rank. Since G is an inverse limit of tori and Čech cohomology commutes with inverse limits, $H^2(G,\mathbf{Z}) \cong \hat{G} \cong H^2(BG,\mathbf{Z})$. In fact, since the cohomology ring of BT is a polynomial ring on a 2-dimensional generator, $H^*(BG,\mathbf{Z})$ is all concentrated in even degrees and is isomorphic to the symmetric algebra on $\hat{G}$. Note, too, that $H^0(\Omega,\mathbf{Z}) \cong H^0(X,\mathbf{Z})$ and that $H^1(\Omega,\mathbf{Z})$ is related to $H^1(X,\mathbf{Z})$ by the following exact Gysin sequence (the sequence of edge terms of the Leray-Serre spectral sequence for p):

$$0 \to H^1(X,\mathbf{Z}) \xrightarrow{p^*} H^1(\Omega,\mathbf{Z}) \xrightarrow{p!} H^0(X,\hat{G}) \to H^2(X,\mathbf{Z}).$$

Now consider the following commutative diagrams of short exact sequences:

$$\text{(A)}\quad \begin{array}{ccccccccc} 0 \to & C(X,\mathbf{T})_0 & \to & C(X,\mathbf{T}) & \to & H^1(X,\mathbf{T}) & \to 0, \\ & \downarrow p^* & & \downarrow p^* & & \downarrow p^* & \\ 0 \to & C(\Omega,\mathbf{T})_0 & \to & C(\Omega,\mathbf{T}) & \to & H^1(\Omega,\mathbf{T}) & \to 0, \end{array}$$

$$\text{(B)}\quad \begin{array}{ccccccccc} 0 \to & H^0(X,\mathbf{Z}) & \to & C(X,\mathbf{R}) & \to & C(X,\mathbf{T})_0 & \to 0 \\ & \cong \downarrow p^* & & \downarrow p^* & & \downarrow p^* & \\ 0 \to & H^0(\Omega,\mathbf{Z}) & \to & C(\Omega,\mathbf{R}) & \to & C(\Omega,\mathbf{T})_0 & \to 0. \end{array}$$

As all maps are equivariant for the action of G and all groups are Polish by the countability assumptions, we can consider the associated diagrams of long exact sequences in G-cohomology. As in the proof of Theorem 3.9, all higher cohomology vanishes for $C(X,\mathbf{R})$ and $C(\Omega,\mathbf{R})$. Furthermore, since $H^0(X,\mathbf{Z})$, $H^1(X,\mathbf{Z})$ and $H^1(\Omega,\mathbf{Z})$ are all trivial discrete G-modules, we may use the fact that by [17], Theorem 4, $H^*(G,M) \simeq H^*(BG,M) \simeq H^*(BG,\mathbf{Z})\otimes M$ for such a module. (The last equality follows from the universal coefficient theorem.) Of course the calculation of H^0 is obvious. From (B) we obtain the commutative diagram

$$\begin{array}{ccc}
H^n(G,C(X,\mathbf{T})_0) & \xrightarrow{\simeq} & H^{n+1}(BG,\mathbf{Z})\otimes H^0(X,\mathbf{Z}) \\
\simeq\downarrow p^* & & \simeq\downarrow p^* \\
H^n(G,C(\Omega,\mathbf{T})_0) & \xrightarrow{\simeq} & H^{n+1}(BG,\mathbf{Z})\otimes H^0(X,\mathbf{Z})
\end{array}$$

for any $n \geq 1$; note in particular that $H^n(G,C(\Omega,\mathbf{T})_0) = 0$ for n even. Then from (A) we obtain commutative diagrams of exact sequences

$$\begin{array}{ccccccccccc}
0 \to & C(X,\mathbf{T})_0 & \to & C(X,\mathbf{T}) & \to & H^1(X,\mathbf{Z}) & \xrightarrow{0} & H^1(G,C(X,\mathbf{T})_0) & \to & H^1(G,C(X,\mathbf{T})) & \to 0 \\
 & \downarrow\simeq & & \downarrow\simeq & & \downarrow p^* & & \simeq\downarrow p^* & & \downarrow p^* & \\
0 \to & C(X,\mathbf{T})_0 & \to & C(X,\mathbf{T}) & \to & H^1(\Omega,\mathbf{Z}) & \to & H^1(G,C(\Omega,\mathbf{T})_0) & \to & H^1(G,C(\Omega,\mathbf{T})) & \to 0
\end{array}$$

and (for $n \geq 1$)

$$\begin{array}{ccccccccc}
0 \to & H^{2n}(G,C(X,\mathbf{T})) & \to & H^{2n}(BG,\mathbf{Z})\otimes H^1(X,\mathbf{Z}) & \to & H^{2n+1}(G,C(X,\mathbf{T})_0) & \to & H^{2n+1}(G,C(X,\mathbf{T})) & \to 0 \\
 & \downarrow p^* & & \downarrow \mathrm{id}\otimes p^* & & \simeq\downarrow p^* & & \downarrow p^* & \\
0 \to & H^{2n}(G,C(\Omega,\mathbf{T})) & \to & H^{2n}(BG,\mathbf{Z})\otimes H^1(\Omega,\mathbf{Z}) & \to & H^{2n+1}(G,C(\Omega,\mathbf{T})_0) & \to & H^{2n+1}(G,C(\Omega,\mathbf{T})) & \to 0
\end{array}$$

from which we can read off almost all we want. In particular,

$$p^* : H^{2n}(G,C(X,\mathbf{T})) \to H^{2n}(G,C(\Omega,\mathbf{T}))$$

is injective and

$$p^* : H^{2n+1}(G,C(X,\mathbf{T})) \to H^{2n+1}(G,C(\Omega,\mathbf{T}))$$

is surjective. Though this would suffice for most of our purposes, we would like to know that $p^* : H^2(G,C(X,\mathbf{T})) \to H^2(G,C(\Omega,\mathbf{T}))$ is actually an isomorphism, when $G = \mathbf{T}$, and then say something about p^* for general G.

When $G = \mathbf{T}$, first note that the exact sequences up to H^1 reduce to

$$\begin{array}{ccccccccc} & & & & H^0(X,\mathbf{Z}) & \xrightarrow{\cong} & H^1(G,C(X,\mathbf{T})) & & \\ & & & & \downarrow \cong & & \downarrow & & \\ 0 \to H^1(X,\mathbf{Z}) & \xrightarrow{p^*} & H^1(\Omega,\mathbf{Z}) & \to & H^0(X,\mathbf{Z}) & \to & H^1(G,C(\Omega,\mathbf{T})) & \to & 0. \end{array}$$

We would like to identify the map $H^1(\Omega,\mathbf{Z}) \to H^0(X,\mathbf{Z})$ as the Gysin map $p_!$. This is not hard to do, modulo a sign which depends on a choice of orientation conventions, since the map must be natural for all $\mathbf{T}$-bundles and have the correct kernel, hence must be of the form $p_!$ followed by multiplication by an integer. To check that the integer is $(\pm)1$, it's enough to know that when p is a trivial bundle (so $p_!$ is surjective), $H^1(G,C(\Omega,\mathbf{T})) = 0$. But in this case, $C(\Omega,\mathbf{T}) \cong \mathrm{Ind}_{1\uparrow\mathbf{T}} C(X,\mathbf{T})$, so the vanishing of H^1 follows from Corollary 3.6. In fact, 3.6 tells us that in this case, $H^{2n}(G,C(\Omega,\mathbf{T})) \cong H^1(X,\mathbf{Z})$ and $H^{2n+1}(G,C(\Omega,\mathbf{T})) = 0$ for $n \geq 1$.

Now (going back to the case of general p, but still with $G = \mathbf{T}$), once we've computed $H^1(G,C(\Omega,\mathbf{T}))$, everything else follows by "periodicity". For we have (by [8], Theorem 9) spectral sequences

$$\begin{array}{ccc} H^p(\mathbf{T},H^q(\mathbf{Z},C(X,\mathbf{T}))) & \Longrightarrow & H^{p+q}(\mathbf{R},C(X,\mathbf{T})) \\ \downarrow p^* & & \downarrow p^* \\ H^p(\mathbf{T},H^q(\mathbf{Z},C(\Omega,\mathbf{T}))) & \Longrightarrow & H^{p+q}(\mathbf{R},C(\Omega,\mathbf{T})), \end{array}$$

and since the $\mathbf{R}$-cohomology vanishes for $p + q \geq 2$ (by [13], Theorem 4.1), we must have a commutative diagram of periodicity isomorphisms

$$\begin{array}{ccc} H^p(\mathbf{T},C(X,\mathbf{T})) & \xrightarrow{\cong} & H^{p+2}(\mathbf{T},C(X,\mathbf{T})) \\ \downarrow p^* & & \downarrow p^* \\ H^p(\mathbf{T},C(\Omega,\mathbf{T})) & \xrightarrow{\cong} & H^{p+2}(\mathbf{T},C(\Omega,\mathbf{T})) \end{array}$$

for $p \geq 1$. This means the maps

$$H^{2n}(BG,\mathbf{Z}) \otimes H^1(\Omega,\mathbf{Z}) \to H^{2n+1}(G,C(\Omega,\mathbf{T})_0) \to H^{2n+1}(G,C(\Omega,\mathbf{T}))$$

must reduce to

$$H^1(\Omega,\mathbf{Z}) \xrightarrow{p_!} H^0(X,\mathbf{Z}) \to H^{2n+1}(G,C(\Omega,\mathbf{T})) \to 0$$

for all $n \geq 1$, and since the kernel of $p_!$ is $p^*(H^1(X,\mathbf{Z}))$,

$$H^1(X,\mathbf{Z}) \simeq H^{2n}(G,C(X,\mathbf{T})) \xrightarrow{p^*} H^{2n}(G,C(\Omega,\mathbf{T}))$$

is in fact an isomorphism, while $H^{2n+1}(G,C(\Omega,\mathbf{T})) \simeq \mathrm{coker}(p_!)$.

The case of general G is similar, but for simplicity we concentrate only on the calculation of H^1 and H^2. (In fact, noting that $H^2(BG,\mathbf{Z}) \simeq \hat{G}$ and $H^3(G,C(\Omega,\mathbf{T})_0) \simeq$ $\simeq H^4(BG,\mathbf{Z})\otimes H^0(X,\mathbf{Z}) \simeq S^2(\hat{G})\otimes H^0(X,\mathbf{Z})$, one may identify the map

$$H^2(BG,\mathbf{Z})\otimes H^1(\Omega,\mathbf{Z}) \simeq \hat{G}\otimes H^1(\Omega,\mathbf{Z}) \longrightarrow H^{2+1}(G,C(\Omega,\mathbf{T})_0) \simeq S^2(\hat{G})\otimes H^0(X,\mathbf{Z})$$

with the Gysin map

$$p_! : H^1(\Omega,\hat{G}) \longrightarrow H^0(X,\hat{G}\otimes\hat{G}),$$

which has kernel $p^*(H^1(X,\hat{G}))$, followed by the map induced by the projection $\hat{G}\otimes\hat{G} \longrightarrow$ $\longrightarrow S^2(\hat{G})$.) By Theorem 2.5, $H^2(G,C(X,\mathbf{T})) \simeq H^1(X,\hat{G})$, and $H^1(G,C(X,\mathbf{T})) \simeq H^2(BG,\mathbf{Z})\otimes$ $\otimes H^0(X,\mathbf{Z}) \simeq H^0(X,\hat{G})$ by our exact sequences. The first step in computing $H^n(G,C(\Omega,\mathbf{T}))$ $(n = 1,2)$ is to compute them when $p : \Omega \longrightarrow X$ is a trivial bundle. For this we need a substitute for Corollary 3.6. For any A and any G, $H^1(G,\mathrm{Ind}_{1\uparrow G}A) \simeq H^1_{cont}(G,\mathrm{Ind}_{1\uparrow G}A) = 0$ as in Proposition 3.3. Furthermore, if G is as in our theorem and A has property F, we may form $EA = U([0,1],A)$ as in [8], which is contractible by Lemma 2.3. Then $\mathrm{Ind}_{1\uparrow G}EA$ is also contractible, and by the argument of [17], p.91, $H^n(G,\mathrm{Ind}_{1\uparrow G}EA) = H^n_{cont}(G,$ $,\mathrm{Ind}_{1\uparrow G}EA) = 0$ for $n \geq 1$. As in the proof of Proposition 2.4 (c), one can see that

$$0 \longrightarrow \mathrm{Ind}_{1\uparrow G}A \longrightarrow \mathrm{Ind}_{1\uparrow G}EA \longrightarrow \mathrm{Ind}_{1\uparrow G}BA \longrightarrow [G,BA] \longrightarrow 0$$

is exact (the non-trivial part being exactness on the right). Applying the long exact cohomology sequence together with the vanishing of $H^1(G,\mathrm{Ind}_{1\uparrow G}\text{--})$, we conclude that

$$H^2(G,\mathrm{Ind}_{1\uparrow G}A) \simeq H^1(G,(\mathrm{Ind}_{1\uparrow G}BA)_0) \simeq [G,BA].$$

Taking $A = C(X,\mathbf{T})$, this says $H^1(G,C(G\times X,\mathbf{T})) = 0$ and gives an explicit calculation of $H^2(G,C(G\times X,\mathbf{T}))$. In other words, when $p : \Omega \longrightarrow X$ is trivial,

$p^* : H^1(G,C(X,\mathbf{T})) \longrightarrow H^1(G,C(\Omega,\mathbf{T}))$ is zero

and

$p^* : H^2(G,C(X,\mathbf{T})) \longrightarrow H^2(G,C(\Omega,\mathbf{T}))$ is an injection, but not always an isomorphism.

Then by naturality (in X) of the diagrams

$$\begin{array}{ccccccccc} & & & & H^0(X,\hat{G}) & \simeq & H^1(G,C(X,\mathbf{T})) & & \\ & & & & \downarrow \simeq & & \downarrow p^* & & \\ 0 \rightarrow H^1(X,\mathbf{Z}) & \xrightarrow{p^*} & H^1(\Omega,\mathbf{Z}) & \rightarrow & H^0(X,\hat{G}) & \rightarrow & H^1(G,C(\Omega,\mathbf{T})) & \rightarrow & 0 \end{array}$$

and

$$\begin{array}{ccc} H^2(G,C(X,\mathbf{T})) & \simeq & H^1(X,\hat{G}) \\ \downarrow p^* & & \downarrow p^* \\ 0 \to H^2(G,C(\Omega,\mathbf{T})) & \to & H^1(\Omega,\hat{G}) \to S^2(\hat{G})\otimes H^0(X,\mathbf{Z}) \end{array}$$

it is easy to see that in general,

$$H^1(G,C(\Omega,\mathbf{T})) \simeq \operatorname{coker}(p_! : H^1(\Omega,\mathbf{Z}) \to H^0(X,\hat{G}))$$

and

$$p^* : H^1(X,\hat{G}) \simeq H^2(G,C(X,\mathbf{T})) \hookrightarrow H^2(G,C(\Omega,\mathbf{T})).$$ ■

Now we apply the theorem to the situation of [13], §1. Let $p : \Omega \to X$ be a principal G-bundle as in the theorem, and B a stable separable continuous-trace algebra with spectrum Ω. Then if $\beta : G \to \mathrm{Aut}(B)$ is an action of G on B inducing the G-action on Ω (free with quotient X), by Theorem 1.1 of [13], there is an isomorphism of B with p^*A, where A is a continuous-trace algebra with spectrum A and p^*A means the pull-back of A in the sense of [14], i.e., $C_0(\Omega)\otimes_{C_0(X)}A$. Furthermore, by Corollary 1.3 (loc.cit.), one may arrange this isomorphism so that β is exterior equivalent to $p^*\mathrm{id}$, the action coming from the tensor product of the translation action τ of G on $C_0(\Omega)$ and the trivial action of G on A. Using Theorem 4.1, we can obtain an interesting complement to the results of [14] and [13] concerning pull-back actions.

PROPOSITION 4.2. *Let* G *be a compact, connected, finite-dimensional, metrizable, abelian group, and let* $p : \Omega \to X$ *be a principal* G*-bundle with* $H^n(X,\mathbf{Z})$ *countable for* $n \le 2$. *Let* A *be any separable continuous-trace algebra with spectrum* X, *and let* $B = p^*A$. *Then if* α_1 *and* α_2 *are two locally unitary actions of* G *on* A *(recall that by Theorem 2.5, this is automatic if* α_1 *and* α_2 *are any actions fixing* X *pointwise), then* $p^*\alpha_1$ *and* $p^*\alpha_2$ *are exterior equivalent as actions of* G *on* B *if and only if* α_1 *and* α_2 *are exterior equivalent as actions of* G *on* A.

PROOF. By Remark 1.2 and Theorem 2.5, the obstruction to exterior equivalence of α_1 and α_2 is the class $\zeta(\alpha_1) - \zeta(\alpha_2) \in H^2(G,C(X,\mathbf{T})) \simeq H^1(X,\hat{G})$, and the obstruction to exterior equivalence of $p^*\alpha_1$ and $p^*\alpha_2$ is a class in $H^2(G,C(\Omega,\mathbf{T}))$, which a simple calculation shows is just $p^*(\zeta(\alpha_1) - \zeta(\alpha_2))$. So the result follows from injectivity of $p^* : H^2(G,C(X,\mathbf{T})) \to H^2(G,C(\Omega,\mathbf{T}))$. ■

REMARK 4.3. One might think, since we showed that $p^* : H^2(G,C(X,\mathbf{T})) \to H^2(G,C(\Omega,\mathbf{T}))$ is often surjective, that every G-action β on p^*A, inducing the given G-action on Ω, is exterior equivalent to some $p^*\alpha$, α a locally unitary action on A. However, as was shown in Example 1.9 of [13], this is false in general. The reason is the following. Since G acts trivially on X, there is a natural map ζ that to any G-action trivial on X associates a class in $H^2(G,C(X,\mathbf{T}))$. However, there is no such map sending a G-action on p^*A inducing the given G-action on Ω to a class in $H^2(G,C(\Omega,\mathbf{T}))$. Instead, an obstruction in $H^2(G,C(\Omega,\mathbf{T}))$ is only defined given a pair of such actions, and the obstruction associated to a pair (α,γ) is not necessarily the sum of the obstructions for (α,β) and (β,γ), because of non-commutativity of the unitaries that appear in the relevant formulae. While one might be tempted to think that the map

$$\beta \mapsto (\text{obstruction to exterior equivalence of } \beta \text{ and } p^*\,\mathrm{id})$$

would have good additivity properties, it does not. In fact, the whole notion of the "basepoint" $p^*\mathrm{id}$ for the G-actions on p^*A inducing the G-action on Ω can be non-canonical, since it might be that $p^*A = p^*C$, and yet $p^*(\mathrm{id}_A)$ and $p^*(\mathrm{id}_C)$ are not exterior equivalent. (This is what happens in the example given in [13].) ∎

As we mentioned in §1, conjugacy together with exterior equivalence defines an equivalence relation on the G-actions on B which is weaker than exterior equivalence but still implies $\hat{G}$-equivariant isomorphism of the crossed products. We studied this phenomenon to some extent in Theorem 1.5 of [13], where we saw that (in the situation of Proposition 4.2) the following conditions are equivalent (assuming A is stable):

(a) $G \ltimes_{p^*\alpha} B$ is $\hat{G}$- and $C_0(X)$-equivariantly isomorphic to $G \ltimes_{p^*\mathrm{id}} B$;

(b) $\langle \zeta(\alpha),[p] \rangle = 0$, where the pairing is the cup product between $H^1(X,\hat{G})$ and $H^1(X,\underline{G})$, with values in $H^2(X,\underline{\mathbf{T}}) \cong H^3(X,\mathbf{Z})$;

(c) $p^*\alpha$ is exterior equivalent to $\gamma(p^*\mathrm{id})\gamma^{-1}$ for some $\gamma \in \mathrm{Aut}_{C_0(\Omega)}B$.

This suggests a natural question: how unique is the γ in (c)? For instance, when can one take it to be the identity? The following answer (when $G = \mathbf{T}$) was conjectured by Iain Raeburn. With more work (which we leave to the reader), it could probably be adapted to the generality of Proposition 4.2.

THEOREM 4.4. *Let $p : \Omega \to X$ be a principal $\mathbf{T}$-bundle, with Ω and X second-countable, locally compact, and with the homotopy type of finite CW-complexes. Let A be a stable separable continuous-trace algebra with spectrum X, and let $\alpha : \mathbf{T} \to \mathrm{Aut}\,A$*

be an action which fixes X *pointwise, such that* $\zeta(\alpha)\cup[p] = 0$. *Then* $p^*\alpha$ *(on* $p^*A = B$*) is exterior equivalent to* $\gamma(p^*\mathrm{id})\gamma^{-1}$ *for* $\gamma \in \mathrm{Aut}_{C_0(\Omega)}B$, *and the class* $[\gamma] \in H^2(\Omega, \mathbf{Z}) = \mathrm{Aut}_{C_0(\Omega)}B/\mathrm{Inn}\,B$ *satisfies* $p_![\gamma] = \zeta(\alpha) \in H^1(X, \mathbf{Z})$. *Here* $p_!$ *is the Gysin map. The class* $[\gamma]$ *is uniquely determined modulo* $p^*(H^2(X, \mathbf{Z}))$.

PROOF. As we said above, it was shown in [13] that if $\zeta(\alpha)\cup[p] = 0$ in $H^3(X, \mathbf{Z})$, then $p^*\alpha$ is exterior equivalent to some $\gamma(p^*\mathrm{id})\gamma^{-1}$. Conversely, for every $\gamma \in \mathrm{Aut}_{C_0(\Omega)}B$, $\gamma(p^*\mathrm{id})\gamma^{-1}$ is an action of $\mathbf{T}$ on B inducing the standard action of $\mathbf{T}$ on Ω, hence there is a well-defined class $\phi(\gamma) \in H^2(\mathbf{T}, C(\Omega,\mathbf{T})) \simeq H^1(X, \mathbf{Z})$ which measures the obstruction to exterior equivalence between $\gamma(p^*\mathrm{id})\gamma^{-1}$ and $p^*\mathrm{id}$. Since exterior equivalence classes don't change under conjugation by inner automorphisms, $\phi(\gamma)$ only depends on $[\gamma] \in H^2(\Omega, \mathbf{Z})$. We shall show ϕ induces a homomorphism $H^2(\Omega, \mathbf{Z}) \to H^1(X, \mathbf{Z})$ with the same formal properties as $p_!$, then deduce that the two coincide.

First we show that ϕ is a homomorphism. If γ is inner, then $\gamma(p^*\mathrm{id})\gamma^{-1}$ and $p^*\mathrm{id}$ are exterior equivalent, so ϕ sends $0 \in H^2(\Omega, \mathbf{Z})$ to $0 \in H^1(X, \mathbf{Z})$. We must check that $\phi(\gamma_1\gamma_2) = \phi(\gamma_1) + \phi(\gamma_2)$. To see this, recall $\phi(\gamma)$ is the coboundary in $H^2(\mathbf{T}, C(\Omega,\mathbf{T}))$ of the 1-cocycle

$$t \mapsto \gamma(p^*\mathrm{id})_t\gamma^{-1}(p^*\mathrm{id})_{t^{-1}} : \mathbf{T} \to U(M(B))/C(\Omega,\mathbf{T}).$$

In other words, $\phi(\gamma)$ is defined by choosing (in a Borel fashion) unitaries v_t implementing $\gamma(p^*\mathrm{id})_t\gamma^{-1}(p^*\mathrm{id})_{t^{-1}}$, then defining a 2-cocycle by

$$(t,s) \mapsto v_{ts}(v_t(p^*\mathrm{id})_t(v_s))^{-1}$$

(which takes values in $U(Z(M(A))) \simeq C(\Omega,\mathbf{T})$ since v_{ts} and $v_t(p^*\mathrm{id})_t(v_s)$ implement the same automorphism). Now given γ_1 and γ_2, choose in this way $\{v_t\}$ corresponding to γ_1 and $\{w_t\}$ corresponding to γ_2, and let $w'_t = \gamma_1(w_t)$. Then

$$\gamma_1\gamma_2(p^*\mathrm{id})_t\gamma_2^{-1}\gamma_1^{-1}(p^*\mathrm{id})_{t^{-1}} = \gamma_1(\gamma_2(p^*\mathrm{id})_t\gamma_2^{-1}(p^*\mathrm{id})_{t^{-1}}(p^*\mathrm{id})_t)\gamma_1^{-1}(p^*\mathrm{id})_{t^{-1}} =$$

$$= \gamma_1(\mathrm{Ad}\,w_t)(p^*\mathrm{id})_t\gamma_1^{-1}(p^*\mathrm{id})_{t^{-1}} = \mathrm{Ad}\,\gamma_1(w_t)\mathrm{Ad}\,v_t = \mathrm{Ad}(w'_tv_t).$$

Thus the 2-cocycle defining $\phi(\gamma_1\gamma_2)$ is

$$(t,s) \mapsto w'_{ts}v_{ts}(w'_tv_t(p^*\mathrm{id})_t(w'_sv_s))^{-1} =$$

$$= w'_{ts}[v_{ts}(p^*\mathrm{id})_t(v_s)^{-1}v_t^{-1}v_t](p^*\mathrm{id})_t(w'_s)^{-1}v_t^{-1}(w'_t)^{-1} =$$

$$= [v_{ts}(p^* id)_t(v_s^{-1})v_t^{-1}]w'_{ts}v_t(p^* id)_t(w'_s)^{-1}v_t^{-1}(w'_t)^{-1} =$$

$$= [v_{ts}(p^* id)_t(v_s^{-1})v_t^{-1}]\gamma_1(w_{ts})\gamma_1(p^* id)_t\gamma_1^{-1}(\gamma_1(w_s))^{-1}(\gamma(w_t))^{-1} =$$

$$= [v_{ts}(p^* id)_t(v_s^{-1})v_t^{-1}]\gamma_1[w_{ts}(p^* id)_t(w_s)^{-1}w_t^{-1}].$$

Since γ_1 was assumed to act trivially on the center of M(B), we see $\phi(\gamma_1\gamma_2) = \phi(\gamma_1) + \phi(\gamma_2)$, and ϕ is a homomorphism $: H^2(\Omega, \mathbf{Z}) \to H^1(X, \mathbf{Z})$. By [13] and our previous remarks, ϕ also has the property that $\phi(\gamma) \cup [p] = 0$ for all γ. Furthermore, ϕ vanishes on the image of $p^* : H^2(X, \mathbf{Z}) \to H^2(\Omega, \mathbf{Z})$, since if γ is pulled back from an automorphism $\delta \in \mathrm{Aut}_{C_0(X)}A$, i.e., γ comes from the automorphism $id \otimes \delta$ of $C_0(\Omega) \otimes_{C_0(X)} A$, then obviously γ commutes with $p^* id = \tau \otimes id$. Thus ϕ has all the formal properties of the Gysin map $p_! : H^2(\Omega, \mathbf{Z}) \to H^1(X, \mathbf{Z})$.

To check that $\phi = p_!$, we should note that ϕ does not depend on the choice of the continuous-trace algebra A, or equivalently, of its Dixmier-Douday invariant in $H^3(X, \mathbf{Z})$. This can be seen as follows. Let $C = C_0(X, \mathcal{K})$ be the stable, stably commutative continuous-trace algebra over X. Then $A \simeq A \otimes_{C_0(X)} C$ and $B \simeq B \otimes_{C_0(\Omega)} p^* C$. Since every class in $H^2(\Omega, \mathbf{Z})$ arises from a locally inner automorphism of $p^* C$, we may assume our automorphism is of the form $id \otimes \delta$, $\delta \in \mathrm{Aut}_{C_0(\Omega)} p^* C$, with respect to this decomposition, and then clearly $\phi(\gamma) = \phi(\delta)$. Thus, without loss of generality, we may assume A and B are stably commutative.

To show that ϕ and $p_!$ coincide, at least when X has the homotopy type of a finite complex, observe that both maps are natural, in the sense that if $(p : \Omega \to X, x \in H^2(\Omega, \mathbf{Z}))$ is the pull-back of $(p' : \Omega' \to X', y \in H^2(\Omega', \mathbf{Z}))$ under some map of $\mathbf{T}$-bundles

$$\begin{array}{ccc} \Omega & \longrightarrow & \Omega' \\ {\scriptstyle p = f^* p'} \downarrow & & \downarrow {\scriptstyle p'} \\ X & \xrightarrow{\ f\ } & X' \end{array},$$

then $\phi(x) = f^*(\phi'(y))$ and $p_!(x) = f^*(p'_!(y))$. Thus it is enough to show that $\phi = p_!$ on "universal examples". But given $p : \Omega \to X$ and $x \in H^2(X, \mathbf{Z})$, the pair $(p_!(x) \in H^1(X, \mathbf{Z}), [p] \in H^2(X, \mathbf{Z}))$ is pulled back from a map

$$X \to K(\mathbf{Z},1) \times K(\mathbf{Z},2) \simeq S^1 \times \mathbf{CP}^\infty,$$

and since $p_!(x) \cup [p] = 0$, the map can be lifted to a map $f : X \to Y$, where Y is the homotopy fiber of the map

$$K(\mathbf{Z},1) \times K(\mathbf{Z},2) \to K(\mathbf{Z},3)$$

corresponding to the cup product. The composite

$$Y \xrightarrow{\text{fiber inclusion}} K(\mathbf{Z},1) \times K(\mathbf{Z},2) \xrightarrow{\text{2nd projection}} K(\mathbf{Z},2)$$

induces a principal $\mathbf{T}$-bundle $\pi : W \to Y$, and from the Serre spectral sequences for the fibrations

$$\begin{array}{ccc} Y \to K(\mathbf{Z},1) \times K(\mathbf{Z},2) & \qquad & S^1 \to W \\ \downarrow & & \downarrow \pi \\ K(\mathbf{Z},3) & & Y \end{array}$$

we see that $H^n(Y, \mathbf{Z}) = \mathbf{Z}$ for $n = 0, 1, 2$ and that $H^2(W, \mathbf{Z})$ is infinite cyclic, say with generator w. Furthermore,

$$\pi^* : H^2(Y, \mathbf{Z}) \to H^2(W, \mathbf{Z})$$

is the zero map, and $\pi_!(w)$ is a generator z of $H^1(Y, \mathbf{Z})$.

Now by our construction, $(p : \Omega \to X) = f^*(\pi : W \to Y)$, and $p_!(x) = f^*(z) = f^*(\pi_!(w)) = p_!(f^*(w))$. Thus $x - f^*(w) \in \ker p_! = p^*(H^2(X, \mathbf{Z}))$. Suppose we could ignore the difficulty that Y isn't locally compact, so that our definition of ϕ doesn't quite make sense for the bundle $\pi : W \to Y$. If we could make sense of $\phi(w)$, it would have to be $\pm z$ (since ϕ is supposed to <u>sur</u>ject onto the kernel of $\bigcup[\pi]$). Adjusting if necessary the choice of sign in the definition of the Gysin map, we can suppose $\phi(w) = z = \pi_!(w)$, and then since ϕ vanishes on the image of p^*,

$$\phi(x - f^*(w)) = 0,$$

hence

$$\phi(x) = \phi(f^*(w)) = f^*(\phi(w)) = f^*(z) = p_!(x).$$

But if X has the homotopy type of a finite complex, then by cellular approximation the map $f : X \to Y$ can be chosen to factor through some finite skeleton Y_n of Y (which is compact metric, so that $C(Y_n, K)$ makes sense). For n large enough, $Y_n \hookrightarrow Y$ induces isomorphisms on cohomology through degree 3, and the above argument works with Y_n in place of Y. ■

REFERENCES

1. **Borel, A. ; Wallach, N. :** *Continuous cohomology, discrete subgroups, and representations of reductive groups*, Annals of Math. Studies, Vol.**94**, Princeton University Press, Princeton, 1980.

2. **Dixmier, J. ; Douady, A. :** Champs continus d'espaces hilbertiens et de C*-algèbres, *Bull. Soc. Math. France* **91**(1963), 227-284.

3. **Dold, A. :** Partitions of unity in the theory of fibrations, *Ann. of Math.* **78**(1963), 223-225.

4. **Guichardet, A. :** *Cohomologie des groupes topologiques et des algèbres de Lie*, Textes Mathématiques, vol.**2**, CEDIC/Nathan, Paris, 1980.

5. **Herman, R.H. ; Rosenberg, J. :** Norm-close group actions on C*-algebras, *J. Operator Theory* **6**(1981), 25-37.

6. **Michael, E. :** Convex structures and continuous selections, *Canad. J. Math.* **11** (1959), 556-575.

7. **Moore, C.C. :** Extensions and low dimensional cohomology theory of locally compact groups. I, *Trans. Amer. Math. Soc.* **113**(1964), 40-63.

8. **Moore, C.C. :** Group extensions and cohomology for locally compact groups. III, *Trans. Amer. Math. Soc.* **221**(1976), 1-33.

9. **Moore, C.C. :** Group extentions and cohomology for locally compact groups. IV, *Trans. Amer. Math. Soc.* **221**(1976), 35-58.

10. **Palais, R. :** On the existence of slices for actions of non-compact Lie groups, *Ann. of Math.* **73**(1961), 295-323.

11. **Pedersen, G.K. :** *C*-algebras and their automorphism groups*, London Math. Soc. Monographs, vol.**14**, Academic Press, London, 1979.

12. **Phillips, J. ; Raeburn, I. :** Crossed products by locally unitary automorphism groups and principal bundles, *J. Operator Theory* **11**(1984), 215-241.

13. **Raeburn, I. ; Rosenberg, J. :** Crossed products of continuous-trace C*-algebras by smooth actions, Preprint 05711-85, MSRI, Berkeley, 1985.

14. **Raeburn, I. ; Williams, D.P. :** Pull-backs of C*-algebras and crossed products by certain diagonal actions, *Trans. Amer. Math. Soc.*, **287**(1985), 755-777.

15. **Vogan Jr., D.A. :** *Representations of real reductive groups*, Progress in Math., vol. **15**, Birkhäuser, Boston, 1981.

16. **Whitehead, G.W. :** *Elements of homotopy theory*, Graduate Texts in Math., vol. **61**, Springer, New York, 1978.

17. **Wigner, D. :** Algebraic cohomology of topological groups, *Trans. Amer. Math. Soc.* **178**(1973), 83-93.

J.Rosenberg
Department of Mathematics
University of Maryland
College Park, MD 20742
U.S.A.

Operator Theory:
Advances and Applications, Vol.17

SPECTRAL MAPPING THEOREMS FOR ANALYTIC FUNCTIONAL CALCULI

K. Rudol

1. INTRODUCTION

The recent decade has brought several new important constructions and applications of analytic functional calculi (mainly related to S.Brown's technique, employing such calculi based on the algebra $H^\infty(G)$ of all bounded analytic functions on a domain $G\subset\mathbf{C}$). One may hope that similar advances are to come in several variables spectral theory, in dealing with an n-tuple τ of commuting operators. Perhaps other algebras $A\subset H^\infty(G)$ will be more suitable when $n > 1$.

Here we want to present what happens at the very beginning of this theory, after a functional calculus $u \mapsto u(\tau)$ based on one of these algebras A is somehow constructed. Then one of the first major tasks is to characterize $\sigma(u(\tau))$, the spectrum of $u(\tau)$ in terms of u and $\sigma(\tau)$, i.e. to establish a spectral mapping theorem. But in order to obtain a sufficiently rich functional calculus, we consider domains G such that their closures, $\bar{G}$ contain $\sigma(\tau)$ - a joint spectrum of τ (specified later on), without assuming that $\sigma(\tau)\subset G$.

Any spectral mapping theorem in the *"strict form"* : $\sigma(u(\tau)) = u(\sigma(\tau))$ is therefore difficult to state and it has no sense unless one adopts a "reasonable" definition of $u(\sigma(\tau))$. Even in the classical setting of the H^∞-functional calculus of a c.n.u. contraction T ($n = 1$, $\tau = T$, $G = \mathbf{D}$, the unit disc), the difficulty is essential. The example from [2] shows that in this case we cannot take $u(\sigma(\tau))$ as the union: $\bigcup\{cl(u;t)\ ;\ t\in\sigma(\tau)\}$ of cluster sets of u at points $t\in\sigma(\tau)$. Our Example 1 will even show that the above "*strict*" formula cannot hold true regardless what would be the definition of $u(\sigma(\tau))$ for $u\in H^\infty$ ($:= H^\infty(\mathbf{D})$).

Because of these difficulties the following "weaker" form of the spectral mapping theorem due to Foias and Mlak [2] seems to be the best possible result as far as the algebra H^∞ is concerned.

THEOREM FM. (a) *For any* c.n.u. *contraction* T *and any function* $u\in H^\infty$ *we have*

$$u(\sigma(T)\cap \mathrm{Cont}\ (u))\subset\sigma(u(T))\subset\bigcup\{cl(u;t)\ ;\ t\in\sigma(T)\},$$

where Cont(u) := $\{t \in \bar{\mathbf{D}}$; *there exists* $u(t) := \lim u(s)$ *as* $s \to t$ *with* $s \in \mathbf{D}\}$ (= $\{t \in \bar{\mathbf{D}}$; u *may be extended to a function continuous on* $\mathbf{D} \cup \{t\}$ $\}$).

(b) *In particular, if* u *is continuously extendible at each point* t *from* $\sigma(T)$ *with* $|t| = 1$, *then* $\sigma(u(T)) = u(\sigma(T))$.

REMARK. In [2], the statement (a) is slightly different (with $\sigma(T) \cap \mathbf{D}$ instead of $\sigma(T) \cap \mathrm{Cont}(u)$); however the proof implies this better estimate.

We shall end this section with the promised example:

EXAMPLE 1. Let $v(z) := \exp((z+1)/(z-1))$. Then $v \in H^{\infty}$ and $\mathrm{cl}(v;1) = \bar{\mathbf{D}}$, the closed unit disc. The example from [2] gives a c.n.u. contraction S with $\sigma(S) = \{1\}$ such that $v(S) = 0$. Now, if $G := \{z \in \mathbf{D};\ |z - \frac{1}{2}| > \frac{1}{2}\}$, then $|v(z)| > \exp(-1)$ for all $z \in G$. Therefore, if N denotes the multiplication by z on L^2(G, planar measure), then $0 \notin \sigma(v(N)) = v(\bar{G})$. The c.n.u. contractions N and $T := S \oplus N$ have equal (and connected) spectra, but $\sigma(v(T)) = \{0\} \cup \sigma(v(N)) \neq \sigma(v(N))$.

REMARKS. 1) This example provides a negative answer to a question posed by C.M. Pearcy, whether the Sz.-Nagy – Foiaş functional calculus preserves the connectedness of spectre. Indeed, $\sigma(v(T))$ is disconnected.

2) The above kind of "spectral misbehaviour" of v may be avoided at the expense of imposing additional assumptions. The following result of [3] seems to be so far the deepest:

If $w \in H^{\infty}$ has finite Dirichlet integral and if T is an absolutely continuous (e.g., c.n.u.) contraction with $\sigma(T)$ connected, then $\sigma(w(T))$ contains a nontangential cluster set of w at t provided that the set $(\partial\mathbf{D} \cap \mathrm{Cont}(w)) \cup (\partial\mathbf{D} \setminus \sigma(T))$ has positive lower density at t.

2. THE SCALAR CASE

In this section we shall generalize Theorem FM by considering: 1) a more general algebra A than $H^{\infty}(\mathbf{D})$; 2) a general, perhaps non-continuous functional calculus based on A; 3) n-tuples rather than single operators and various types of joint spectra.

Let G be a bounded open subset of $\mathbf{C}^n$, fixed throughout the paper, and let A be a Banach algebra consisting of certain functions continuous on G, containing all polynomials in $(z_1, \dots, z_n)$-complex variables (on G). By M(A) we shall denote the space of maximal ideals (spectrum) of A.

2.1. DEFINITION. For $t \in \mathbf{C}^n$ the *fibre* M_t of M(A) over the point t is defined as the set of all characters $m \in M(A)$ satysfying $m(z_j) = t_j$ for $j = 1, \dots, n$. Then A is said to have the *cluster value property*, if for any $t \in \bar{G}$, $w \in A$ and $s \in \mathbf{C}^n \setminus \bar{G}$ we have $M_s = \emptyset$ and $\hat{w}(M_t) = cl(w;t)$.

Here $\hat{w}$ denotes the Gelfand transform of w, cl(w;t) - the cluster set of w at t, defined as $\{\lim_{k\to\infty} w(s_k) ; s_k \to t, \ s_k \in G, \lim w(s_k) \text{ exists}\}$.

We shall discuss this cluster value property for certain algebras in Remark 2.7, after giving a generalization of the FM Theorem (which heavily exploits this property). Our theorem may be stated for various types of joint spectrum, as that in the sense of J.L. Taylor, approximate point spectrum, essential spectrum and so on. To handle all these cases in a single statement, it is convenient to use (as in [14]) the axiomatic definition of the so called subspectrum, introduced by W. Żelazko in [19].

Let B be a Banach algebra with the unit 1. Let us assume that to each commutative tuple $\beta = (b_1, \dots, b_k) \in B^k$ there corresponds a compact nonempty set $\sigma(\beta) \subset \mathbf{C}^k$ in a way compatible with the natural projections: $B^k \to B^p$ and $\mathbf{C}^k \to \mathbf{C}^p$ $(1 \le p \le k)$ onto fewer variables. If, moreover $\sigma(b_1) \subset \sigma_B(b_1) := \{\lambda \in \mathbf{C} ; b_1 - \lambda 1$ is not invertible in B$\}$ (k = 1) for all ($\beta =$) $b_1 \in B$, then σ is called a spectral system on B having the projection property (cf. [18], [19]).

2.2. DEFINITION. Under the above assumptions, the system σ is referred to as *a subspectrum on* B, if in addition for any k-tuple β of commuting elements of B and for any polynomial $p : \mathbf{C}^k \to \mathbf{C}$ we have $\sigma(P(\beta)) = P(\sigma(\beta))$. (For other characterizations of subspectra see [19], [14].)

Here we shall consider mainly commutative algebras B. This assumption allows us to characterize subspectra on B as spectral systems arising from compact subsets $\Delta = \Delta(\sigma, B)$ of M(B), the maximal ideal space, by means of the following "Gelfand formula for σ":

$$\sigma(\beta) = \hat{\beta}(\Delta) \qquad \text{for all } \beta = (b_1, \dots, b_k) \in B^k.$$

Here $\hat{\beta}$ stands for the Gelfand transform (of the k-tuple $\hat{\beta}(m) = (\hat{b}_1(m), \dots, \hat{b}_k(m))$).

This "representing for σ" set $\Delta(\sigma, B)$ (denoted in [12], [18] as $\sigma(B)$) is determined uniquely by the above formulae as β runs through an arbitrary set Γ of tuples of elements of B which separates compact subsets H of M(B) from points $m \in M(B) \setminus H$ (so that there always exists $\beta \in \Gamma$ such that $\hat{\beta}(m) \notin \hat{\beta}(H)$). Indeed, we have the equality:

$$\Delta(\sigma, B) = \{m \in M(B) ; \hat{\beta}(m) \in \sigma(\beta) \text{ for any tuple } \beta \in \Gamma\}.$$

We may take as Γ the set of all (finite) tuples from B, but in many cases the set Γ may be much smaller. From this easy remark follows the uniqueness theorem for subspectra on completely regular Banach algebras (e.g. for algebras consisting of normal operators, treated differently in [19]).

Here we list some other immediate but usefull consequences of the Gelfand formula for a subspectrum $\tilde{\sigma}$ on B:

(2.3) $\tilde{\sigma}(\beta) \subset \sigma_B(\beta) := \hat{\beta}(M(B))$, the joint spectrum of β in B.

(2.4) Spectral Mapping Theorem: $\tilde{\sigma}(f(\beta)) = f(\tilde{\sigma}(\beta))$ holds true for any n-tuple f of functions analytic in a neighborhod of $\sigma_B(\beta)$, if $f(\beta)$ is given by the functional calculus of Shilov–Arens–Calderòn. For B being a Banach subalgebra of $L(X)$, where X is a Banach space, the same holds true for the functional calculus of J.L. Taylor (for f holomorphic in a neighborhood of $\sigma_T(\beta)$, Taylor's spectrum of β), provided that $\Delta(\tilde{\sigma}, B) \subset \Delta(\sigma_T, B)$ (see [12] and Remark 2 in Section 2 of [14]).

(2.5) The subspectrum $\tilde{\sigma}$ is upper and lower semicontinuous. The notion and the proof are similar to that given by J.D. Newbourgh in [9]. If one drops here the commutativity assumption on B, then $\tilde{\sigma}$ will be upper semicontinuous on commuting k-tuples iff $\{\beta \in B^k ; 0 \in \tilde{\sigma}(\beta), \beta\text{-commutative}\}$ is closed in B^k (with its Cartesian product topology, k fixed).

Now we shall state the main result of this section. Let us consider a commutative n-tuple $\tau = (T_1, \dots, T_n)$ of bounded linear operators on a Banach space X (i.e. $T_j \in L(X)$). Given a bounded domain $G \subset \mathbf{C}^n$ and a Banach algebra A of certain functions continuous on G which contains $1, z_1, \dots, z_n$, we shall call an *A-functional calculus in* τ any algebraic homomorphism: $A \ni u \mapsto u(\tau) \in L(X)$, such that $u(1) = I_X$ (identity on X) and $u(z_j) = T_j$ for any $j = 1, \dots, n$.

For a k-tuple $U = (u_1, \dots, u_k)$ in A let $U(\tau) := (u_1(\tau), \dots, u_k(\tau))$ and let $\mathrm{Cont}(U) := \{t \in \bar{G} ; \lim U(s)$ (called $U(t)$) exists in $\mathbf{C}^k$ as $s \to t$, $s \in G\}$. Then for any subspectrum $\tilde{\sigma}$ defined on an arbitrary Banach subalgebra of $L(X)$ which contains each $u(\tau)$ for $u \in A$, Theorem FM may be set as follows:

2.6.THEOREM. *If the algebra A has the cluster value property* (cf. 2.1) *then :*

(i) $\tilde{\sigma}(\tau) \subset \bar{G}$;

(ii) $u(\tilde{\sigma}(\tau) \cap \mathrm{Cont}(u)) \subset \tilde{\sigma}(u(\tau)) \subset \bigcup \{\mathrm{cl}(u;t) ; t \in \tilde{\sigma}(\tau)\}$ *for any* $u \in A$;

(iii) *For any n-tuple* U *in* A *we have* $U(\tilde{\sigma}(\tau) \cap \mathrm{Cont}(U)) \subset \tilde{\sigma}(U(\tau))$, *and if moreover* $\tilde{\sigma}(\tau) \subset \mathrm{Cont}(U)$, *then* $U(\tilde{\sigma}(\tau)) = \tilde{\sigma}(U(\tau))$;

(iv) *The validity of the second containment in* (ii) *for all tuples* U *(in place of* u) *and for all* A*-functional calculi is equivalent to the density of* G *in* M(A) *in the Gelfand topology, i.e. to the Corona Theorem in* A.

Here G is also considered as a subset of M(A) by means of evaluation functionals. Of course, this theorem deserves an indication to which algebras A it applies (which will be given in Remark 2.8).

PROOF. For any tuple U in A let $\overline{\sigma}(U) := \tilde{\sigma}(U(\tau))$. Then the so defined $\overline{\sigma}$ is a subspectrum on A and $\overline{\sigma}(Z) = \tilde{\sigma}(\tau)$, where Z stands for the n-tuple $(z_1, \ldots, z_n)$ of coordinate functions. Let $\Delta := \Delta(\overline{\sigma}, A)$ be the related representing set for $\overline{\sigma}$. Since $\hat{Z}(\Delta) = \overline{\sigma}(Z)$, the assumption $M_s = \emptyset$ if $s \notin \overline{G}$ implies (i).

Therefore $\hat{Z}(M_t) = \{t\}$ if $t \in \tilde{\sigma}(\tau)$, so finally $\Delta \cap M_t \neq \emptyset$ iff $t \in \tilde{\sigma}(\tau)$. Using this we may decompose Δ into a union of fibres: $\Delta = \bigcup\{\Delta \cap M_t ; t \in \tilde{\sigma}(\tau)\}$. For any $u \in A$ the Gelfand formula for $\tilde{\sigma}$ yields

$$\tilde{\sigma}(u(\tau)) = \hat{u}(\Delta) = \hat{u}(\bigcup\{\Delta \cap M_t ; t \in \tilde{\sigma}(\tau)\}) = \bigcup \hat{u}(\Delta \cap M_t) \subset \bigcup\{cl(u;t) ; t \in \tilde{\sigma}(\tau)\} .$$

If $t \in \tilde{\sigma}(\tau) \cap Cont(u)$, then $\Delta \cap M_t \neq \emptyset$, so $\{u(t)\} = cl(u;t) = \hat{u}(\Delta \cap M_t) \subset \tilde{\sigma}(u(\tau))$. The same argument proves (iii), since for a system U in A and for $t \in Cont(U)$ we have $\hat{U}(M_t) = \{U(t)\}$.

Finally the Corona Theorem is known [17] to be equivalent to the validity of the cluster value property for all tuples U (i.e., to equalities: $\hat{U}(M_t) = cl(U,t)$ if $t \in \overline{G}$, $M_s = \emptyset$, $s \notin \overline{G}$).

Similarily one can see that (i) and (ii) (jointly) imply the cluster value property for A (consider T_j (= multiplication by z_j on A) $\in L(A)$ for one implication and the proof of (ii) for the other one).

Now we shall use the notion of strict pseudoconvexity (**spsc**) (see [5], [13]).

2.7.REMARK. The statements (i) and (iii) of Theorem 2.6 hold still true if instead of the cluster value property of A we assume that G is **spsc** and that:

(a) A contains $H(\overline{G}) := \{f : G \to \mathbf{C} ; f \text{ is holomorphic on a neighborhood of } \overline{G}\}$;

(b) The characters $m \in M(A)$ are continuous in the sup-norm $\| \ \|_G$ (over G) (e.g. if for $u \in A$, $\sup\{|u(s)|^{-1} ; s \in G\} < \infty$ implies that u is invertible in A);

(c) If $t \in G$ then any $u \in A$ has a decomposition $u(z) - u(t) = \sum(z_j - t_j)w_j(z)$ where z runs G, for certain $w_1, \dots, w_n \in A$.

Indeed, (i) follows easily from (a) and (b): If $H(\bar{G})$ stands for the $\| \|_G$-closure of $H(\bar{G})$, then its spectrum coincides with $\bar{G}$ by [13], since $\bar{G}$ has a basis of neighborhoods consisting of domains of holomorphy as Theorem 1.5.21 in [5] says. To see (iii) it suffices to prove that for $t \in \bar{G} \cap \mathrm{Cont}(u)$, $u \in A$ we have $\hat{u}(M_t) = \{t\}$. If $t \in G$, it follows from (c), while for $t \in \partial G$ (boundary), the standard use of a peak function at t (see [8]), which exists since G is spsc [13], does the job.

This remark applies to the algebras $H^{\infty,k}(G)$, $A^k(G)$ considered in [6], as well as for the algebra $BD(G) := \{f \in H^\infty(G)\ ;\ f$ has finite Dirichlet integral$\}$ if $n = 1$.

2.8.REMARK. The cluster value property (although generally difficult to prove, especially when $n > 1$) is enjoyed by many algebras. The most interesting cases are: $H^\infty(G)$, $A(G) := H^\infty(G) \cap C(\bar{G})$, and certain algebras of $H^\infty + C$ – type (cf. [8]). A warning is provided by [16]: there is a Runge domain $G_0 \subset\subset \mathbf{C}^2$ for which certain fibres over points $s \notin \bar{G}_0$ are nonempty – so that the cluster value and Corona theorems fail. (This domain can be used to show that certain joint spectra of subnormal systems misbehave while taking infinite direct sum [15].)

Let us consider the cluster value property of $H^\infty(G)$ and of $A(G)$. For a polydomain ($G = G_1 \times \dots \times G_n$, $G_j \subset\subset \mathbf{C}$) it was proved by Gamelin in [4] and for a strictly pseudoconvex (**spsc**) domain G with "$\partial G \in C^2$" – essentially in [8] and in [14]. Let us briefly review the problem both to get rid of the "$\partial G \in C^2$" requirement and to provide a base for Theorem 3.1. For $t \in G$ it suffices to prove the decomposition property 2.7 (c) by adapting the argument from [10] (cf.[1], [6], [14]). What one needs here is only the L^∞-regularity (resp., the strong L^∞-regularity if $A = A(G)$) of the $\bar{\partial}$-problem on G. It means that for any (0, p)-form $\phi = \sum \phi_\alpha d\bar{z}_\alpha$ with $1 \le \alpha_1 < \dots < \alpha_p \le n$, such that $\bar{\partial}\phi = 0$ and $\|\phi\| := \sum \sup\{|\phi_\alpha(s)|\ ;\ s \in G\} < +\infty$, there exists a (0,p - 1)-form ψ satysfying $\bar{\partial}\psi = \phi$ and $\|\psi\| \le C\|\phi\|$, with $C > 0$ depending only on G (and ψ has coefficients continuous on $\bar{G}$, resp. for strong regularity).

The L^∞-regularity takes place in many polydomains (e.g. in $\mathbf{D}^n$) and in **spsc** polyhedra, while the strong regularity is proved in [5] for all **spsc** domains even without assuming that ∂G is C^2-smooth.

Since the case $t \notin \bar{G}$ is settled by the Remark 2.7, it suffices now to consider $t \in \partial G$. The method of [8] is then based on the strong L^∞-regularity and uses a peak function at t. Here is a short proof from [1], simplified in one place: Let W be a neighborhood of t in $\mathbf{C}^n$. If for any $f \in H^\infty(G \cap W)$ we find $F \in H^\infty(G)$ such that F - f may be extended as analytic in a neighborhood W_1 of t (say, (F - f)(t) = 0), then any $m \in M_t$ (i.e., $m \in M(H^\infty(G))$ with $\hat{Z}(m) = t$) may be extended to $\mathfrak{m} \in M(H^\infty(G \cap W))$ by putting $\mathfrak{m}(f) := m(F)$. If F_1 is any other function having these properties, then $F - F_1 \in H^\infty(G_0)$, where G_0 is a **spsc** domain containing $G \cup \{t\}$ and such that $G_0 \setminus G \subset W_1$. Such a G_0 may be easily constructed and now t becomes an interior point, so that 2.7(c) implies that $m(F - F_1) = 0$. The definition of $\mathfrak{m}$ will then be correct and, as W varies, this will prove the cluster value property at t. To construct F, let us take a **spsc** domain $\tilde{G}$ with $G \cup \{t\} \subset \tilde{G} \subset G \cup W$, and a C^∞-smooth function χ supported by W, such that $0 \leq \chi \leq 1$, $\chi \equiv 1$ in a neighborhood of $\tilde{G} \setminus G$. Then $\phi := -f\,\bar{\partial}\chi$ is a $\bar{\partial}$-closed (0,1)-form on $\tilde{G}$, bounded there. Since $\bar{\partial}$ is here L^∞-regular, one can find a bounded (0,0)-form, i.e. a function $u \in L^\infty(\tilde{G})$ such that $\bar{\partial}u = \phi$. Obviously, $F := u + f\chi$ satisfies the requirements (precisely, F - F(t) does), since $\bar{\partial}F = 0$ on G and $\bar{\partial}(F - f) = \bar{\partial}u = 0$ on a neighborhood of $\tilde{G} \setminus G$ (and of t).

(The above method originates from the work of Gamelin and Garnett in the case n = 1.)

To end this section, let us see what happens when A is one of the folowing algebras: R(X) (= the $\| \ \|_X$-closure of the set of rational functions regular on X, a compact subset of $\mathbf{C}^n$) or C(K) (= all continuous functions on a compact topological space K).

In the first case M(A) = (the rational hull of X) (= X, when X is rationally convex, e.g. if n = 1). We shall assume that convexity. Then the spectral mapping theorem: $\tilde{\sigma}(U(\tau)) = U(\tilde{\sigma}(\tau))$ for any system U in R(X) (with $\tilde{\sigma}$, τ, $U(\tau)$ as previously) is an easy consequence of the Gelfand formula for $\tilde{\sigma}$, since $\hat{Z}$ maps $\Delta(\bar{\sigma}, A)$ onto $\tilde{\sigma}(\tau)$ in a bijective way. The same holds true for A = C(K) and this may be used to simplify the proof of the spectral mapping theorem for compactly supported spectral measures P in [11]. It sufficies to know that $C(K) \ni f \mapsto \int f\, dP \in L(X)$ is multiplicative and unit-preserving. Then, if $\tilde{\sigma}$ is any subspectrum on $L(X)$ such that $\tilde{\sigma}(T) = \sigma(T)$ for all $T \in L(X)$, then for $\bar{\sigma}(U) := \tilde{\sigma}(\int U dP)$ where U is a tuple in C(K), one has $\Delta(\bar{\sigma}, C(K)) = K$. The statement of Theorem 2 in [11] may be extended to cover tuples of functions (cf. the uniqueness of $\bar{\sigma}$, treated at the beginning of this section).

3. OPERATOR - VALUED CASE

The notion of functional calculus may be easily extended to a more general setting, where $F(\tau)$ is defined for certain operator-valued functions F (e.g. for polynomials in $Z := (z_1, \ldots, z_n)$ whose coefficients are elements of $L(X)$, commuting with τ).

Let us consider two commutative Banach subalgebras $A \subset A_1$ of $L(X)$ with $I_X \in A$ and a subspectrum $\tilde{\sigma}$ on A_1. By A we shall denote the Banach algebra $H^\infty(G,A)$ of operator-valued, analytic functions $F : G \to A$ such that $\|F\|_G := \sup\{\|F(t)\| \; ; \; t \in G\} < \infty$. Here G stands for a bounded domain in $\mathbf{C}^n$ which is **spsc** or a polydomain. For $F \in A$, $t \in \bar{G}$ let us intruduce the notion of spectral cluster set $\mathbf{scl}(F,t)$ of F at t:

$$\mathbf{scl}(F;t) := \{\lambda \in \mathbf{C} \; ; \text{ the function } G \cap W \ni s \mapsto F(s) - \lambda I_X \text{ is non-invertible in } H^\infty(G \cap W, A) \text{ for any neighborhood } W \text{ of } t \text{ in } \bar{G}\}.$$

Let $\tau = (T_1, \ldots, T_n)$ be an n-tuple in A_1. We shall consider a homomorphism $A \ni F \mapsto F(\tau) \in A_1$ which is a functional calculus in τ, i.e. such that $\Phi(\tau) = S$ if $S \in A$ and $\Phi(z) \equiv S$, $\psi_j(\tau) = T_j$, if $\psi_j(z) = z_j I_X$, $j = 1, \ldots, n$.

Then, in particular, $H^\infty(G) \ni u \mapsto (uI_X)(\tau)$ is a (scalar) functional calculus in τ, so that $\tilde{\sigma}(\tau)$ has to be contained in $\bar{G}$, by 2.6. Moreover we have the following.

3.1. THEOREM. (i) $\tilde{\sigma}(F(\tau)) \subset \bigcup \{\mathbf{scl}(F;t) \; ; \; t \in \tilde{\sigma}(\tau)\}$ *for all* $F \in A$.

(ii) *If* $F_1, \ldots, F_k \in A$ *are continuously (in the norm topology of* $L(X)$*) extendible to* $G \; \{t\}$ *for all* $t \in \tilde{\sigma}(\tau)$, *then*

$$\tilde{\sigma}(F_1(\tau), \ldots, F_k(\tau)) \subset \bigcup \{\tilde{\sigma}(F_1(t), \ldots, F_k(t)) \; ; \; t \in \tilde{\sigma}(\tau)\} .$$

REMARKS: 1) This theorem extends a result of [7], where the union was performed over $\sigma(T_1) \times \ldots \times \sigma(T_n)$ instead of $\tilde{\sigma}(\tau)$, $k = 1$, F was a function analytic in a neighborhood of this Cartesian product.

2) For a function $F \in A$ which is continuous on $G \cup \{t\}$ its spectral cluster set at t equals $\sigma_A(F(t))$.

3) Even on a 2-dimensional Hilbert space X and for F being the polynomial $F(z) := Qz$, where $Q \in L(X)$, the strict form of the spectral mapping theorem $\sigma(F(T)) = \bigcup \{\sigma(F(t)) \; ; \; t \in \sigma(T)\}$ is impossible (take Q, T as orthogonal projections onto mutually orthogonal complex lines in X).

Idea of PROOF: As in the scalar case, define the subspectrum $\bar{\sigma}$ on A and split its representing set Δ into a union of fibres $\Delta \cap M_t$, $t \in \tilde{\sigma}(\tau)$, where $M_t := \{m \in M(A) \; ; \; m(\psi_j) = t_j \text{ if } j = 1, \ldots, n\}$ (ψ_j as above). Now fix $m \in \Delta \cap M_t$. It sufficies to show that $m(F) \in \mathbf{scl}(F;t)$ in the case (i) and that $m(F_j) = m(F_j(t) I_X)$ if F_j is continuous on $G \cup \{t\}$.

The result will then follow from the formulae:

$$\tilde{\sigma}(F_1(\tau),\dots,F_k(\tau)) = (F_1,\dots,F_k)^\wedge(\Delta) \text{ and } \tilde{\sigma}(F(\tau)) \subset \sigma_A(F(\tau)).$$

But these both facts may be proved just in the same way as in the scalar case (see Remark 2.7). The point is that the L^∞-regularity of $\bar{\partial}$-problem on G may be easily extended to (0,q)-forms with bounded, A-valued coefficients by using integral formulae form [6], [5].

Acknowledgements. The author wishes to express his gratitude to Professors W. Mlak and J. Janas for introducing him to these problems and to Professors M. Putinar and H. Bercovici for letting him know about their results prior to publication.

REFERENCES

1. **Cufi, J.** : $H^\infty + L_E$ in several variables, *Collect.Math.* **23**(1982), 109-123.
2. **Foiaş, C. ; Mlak, W.** : The extended spectrum of completely non-unitary contractions and the spectral mapping theorem, *Stud. Math.* **26**(1966), 239-245.
3. **Foiaş, C. ; Pearcy, C.M. ; Bercovici, H.** : A spectral mapping theorem for functions with finite Dirichlet integrals, to appear.
4. **Gamelin, T.M.** : Inversens theorem and fiber algebras, *Pacific J. Math.* **46**(1973), 389-414.
5. **Henkin, G.M. ; Leiterer, J.** : *Theory of functions on complex manifolds*, Birkhäuser Verlag, 1984, Basel, Boston, Stuttgart.
6. **Jakóbczak, P.** : Approximation and decomposition theorems for algebras of analytic functions in strictly psc. domains, *Zeszyty naukowe, Universytetu Jagiellenskiege, Prace Mat.* **22**(1981), 95-109.
7. **Marmestein, I.I.** : A property of the spectrum of a family of linear operators (Russian), *Teor.Funkcij, Funct. Analiz i Prilož.* **32**(1979), 53-61.
8. **McDonald, G.** : The maximal ideal space of $H^\infty + \mathbf{C}$ on the ball in $\mathbf{C}^n$, *Canad. J. Math.* **31**(1979), 79-86.
9. **Newbourgh, J.D.** : The variation of spectra, *Duke Math. J* **18**(1951), 165-176.
10. **Øvrelid, N.** : Generators of the maximal ideals of A(**D**), *Pacific J. Math.* **39**(1971), 219-224.
11. **Panchapagesan, T.V. ; Palled, S.V.** : A generalized spectral mapping theorem, *J. Madras Univ. B.* **41**(1978), 46-53.
12. **Putinar, M.** : Functional calculus and the Gelfand transformation, *Stud. Math.* **79**(1984), 83-86.
13. **Rossi, H.** : Holomorphically convex sets in several complex variables, *Ann. of Math.* **74**(1961), 470-493.
14. **Rudol, K.** : On spectral mapping theorems, *J. Math. Anal. Appl.* **97**(1983), 131-139.

15. **Rudol, K.** : Extended spectrum of subnormal representations, *Bull. Polish Acad. Sci., Math.* **31**(1983), 361-368.

16. **Sibony, N.** : Prolongement des fonctions holomorphes bornées et metrique de Carathéodory, *Invent. Math.* **29**(1975), 206-244.

17. **Scheinberg, S.** : *Cluster sets and corona theorems*, Lect.Notes in Math. **604** (1977).

18. **Słodkowski, Z.** ; **Zelazko, W.** : On spectra of commuting families of operators, *Stud. Math.* **50**(1974), 127-148.

19. **Żelazko, W.** : Axiomatic approach to joint spectra. I, *Stud. Math.* **64**(1971), 250-261.

K. Rudol

Instytut Matematyczny PAN
Solskiego 30, Krakow
Poland.

Operator Theory:
Advances and Applications, Vol.17

PREDICTION THEORY AND CHOICE SEQUENCES: AN ALTERNATE APPROACH

Dan Timotin

Beginning with the series of papers [6], [7], [2], choice sequences have been developed as an interesting object of study, which makes its appearance in many instances when it is the case to parametrize certain classes of objects. The first (historical!) reference is the work of Schur ([16], see also [11]); after the papers [15] and [1] it became clear that one may subsume many "classical" function theory problems to a single operatorial frame, which has been treated in its most general form in [2]. The whole algorithm in [2] is rather intricate; subsequently, the papers of Constantinescu [8], [9], [10] concerning the structure of positive Toeplitz (and, recently, non-Toeplitz) matrices showed a more direct way of applying choice sequences (respectively, generalized choice sequences). There are extensive ramifications of this work, touching problems as Naimark dilation, estimation of spectra, Szegö polynomials, Cholesky factorizations, etc. (see [3], [4], [5], [9]).

On the other hand, the approach in [8], [9] or [10] is quite computational. This paper tries to provide a new insight, by showing the construction of Constantinescu as following from a simple geometric frame, suggested by the objects of prediction theory. The choice sequence is thus seen to emerge from a direct generalization of the classical Gram-Schmidt procedure of orthogonalization. Note that already in [2], VII, Corollary 4.1, a choice operator was seen to be the angle operator between two subspaces.

Thus (with the possible exception of Remark 2), there are almost no new results in the sequel, but an alternate (more transparent, we hope) approach to the results of [10], which might prove fruitful also for the related subjects quoted above.

We start with the problem of describing all positive matrices with entries operators on a given Hilbert space H. The connection with the frame of prediction theory is made by the following standard dilation theorem, whose scalar case may be traced to Kolmogorov (see [14]).

THEOREM A. *For any positive matrix* $\{S_{ij}\}_{i,j\in\mathbf{Z}}$, $S_{ij}\in L(H)$ *there exists a Hilbert space* K *and operators* $V_i\in L(H,K)$, *such that* $S_{ij}=V_j^*V_i$. *The natural condition* *of minimality determines K and* $\{V_i\}_{i\in\mathbf{Z}}$ *up to unitary equivalence.*

To keep the computation simpler (and to stick close to [10]) we will consider hereafter only matrices $\{S_{ij}\}$ with $S_{ii} = I$; that corresponds to the V_i's being isometries. Also, obviously $S_{ij} = S_{ji}^*$.

In prediction theory (see [17]) it is the case to consider the space $L(E,K)$ (E,K Hilbert spaces) as a generalization of a Hilbert space; instead of multiplication by scalars we have a right action of $L(E)$, while the role of the scalar product is played by the *correlation* $[T,S] = S^* T$, which has also values in $L(E)$. We will pursue this analogy in the sequel, but using occasionally different spaces for E. Thus, any $T \in L(E,K)$ can be written as $T = VA$, where $V \in L(E,K)$ is an isometry and $A \in L(E)$ (the polar decomposition gives one way of doing it, but it is not the single one that will appear). If $K' \subset K$ is a subspace, then $P_{K'}T$ will be called the projection of T onto K'.

To develop our construction, we will rely on a few elementary lemmas, which are stated below. Their proofs, which we omit, follow more or less immediately from the well known relation $\|h\|^2 = \|Th\|^2 + \|D_T h\|^2$. Here $D_T = (I - T^*T)^{\frac{1}{2}}$ is the defect operator of the contraction T; it will often appear in the sequel, together with the associated defect space $\mathcal{D}_T$ = closed range of D_T.

LEMMA 1. *Suppose that, for* $i = 1, 2$, $V_i : E_i \to K$ *are two isometries. The general form of the projection of* V_1 *onto* V_2E_2 *is*

$$P_{V_2E_2} = V_2\Gamma$$

where $\Gamma = V_2^*V_1$ *is an arbitrary contraction in* $L(E_1,E_2)$. *If* Γ *is fixed, then* $V_1 = V_2\Gamma + V_2'D_\Gamma$, *where* $V_2' : E \to K \ominus V_2E_2$ *is an isometry.*

We may say that Γ measures the "angle" between the subspaces V_1E_1 and V_2E_2.

LEMMA 2. *As a consequence, if* $V_i : E_i \to K$ *are isometries with orthogonal ranges* ($i = 1,\dots,N$), *any isometry* $V : E \to K$ *has the form*

$$(1)\qquad Vh = V_1\Gamma_1 h + V_2\Gamma_2 D_{\Gamma_1} h + \dots + V_N\Gamma_N D_{\Gamma_{N-1}} \cdots D_{\Gamma_1} h + V'D_{\Gamma_N} \cdots D_{\Gamma_1} h$$

where $\Gamma_1 : E \to E_1$, $\Gamma_{i+1} : \mathcal{D}_{\Gamma_i} \to E_i$ ($i = 1,\dots,N-1$) *are contractions and* $V' : \mathcal{D}_{\Gamma_N} \to K$ *is an isometry with range orthogonal to all* V_iE_i ($i = 1,\dots,N$).

Conversely, any N-tuple of contractions $\{\Gamma_1,\dots,\Gamma_N\}$ *may occur in a decomposition of some* V.

Note that the N-tuple $\{\Gamma_1,\ldots,\Gamma_N\}$ depends on the order in which we have considered the V_i's. The decomposition (1) is the "orthogonal decomposition" of V corresponding to $V_1,\ldots,V_N$. It may be used to calculate easily correlations as follows:

LEMMA 3. *If* $V_i : E_i \to K$ *are isometries with orthogonal ranges,* A_i, $B_i \in L(E_i)$ *(i = 1,...,N), and* $T = \sum_{i=1}^{N} V_iA_i$, $S = \sum_{i=1}^{N} V_iB_i$, *then*

$$[T,S] = \sum_{i=1}^{N} [A_i,B_i].$$

Our main task in the sequel will be to construct orthogonal decompositions of the type that appear in the statement of Lemmas 2 and 3. The basic step is given by the following lemma.

LEMMA 4. *Let* $V_i : E_i \to K$ *be isometries,* i = 1, 2 *and* $K'\subset K$, $K' = V_1E_1 \vee V_2E_2$, $\Gamma = V_2^*V_1$. *We have then the following isometries:*

$$\Omega : D_\Gamma \to K' \ominus V_2E_2$$

$$\Omega(D_\Gamma h) = (I - P_{V_2E_2})V_1h = (I - V_2V_2^*)V_1h$$

$$\Omega_* : D_{\Gamma^*} \to K' \ominus V_1E_1$$

$$\Omega_*(D_{\Gamma^*}h) = (I - P_{V_1E_1})V_2h = (I - V_1V_1^*)V_2h .$$

Therefore

(2) $$K' = \Omega D_\Gamma \oplus V_2E_2 = V_1E_1 \oplus \Omega_* D_{\Gamma^*}$$

and

$$V_1h = \Omega D_\Gamma h + V_2\Gamma h$$

$$V_2h = V_1\Gamma^*h + \Omega_* D_{\Gamma^*}h.$$

Obviously, Ω and Ω_*, as well as the two orthogonal decompositions of K' in (2) are in some sense dual one another. Finally, let us remark that, as a consequence of (2), we have a unitary operator

$$(\Omega, V_2)^*(V_1, \Omega_*) : E_1 \oplus D_{\Gamma^*} \to D_\Gamma \oplus E_2$$

which is nothing else than the "Julia operator" ([2]) or "elementary rotation" (in [4])

$$J(\Gamma) = \begin{bmatrix} D_\Gamma & -\Gamma^* \\ \Gamma & D_{\Gamma^*} \end{bmatrix}.$$

Let us pass now to the actual construction. By Theorem A, in a positive matrix $\{S_{ij}\}$ the entries are viewed as correlations $[V_i,V_j] = V_j^* V_i$ between elements $V_i \in L(H,K)$. We will develop a description of all such possible sequences of V_i's (up to unitary equivalence) in a manner suggested by Lemma 2 above. By using a reccurrent procedure, we find at each step $(n \geq 1)$ the possible values of S_{ij} for $i - j = n$, once S_{ij} is given for $i - j \leq n - 1$.

Let us describe the main idea, before writing down the exact formulas. We will use the following notations:

$$K_{ij} = \bigvee_{r=i}^{j} V_r H, \quad \text{for } i \leq j$$

$$G_{ij} = K_{ij} \ominus K_{i+1,j}\,, \; G_{ij}^* = K_{ij} \ominus K_{i,j-1}, \quad \text{for } i < j \quad .$$

Thus, K_{ij} is the subspace of K spanned by the ranges of a finite number of V_i's, while G_{ij} and G_{ij}^* are, in terms of prediction theory, the "innovation" parts of the process $\{V_i\}$ with respect to a "finite length part"; the "*" corresponds then to reversing the direction of time.

The recursion goes then as follows: suppose that all correlations $[V_i,V_j] = S_{ij}$ are fixed for $i \leq j \leq i + n - 1$ (equivalently, for $|i - j| \leq n - 1$). We must determine the correlation $[V_i, V_{i+n}]$. We have the decomposition

(3) $$K_{i,i+n-1} = G_{i,i+n-1} \oplus K_{i+1,i+n-1}$$

(4) $$K_{i+1,i+n} = K_{i+1,i+n-1} \oplus G_{i+1,i+n}^* \quad .$$

We may then decompose V_i corresponding to formula (3) and V_{i+n} corresponding to (4). The projections of both V_i and V_{i+n} onto $K_{i+1,i+n-1}$ and hence their correlations are already determined by the correlations at distance smaller than n - 1; the freedom we have is choosing the "angle" between $G_{i,i+n-1}$ and $G_{i+1,i+n}^*$. This will be given by an arbitrary contraction $\Gamma_{i,i+n} : G_{i,i+n-1} \to G_{i+1,i+n}^*$.

It thus becomes clear that the actual computations will require a precise identification of the spaces $G_{i,j}$. The dual decompositions

(5) $$K_{i,j} = G_{i,j} \oplus G_{i+1,j} \oplus \cdots \oplus G_{j-1,j} \oplus V_j H$$

(6) $$K_{i,j} = V_iH \oplus G^*_{i,i+1} \oplus \ldots \oplus G^*_{i,j-1} \oplus G^*_{i,j}$$

which are completely analogous to those given by the classical Gram-Schmidt procedure, will be developed step by step. They will be used for $K_{i+1,i+n-1}$ in formulas (3) and (4), and therefore V_i and V_{i+n} will be decomposed accordingly. To obtain the final formula, we will then need (and obtain recurrently) a precise formula for a "transition operator" between the decompositions (5) and (6). A word for the terminology: in connection with (5) and (6), it will often be the case that a certain subspace $\tilde{K} \subset K$ is decomposed in two ways:

(7) $$\tilde{K} = \bigoplus_i \tilde{K}_i = \bigoplus_j \tilde{K}'_j$$

and that we have unitary operators $\Phi_i : E_i \to \tilde{K}_i$, $\Phi'_j : E'_j \to \tilde{K}'_j$. Then we will say that the unitary operator

$$(\bigoplus_i \Phi_i)^{-1}(\bigoplus_j \Phi'_j) : \bigoplus_j E'_j \to \bigoplus_i E_i$$

corresponds to the two decompositions in (7).

We pass now to the details.

Step 1. By Lemma 1, $S_{i,i+1} = \Gamma_{i,i+1}$, where $\Gamma_{i,i+1} \in L(H)$ may be an arbitrary contraction.

Step 2. By Lemma 4, we have

$$V_{i-1}H \vee V_iH = \Omega^i_{i-1} D_{\Gamma_{i-1,i}} \oplus V_iH$$

$$V_iH \vee V_{i+1}H = V_iH \oplus \Omega^{i+1}_{*,i} D_{\Gamma^*_{i,i+1}} .$$

That is, Ω^i_{i-1} and $\Omega^{i+1}_{*,i}$ identify $D_{\Gamma_{i-1,i}}$ and $D_{\Gamma^*_{i,i+1}}$ with $G_{i-1,i}$ and $G^*_{i,i+1}$ respectively.

Also

$$V_{i-1} = \Omega^i_{i-1} D_{\Gamma_{i-1,i}} + V_i \Gamma_{i-1,i}$$

$$V_{i+1} = V_i \Gamma^*_{i,i+1} + \Omega^{i+1}_{*,i} D_{\Gamma^*_{i,i+1}} .$$

The orthogonality of these decompositions yields

$$V^*_{i+1} V_{i-1} = \Gamma_{i,i+1} \Gamma_{i-1,i} + D_{\Gamma^*_{i,i+1}} \Omega^{i+1}_{*,i} \Omega^i_{i-1} D_{\Gamma_{i-1,i}} .$$

The first term on the right corresponds to the projections of V_{i+1} and V_{i-1} onto

$V_i H$; the freedom we have is to choose in $K \ominus V_i H$ the correlation between Ω^{i}_{i-1} and $\Omega^{i+1}_{*,i}$ (respectively, the "angle" between $G_{i-1,i}$ and $G^*_{i,i+1}$). By Lemma 1, this is given by an arbitrary contraction $\Gamma_{i-1,i+1} : D_{\Gamma_{i-1,i}} \to D_{\Gamma^*_{i,i+1}}$.

Having fixed $\Gamma_{i-1,i+1}$, let us apply Lemma 4 to Ω^{i}_{i-1} and $\Omega^{i+1}_{*,i}$. We obtain thus isometries

$$\Omega^{i+1}_{i-1} : D_{\Gamma_{i-1,i+1}} \to (\Omega^{i}_{i-1} D_{\Gamma_{i-1,i}} \vee \Omega^{i+1}_{*,i} D_{\Gamma^*_{i,i+1}}) \ominus \Omega^{i+1}_{*,i} D_{\Gamma^*_{i,i+1}} = G_{i-1,i+1}$$

$$\Omega^{i+1}_{*,i-1} : D_{\Gamma^*_{i-1,i+1}} \to (\Omega^{i}_{i-1} D_{\Gamma_{i-1,i}} \vee \Omega^{i+1}_{i} D_{\Gamma^*_{i,i+1}}) \ominus \Omega^{i}_{i-1} D_{\Gamma_{i-1,i}} = G^*_{i-1,i+1} .$$

There is no need to write explicitely these isometries; the only fact we will retain is that, following Lemma 4, to the two decompositions

$$\Omega^{i}_{i-1} D_{\Gamma_{i-1,i}} \vee \Omega^{i+1}_{*,i} D_{\Gamma^*_{i,i+1}} = G_{i-1,i+1} \oplus G^*_{i,i+1} = G_{i-1,i} \oplus G^*_{i-1,i+1}$$

corresponds the operator

$$J(\Gamma_{i-1,i+1}) : D_{\Gamma_{i-1,i}} \oplus D_{\Gamma^*_{i-1,i+1}} \to D_{\Gamma_{i-1,i+1}} \oplus D_{\Gamma^*_{i,i+1}} .$$

<u>Step n</u>. We state the induction hypothesis. All that follows is valid for $i < j \leq i + n - 1$.

We have fixed the contractions

$$\Gamma_{i,j} : D_{\Gamma_{i,j-1}} \to D_{\Gamma^*_{i+1,j}} .$$

We have determined (in a manner that will be precised in the induction step) unitary operators

$$\Omega^{j}_{i} : D_{\Gamma_{i,j}} \to G_{i,j}$$

$$\Omega^{j}_{*,i} : D_{\Gamma^*_{i,j}} \to G^*_{i,j} .$$

Corresponding to the two decompositions (5), (6) of $K_{i,j}$ we have a unitary operator

$$U_{i,j} : H \oplus D_{\Gamma^*_{i,i+1}} \oplus \cdots \oplus D_{\Gamma^*_{i,j}} \to D_{\Gamma_{i,j}} \oplus \cdots \oplus D_{\Gamma_{j-1,j}} \oplus H .$$

We have

(8) $P_{K_{i+1,j}} V_i = V_{i+1} \Gamma_{i,i+1} + \Omega^{i+2}_{*,i+1} \Gamma_{i,i+2} D_{\Gamma_{i,i+1}} + \ldots + \Omega^{j}_{*,i+1} \Gamma_{i,j} D_{\Gamma_{i,j-1}} D_{\Gamma_{i,j-2}} \cdots D_{\Gamma_{i,i+1}}$

(9) $$P_{G_{i,j}} V_i = \Omega_i^j D_{\Gamma_{i,j}} D_{\Gamma_{i,j-1}} \dots D_{\Gamma_{i,i+1}}$$

(10) $$P_{K_{i,j-1}} V_j = V_{j-1}\Gamma^*_{j-1,j} + \Omega^{j-1}_{j-2}\Gamma^*_{j-2,j} D_{\Gamma^*_{j-1,j}} + \dots + \Omega^{j-1}_i \Gamma^*_{i,j} D_{\Gamma^*_{i+1,j}} \dots D_{\Gamma^*_{j-1,j}}$$

(11) $$P_{G^*_{i,j}} V_j = \Omega^j_{*,i} D_{\Gamma^*_{i,j}} D_{\Gamma^*_{i+1,j}} \dots D_{\Gamma^*_{j-1,j}}$$

((8) and (10) are orthogonal decompositions).

We make now the induction step. By (3) and (4), we must have

(12) $$[V_i, V_{i+n}] = [P_{K_{i+1,i+n-1}} V_i , P_{K_{i+1,i+n-1}} V_{i+n}] + [P_{G_{i,i+n-1}} V_i , P_{G^*_{i+1,i+n}} V_{i+n}].$$

The first term on the right is fixed by (8) and (10); the freedom in the second is to choose the "angle" between $G_{i,i+n-1}$ and $G^*_{i+1,i+n}$, or equivalently, the correlation $[\Omega_i^{i+n-1}, \Omega^{i+n}_{*,i+1}]$. This is given by an arbitrary contraction

$$\Gamma_{i,i+n} : D_{\Gamma_{i,i+n-1}} \to D_{\Gamma^*_{i+1,i+n}} .$$

Once $\Gamma_{i,i+n}$ is fixed, we apply Lemma 4 and obtain the unitary operators

$$\Omega_i^{i+n} : D_{\Gamma_{i,i+n}} \to G_{i,i+n}$$

$$\Omega^{i+n}_{*,i} : D_{\Gamma^*_{i,i+n}} \to G^*_{i,i+n} .$$

We have the two decompositions

$$K_{i,i+n} \ominus K_{i+1,i+n-1} = G_{i,i+n} \oplus G^*_{i+1,i+n} = G_{i,i+n-1} \oplus G^*_{i,i+n}$$

and a corresponding unitary operator

$$J(\Gamma_{i,i+n}) : D_{\Gamma_{i,i+n-1}} \oplus D_{\Gamma^*_{i,i+n}} \to D_{\Gamma_{i,i+n}} \oplus D_{\Gamma^*_{i+1,i+n}} .$$

The definition of $\Gamma_{i,i+n}$ allow us to extend immediately the range of validity of formulas (8)-(11) to the case $j = i + n$. As a consequnce, we may write explicitely formula (12) as

(13) $$S_{i,i+n} = [V_i, V_{i+n}] = Y^*_{i+1,i+n} U_{i+1,i+n-1} X_{i,i+n-1} + (\Delta^{i+n}_{*,i+1})^* \Gamma_{i,i+n} \Delta_i^{i+n-1}$$

where

$$X_{i,j} = (\Gamma_{i,i+1} , \Gamma_{i,i+2} D_{\Gamma_{i,i+1}} , \dots , \Gamma_{i,j} D_{\Gamma_{i,j-1}} \dots D_{\Gamma_{i,i+1}})$$

$$Y_{i,j} = (\Gamma^*_{j-1,j}, \Gamma^*_{j-2,j} D_{\Gamma^*_{j-1,j}}, \ldots, \Gamma^*_{i,j} D_{\Gamma^*_{i+1,j}} \ldots D_{\Gamma^*_{j-1,j}})$$

$$\Delta_i^j = D_{\Gamma_{i,j}} \ldots D_{\Gamma_{i,i+1}}$$

$$\Delta_{*,i}^j = D_{\Gamma^*_{i,j}} \ldots D_{\Gamma^*_{j-1,j}} .$$

The last thing we need to make (13) explicit is a recurrent formula for $U_{i,i+n}$. First of all, we note that

$$K_{i,j} = G_{i,j} \oplus G_{i+1,j} \oplus \ldots \oplus V_j H = G_{i,j} \oplus V_{i+1} H \oplus G^*_{i+1,i+2} \oplus \ldots \oplus G^*_{i+1,j}$$

and the corresponding operator is $I_{D_{\Gamma_{i,j}}} \oplus U_{i+1,j}$.

Then, we use repeatedly the fact that to the decompositions

$$K_{r,s} \ominus K_{r+1,s-1} = G_{r,s-1} \oplus G^*_{r,s} = G_{r,s} \oplus G^*_{r+1,s}$$

corresponds, as seen above, the operator $J(\Gamma_{r,s})$.

Thus, to the chain of decompositions

$$G_{i,j} \oplus V_{i+1} H \oplus \ldots \oplus G^*_{i+1,j} =$$

$$= G_{i,j-1} \oplus V_{i+1} H \oplus \ldots \oplus G^*_{i+1,j-1} \oplus G^*_{i,j} =$$

$$= G_{i,j-2} \oplus V_{i+1} H \oplus \ldots \oplus G^*_{i+1,j-2} \oplus G^*_{i,j-1} \oplus G^*_{i,j} =$$

$$\cdots\cdots\cdots\cdots\cdots\cdots\cdots\cdots\cdots\cdots\cdots\cdots$$

$$= G_{i,i+1} \oplus V_{i+1} H \oplus G^*_{i,i+2} \oplus G^*_{i,i+3} \oplus \ldots \oplus G^*_{i,j} =$$

$$= V_i H \oplus G^*_{i,i+1} \oplus G^*_{i,i+2} \oplus \ldots \oplus G^*_{i,j}$$

corresponds a chain of elementary rotations $J(\Gamma_{i,j})$, $J(\Gamma_{i,j-1})$, $\ldots$, $J(\Gamma_{i,i+1})$.

Thus, if we denote by

$$\tilde{J}_{i,j}(\Gamma_{i,r}) : \overset{*}{D}_{\Gamma_{i,r}} \oplus H \oplus D_{\Gamma^*_{i+1,i+2}} \oplus \ldots \oplus \overset{*}{D}_{\Gamma^*_{i+1,r}} \oplus D_{\Gamma^*_{i,r+1}} \oplus \ldots \oplus D_{\Gamma^*_{i,j}} \rightarrow$$

$$\rightarrow \overset{*}{D}_{\Gamma_{i,r-1}} \oplus H \oplus D_{\Gamma^*_{i+1,i+2}} \oplus \ldots \oplus \overset{*}{D}_{\Gamma^*_{i,r}} \oplus D_{\Gamma^*_{i,r+1}} \oplus \ldots \oplus D_{\Gamma^*_{i,j}}$$

the operator that acts as $J(\Gamma_{i,r})$ on the starred spaces and as the identity on the rest (for $r = i+2, \ldots, j-1$; there is an obvious modification for $r = i+1$ and $r = j$), we have the recurrent formula

(14) $$U_{i,j} = \tilde{\mathfrak{J}}_{i,j}(\Gamma_{i,i+1})\tilde{\mathfrak{J}}_{i,j}(\Gamma_{i,i+2})\dots\tilde{\mathfrak{J}}_{i,j}(\Gamma_{i,j})(I_{\mathcal{D}_{\Gamma_{i,j}}} \oplus U_{i+1,j}).$$

Leaving aside all our geometric construction in K, we state the theorem we have proved, and which is in fact the basic result of [10]:

THEOREM. *There is a one-to-one correspondence between positive matrices* $\{S_{i,j}\}_{i,j\in\mathbf{Z}}$, $S_{i,j} \in L(H)$ *and generalized choice sequences; that is, families of contractions* $\{\Gamma_{i,j}\}_{i<j}$, *where* $\Gamma_{i,i+1} \in L(H)$ *and* $\Gamma_{i,j} \in L(\mathcal{D}_{\Gamma_{i,j-1}}, \mathcal{D}_{\Gamma^*_{i+1,j}})$. *The correspondence is given by formula* (13).

OTHER REMARKS. 1. Toeplitz case. Historically, the first treated was the case of Toeplitz matrices. This corresponds in Theorem A (and in the language of prediction theory) to a *stationary* process $\{V_i\}_{i\in\mathbf{Z}}$; that is, with correlations depending only on the difference of the indices. The inductive procedure above yields then a *single* contraction $\Gamma_i : \mathcal{D}_{\Gamma_{i-1}} \to \mathcal{D}_{\Gamma^*_{i-1}}$ at each step, and we obtain the usual notion of "choice sequence".

2. The Levinson algorithm. In the scalar stationary case of prediction theory (**dim** $H = 1$), there is a classical procedure (the Levinson algorithm [13]) which allows us, once we know the correlations S_{ij}, to write recurrently, in a simple and efficient way, the projection of V_i onto $K_{i+1,i+n}$ as a linear combination of $V_{i+1},\dots,V_{i+n}$ (this is, in fact, the actual problem of prediction theory). It has been extended to multivariate prediction (that is, **dim** $H < \infty$) (see [12]). This would correspond in our case to a formula as

(15) $$P_{K_{i+1,i+n}} V_i = \sum_{j=i+1}^{i+n} V_j A^n_{i,j}, \quad \text{with } A^n_{i,j} \in L(H).$$

Unfortunately, this is not always possible in the general (even stationary) case. The obstruction is the fact that if $T_1, T_2 \in L(H,K)$, and $\overline{T_1 H} = \overline{T_2 H}$, it is not always possible to write $T_1 = T_2 S$, with $S \in L(H)$. Thus, the projection of $T \in L(H,K)$ onto $\overline{T_2 H}$ is not necessarily of the form $T_2 S$, with $S \in L(H)$; consequently, (15) may fail already for n = 1. The general condition on T_2 for such a formula to hold for any T_1 is:

(*) $$0 \text{ is not an accumulation point of } \sigma(|T_2|).$$

If it is satisfied, the generalized (Moore-Penrose) inverse of T_2, $T_2^{(-1)}$ exists as a bounded operator, and we have

(16) $$P_{\overline{T_2 H}} T = T_2(T_2^{(-1)}T)$$

We will now briefly present the Levinson algorithm, assuming condition (*) satisfied whereever it will be necessary.

Suppose (15) is valid, together with

(15') $$P_{K_{i-n,i-1}} V_i = \sum_{j=i-n}^{i-1} V_j A'^n_{i,j}$$

(it is, of course, natural that we should develop the algorithm simultaneously "in both directions").

Then

(17) $$P_{K_{i+1,i+n+1}} V_i = P_{K_{i+1,i+n}} V_i + P_{G^*_{i+1,i+n+1}} V_i = P_{K_{i+1,i+n}} V_i + P_{G^*_{i+1,i+n+1}} P_{G_{i,i+n}} V_i.$$

Then, by (9) and (11)

(18) $$P_{G_{i,i+n}} V_i = \Omega_i^{i+n} \Delta_i^{i+n}$$

(19) $$P_{G^*_{i+j,i+n+1}} V_{i+n+1} = \Omega^{i+n+1}_{*,i+1} \Delta^{i+n+1}_{*,i+1} .$$

Moreover, $G^*_{i+1,i+n+1}$ is spanned by $P_{G^*_{i+1,i+n+1}} V_{i+n+1} H$, while, by (15'),

(20) $$P_{G^*_{i+1,i+n+1}} V_{i+n+1} = V_{i+n+1} - P_{K_{i+1,i+n}} V_{i+n+1} = V_{i+n+1} - \sum_{j=i+1}^{i+n} V_j A'^n_{i+n+1,j} .$$

We may thus use (18), (19) and (20) in formula (16) in order to calculate the second term of (17); we get

$$P_{G^*_{i+1,i+n+1}} P_{G_{i,i+n}} V_i = (V_{i+n+1} - \sum_{j=i+1}^{i+n} V_j A'^n_{i+n+1,j}) E_{i,i+n+1}$$

where

$$E_{i,j} = (\Delta^j_{*,i+1})^{(-1)} \Gamma_{i,j} \Delta_i^{j-1} .$$

Since the first term in (17) is given by (15), we get

$$P_{K_{i+1,i+n+1}} V_i = \sum_{j=i+1}^{i+n} V_j (A^n_{i,j} - A'^n_{i+n+1,j} E_{i,i+n+1}) + V_{i+n+1} E_{i,i+n+1}$$

whence

$$(21)\qquad \begin{cases} A^{n+1}_{i,j} = A^{n}_{i,j} - A'^{n}_{i+n+1,j}E_{i,i+n+1} & \text{for } j = i+1,\dots,i+n \\ A^{n+1}_{i,i+n+1} = E_{i,i+n+1}\,. \end{cases}$$

Similarly,

$$(21')\qquad \begin{cases} A'^{n+1}_{i,j} = A'^{n}_{i,j} - A^{n}_{i-n-1,j}E'_{i-n-1,j} & \text{for } j = i-n,\dots,i-1 \\ A'^{n+1}_{i,i-n-1} = E'_{i-n-1,i} \end{cases}$$

with

$$E'_{i,j} = (\Delta^{j-1}_{i})^{(-1)}\Gamma^{*}_{i,j}(\Delta^{j}_{*,i+1}).$$

To determine $E_{i,j}$ and $E'_{i,j}$, we need not the algorithm of Theorem 1, since there is another simple inductive method based on (15) and (15'). We have (using also (18) and (19)):

$$S_{i,i+n+1} = [V_i,V_{i+n+1}] = [P_{G_{i,i+n}}V_i,P_{G^{*}_{i+1,i+n+1}}V_{i+n+1}] + [P_{K_{i+1,i+n}}V_i,V_{i+n+1}] =$$

$$= (\Delta^{i+n+1}_{*,i+1})^{*}\,\Gamma_{i,i+n+1}\,\Delta^{i+n}_{i} + \sum_{j=i+1}^{i+n} S_{j,i+n+1}A^{n}_{i,j}\,.$$

Thus,

$$(22)\qquad E_{i,i+n+1} = [(\Delta^{i+n+1}_{*,i+1})^{*}\,\Delta^{i+n+1}_{*,i+1}]^{(-1)}[S_{i,i+n+1} - \sum_{j=i+1}^{i+n} S_{j,i+n+1}A^{n}_{i,j}]$$

and, similarly,

$$(22')\qquad E'_{i-n-1,i} = [(\Delta^{i-n}_{i-n-1})^{*}\,\Delta^{i-n}_{i-n-1}]^{(-1)}[S_{i-n-1,i} - \sum_{j=i-n}^{i-1} S_{i-n-1,j}A'^{n}_{i,j}].$$

The formulas (21), (21'), (22), (22') (and the relations between $E_{i,j}$, $E'_{i,j}$ and $\Gamma_{i,j}$) constitute the general, nonstationary form of the Levinson algorithm. For the algorithm to work, it is obviously necessary that all Δ^{j}_{i}, $\Delta^{j}_{*,i}$ satisfy condition (*).

We will not develop all the simplifications of the algorithm in different particular cases. However, note that an important situation in which condition (*) is automatically satisfied is, of course, $\dim H < \infty$. We thus obtain a nonstationary variant of the multivariate Levinson algorithm.

For the stationary case, an alternate condition "a priori" for the functioning of the Levinson algorithm is given in [18].

Finally, we leave to the interested reader the task of recapturing in the geometrical frame presented above several results of [3], [8], [9] (or even [5]).

REFERENCES

1. **Adamjan, V.M. ; Arov, D.Z. ; Krein, M.G. :** Analytic properties of Schmidt pairs for a Hankel operator and the generalized Schur-Takaji problem (Russian), *Mat. Sb.* **86**(1971), 34-75.
2. **Arsene, Gr. ; Ceaușescu, Zoia ; Foiaș , C. :** On intertwining dilations. VII, LNM 747, *Complex Analysis Joensuu 1978*, Springer Verlag, 1979; VIII, *J. Operator Theory* **4**(1980), 55-92.
3. **Arsene, Gr. ; Constantinescu, T. :** The structure of the Naimark dilation and Gaussian stationary processes, *Integral Equations Operator Theory* **8**(1985), 181-204.
4. **Arsene, Gr. ; Constantinescu, T. :** Structure of positive block-matrices and non-stationary prediction, *J. Functional Analysis*, to appear.
5. **Burg, J.P. :** *Maximum entropy spectral analysis*, Ph. D. Thesis, Stanford, 1975.
6. **Ceaușescu, Zoia ; Foiaș, C. :** On intertwining dilations. V, *Acta Sci. Math. (Szeged)* **40**(1978), 9-32.
7. **Ceaușescu, Zoia ; Foiaș, C. :** On intertwining dilations. VI, *Rev. Roumaine Math. Pures Appl.* **23**(1978), 1471-1482.
8. **Constantinescu, T. :** On the structure of positive Toeplitz forms, in *Dilation theory, Toeplitz operators and other topics*, Birkhäuser Verlag (OT-Series 11), 1983, pp.127-149.
9. **Constantinescu, T. :** On the structure of Naimark dilations, *J. Operator Theory* **12**(1984), 159-175.
10. **Constantinescu, T. :** Schur analysis of positive block-matrices, INCREST Preprint **54**(1984).
11. **Geronimus, I.L. :** *Orthogonal polynomials on the circle and on the line* (Russian), Moscow, 1958.
12. **Kailath, T. :** A view of three decades of linear filtering theory, *IEEE Trans. Information Theory* **IT-20**(1974), 146-181.
13. **Levinson, N. :** The Wiener rms error criterion in filter design and prediction, *J. Math. Phys.* **25**(1946), 261-278.
14. **Parthasarathy, K.R. ; Schmidt, K. :** *Positive definite kernels, continuous tensor products and central limit theorems of probability theory*, LNM **272**, Springer Verlag, Berlin – Heidelberg – New York, 1972.
15. **Sarason, D. :** Generalized interpolation in H^{∞}, *Trans. Amer. Math. Soc.* **127** (1967), 179-203.
16. **Schur, I. :** Über Potenzreihen, die im Innern des Einheitskreises beschränkt sind, *J. Reine Angew. Math.* **148**(1918), 122-145.
17. **Suciu, I. ; Valușescu I. :** Factorization theorems and prediction theory, *Rev. Roumaine Math. Pures Appl.* **23**(1978), 1393-1423.
18. **Timotin, D. :** The Levinson algorithm in linear prediction, in *Invariant subspaces and other topics*, Birkhäuser Verlag (OT-Series 8), 1982, pp.217-223.

Dan Timotin

Department of Mathematics, INCREST
Bdul Păcii 220, 79620 Bucharest
Romania.

Operator Theory:
Advances and Applications, Vol.17

IDEAL PROPERTIES OF ORDER BOUNDED OPERATORS ON ORDERED BANACH SPACES WHICH ARE NOT BANACH LATTICES

Dan Tudor Vuza

0. INTRODUCTION

In 1979, P.Dodds and D.H.Fremlin published their famous result asserting that if E, F are Banach lattices such that E' and F have order continuous norms and if U, V : E → F are linear operators such that $0 \leq U \leq V$ then the compacity of V implies the compacity of U.

Since then, many theorems of Dodds-Fremlin type for various classes of operators were given. In particular, the problem of the inclusion of the order ideal generated by an order bounded operator into the closed algebraic ideal generated by it was considered by several authors (C.D.Aliprantis and O.Burkinshaw [1], N.J.Kalton and P.Saab [6], H.Leinfelder [7], B. de Pagter [10], D.Vuza [17], etc.) All this was done assuming that the operators act between Banach lattices (or more generally, vector lattices endowed with locally solid topologies). The present paper considers a variant of the same problem in the case when the operators are defined on an ordered Banach space which might not be a Banach lattice (i.e. it is not a lattice and/or its topology is not locally solid). The importance of this case is due to the existence of an order relation on the Sobolev spaces.

The methods used here are based on principal modules theory. This theory was developed by the author during the years 1980-1981 in a series of papers circulated as INCREST preprints (see [12], [13]). Some applications of it were presented at the First Romanian-GDR Seminar on Banach space theory held in Bucharest, 1981 (see [14]) and at the International Conference on operator algebras and ideals held in Leipzig, 1983 (see [16]).

The theory of principal modules provides an unified framework for the results in the area of Dodds-Fremlin type theorems. Besides its applications we shall present here, we refer the reader to [14] for an application to perfect M-tensor products and to [16] for a proof of Schep's theorem on kernel operators based on principal modules theory as well for various applications to approximable operators and to the characterization of the band generated by the finite rank operators.

1. PRELIMINARIES

1_M will always denote the identity map of a set M.

For an ordered vector space E we shall use the standard notations:

$$E_+ = \{x \mid x \in E, x \geq 0\},$$

$$[x,y] = \{z \mid z \in E, x \leq z \leq y\}.$$

We say that the positive cone E_+ in the ordered vector space E is generating if $E = E_+ - E_+$. If E is an ordered normed vector space we say that E_+ is b-strict if there is $a > 0$ such that for every $x \in E$ there is $y \in E$ verifying $-y \leq x \leq y$ and $\|y\| \leq a\|x\|$. Every b-strict cone is generating. By Corollary 1.28 of [11], if the positive cone of an ordered Banach space is closed and generating, it is also b-strict.

The vector space of all linear maps between two ordered vector spaces is ordered in the usual way: $U \geq 0$ if $U(E_+) \subset F_+$.

For E, F vector lattices with F order complete we denote by $L_r(E,F)$ the vector lattice of all order bounded linear maps $U : E \to F$.

If E, F are Banach spaces, L(E,F) will be the Banach space of all linear continuous maps $U : E \to F$. The dual of a Banach space E will be denoted by E'.

A set M in a Banach lattice E is called L-bounded if for every $\varepsilon > 0$ there is $y \in E_+$ such that $\|(|x| - y)_+\| \leq \varepsilon$ for every $x \in M$. Every compact set is L-bounded; the solid convex hull of an L-bounded set is L-bounded. If E is a Banach space and F a Banach lattice, LW(E,F) will be the space of all linear maps $U : E \to F$ which carry the unit ball of E into an L-bounded subset of F. Every compact linear map from E to F is in LW(E,F).

Let E be an Archimedean vector lattice. The center of E is the set of all linear maps $U : E \to F$ for which there is $a \geq 0$ such that $|U(x)| \leq a|x|$ for every $x \in E$. Denote by C(E) the center of E; it is a subalgebra of the algebra of all linear maps on E and a vector lattice having 1_E as strong order unit. The modulus $|U|$ of $U \in C(E)$ is given by

$$|U|(x) = |U(x)|$$

for every $x \in E_+$.

Let E, F be Archimedean vector lattices. The tensor product $E\bar{\otimes} F$ in the sense of D.H.Fremlin ([3]) is an Archimedean vector lattice and there is a canonical Riesz bimorphism $\psi : E \times F \to E\bar{\otimes} F$; we use the notation $x \otimes y$ for $\psi(x,y)$. The couple $(E\bar{\otimes} F, \psi)$ is universal in the following sense: for every Archimedean vector lattice G and every Riesz bimorphism $\phi : E \times F \to G$ there is an unique Riesz morphism $\bar{\Phi} : E\bar{\otimes} F \to G$ such that $\phi = \overline{\phi}\psi$. We recall that the linear map $U : E \to F$ is a Riesz morphism if $|U(x)| = U(|x|)$ for every $x \in E$; the bilinear map $\phi : E \times F \to G$ is a Riesz bimorphism if $|\phi(x,y)| = \phi(|x|, |y|)$ for every $x \in E$, $y \in F$. The canonical morphism ψ induces an injective map from the algebraic tensor product $E \otimes F$ into $E \overline{\otimes} F$; we shall identify $E \otimes F$ with its image in $E \overline{\otimes} F$.

2. PRINCIPAL MODULES

We collect here the basic definitions and results we shall need from principal modules theory (see [12] and [13]).

A lattice-ordered algebra with unit is an Archimedean vector lattice A with a strong unit e endowed with a bilinear multiplication which is a Riesz bimorphism and admits e as algebraic unit. On every lattice-ordered algebra with unit e we can give a norm by

$$\|x\| = \inf\{a \mid a \in \mathbf{R}_+,\ |x| \leq ae\}.$$

By a lattice-ordered subalgebra of the lattice-ordered algebra A we shall mean a subalgebra which is also a vector sublattice.

Let A be a lattice-ordered algebra with unit. By an A-module we shall mean a vector lattice E which is an algebraic module over A such that the map $(a,x) \mapsto ax$ (from $A \times E$ into E) is a Riesz bimorphism.

The center $C(E)$ of an Archimedean vector lattice E is a lattice-ordered algebra with unit; the map $(U,x) \mapsto U(x)$ defines a structure of $C(E)$-module on E.

A principal A-module E is an A-module endowed with a locally solid topology such that for every $x \in E$ the set $\{ax \mid a \in A\}$ is dense in the principal order ideal generated by x. This is equivalent to require the set $\{ax \mid a \in [0,e]\}$ to be dense in $[0,x]$ for every $x \in E_+$ or to require the set $\{ax \mid a \in [-e,e]\}$ to be dense in $[-|x|,|x|]$ for every $x \in E$.

THEOREM 2.1. ([12]) *Let* E *be an Archimedean* A*-module endowed with a locally solid topology. Then* E *is principal if and only if for every neighborhood* V *of* 0 *and every* $x_1, x_2 \in E$ *such that* $x_1 \wedge x_2 = 0$ *there are* $a_1, a_2 \in A$ *such that* $a_1 \wedge a_2 = 0$ *and* $x_i - a_i x_i \in V$, $i = 1, 2$.

If E is an order complete vector lattice and $x_1, x_2 \in E$, $x_1 \wedge x_2 = 0$ we can find band projections P_1, P_2 such that $P_1 \wedge P_2 = 0$ and $x_i - P_i(x_i) = 0$, $i = 1, 2$. Hence, by Theorem 2.1, E is a principal $C(E)$-module for every locally solid topology on E.

If A, B are lattice-ordered algebras with units e_1, e_2, by the universality property of $A\bar{\otimes}B$ there is a unique structure of lattice-ordered algebra on $A\bar{\otimes}B$ such that $(a_1\otimes b_1)(a_2\otimes b_2) = a_1a_2\otimes b_1b_2$ for $a_1, a_2 \in A$, $b_1, b_2 \in B$; its unit is $e_1\otimes e_2$.

Suppose that E is an Archimedean A-module and also a B-module. Then we can give a unique structure of $A\bar{\otimes}B$-module on E such that $(a\otimes b)x = a(bx)$ for every $a \in A$, $b \in B$, $x \in E$.

Let E be an A-module and let F be a B-module. Suppose that E and F are principal, that F is order complete and its topology is order continuous. Denote by $L_r'(E,F)$ the vector lattice of all continuous linear maps in $L_r(E,F)$. Define structures of A-module and B-module on $L_r'(E,F)$ by

$$(aU)(x) = U(ax),$$

$$(bU)(x) = bU(x)\,.$$

Then $L_r'(E,F)$ becomes an $A\bar{\otimes}B$-module. The solid strong topology on $L_r'(E,F)$ has as a basis of neighborhoods of 0 the sets $\{U \mid U \in L_r'(E,F),\ |U|(x) \in V\}$ for every $x \in E_+$ and every neighborhood V of 0 in F.

THEOREM 2.2. ([13]) *With respect to the solid strong topology,* $L_r'(E,F)$ *is a principal* $A\bar{\otimes}B$*-module.*

3. ADDITIONAL RESULTS ABOUT A-MODULES

It is well known that if B is a subalgebra of a lattice-ordered algebra A and $e \in B$ then its closure $\bar{B}$ is a lattice-ordered subalgebra of A.

DEFINITION 3.1. Let E be an A-module endowed with a locally solid topology and let B be a subalgebra of A. We say that E is B-*principal* if for every $x_1, x_2 \in E$ such that $x_1 \wedge x_2 = 0$ and every neighborhood V of 0 in E there are $b_1, b_2 \in B$ such that $x_1 - bx_1 \in V$ and $bx_2 \in V$.

THEOREM 3.1. *Let* E *be an Archimedean* A*-module endowed with a locally solid topology and let* B *be a subalgebra of* A *containing* e*. Then* E *is* B*-principal if and only if* E *is a principal* $\bar{B}$*-module.*

PROOF. We shall apply Theorem 2.1. Let E be B-principal and let V be a neighbourhood of 0. There is a solid neighborhood W of 0 such that $W + W \subset V$. Take $x_1, x_2 \in E$ such that $x_1 \wedge x_2 = 0$. By the hypothesis there is $b \in B$ such that $x_1 - bx_1 \in W$

and $bx_2 \in W$. Put $b_1 = b$, $b_2 = e - b$. From

$$|x_i - |b_i|x_i| = ||x_i| - |b_ix_i|| \leq |x_i - b_ix_i|$$

it follows that $x_i - |b_i|x_i \in W$. We have

$$(|b_1| \wedge |b_2|)x_1 \leq |b_2|x_1 = |b_2x_1|$$

hence $(|b_1| \wedge |b_2|)x_1 \in W$; similarly $(|b_1| \wedge |b_2|)x_2 \in W$. Consequently, if $c_i = |b_i| - |b_1| \wedge |b_2|$ then $c_i \in \bar{B}$, $c_1 \wedge c_2 = 0$ and

$$|x_i - c_ix_i| \leq |x_i - |b_i|x_i| + (|b_1| \wedge |b_2|)x_i ;$$

therefore $x_i - c_ix_i \in W + W \subset V$. As V is arbitrary, we have obtained that E is $\bar{B}$-principal.

Conversely, suppose that E is $\bar{B}$-principal. Let $x_1, x_2 \in E$ be such that $x_1 \wedge x_2 = 0$ and let V be a neighborhood of 0 in E. There is a solid neighborhood W of 0 such that $W + W \subset V$. By the hypothesis there are $c_1, c_2 \in \bar{B}$ such that $c_1 \wedge c_2 = 0$ and $x_i - c_ix_i \in W$, $i = 1, 2$; we may assume that $c_i \in [0,e]$. Let $\varepsilon > 0$ be such that $\varepsilon(x_1 + x_2) \in W$. There is $b \in B$ with $\|c_1 - b\| \leq \varepsilon$. We have

$$|x_1 - bx_1| \leq |x_1 - c_1x_1| + |c_1 - b|x_1 ,$$

$$|bx_2| \leq |c_1x_2| + |b - c_1|x_2 \leq |c_1(x_2 - c_2x_2)| + |c_1c_2x_2| + |b - c_1|x_2 \leq$$

$$\leq |x_2 - c_2x_2| + |b - c_1|x_2 ,$$

hence $x_1 - bx_1 \in V$ and $bx_2 \in V$. As V is arbitrary, we have obtained that E is B-principal.

THEOREM 3.2. *Let* E *be an Archimedean* A*-module endowed with a locally solid topology and let* B *be a subalgebra of* A *containing* e. *Consider a vector subspace* F *of* E *and a vector subspace* F_0 *dense in* F *with the following properties:*

i) $BF_0 \subset F$;

ii) *For every* $x \in F_0$ *and every neighborhood* V *of* 0 *in* E *there is* $b \in B \cap [0,e]$ *such that* $bx \geq 0$ *and* $(e - b)x_+ \in V$.

Then the following are true:

i) *The closure* $\bar{F}$ *of* F *is a vector sublattice of* E.

ii) $(\bar{F})_+$ *is equal to the closure of* $F \cap E_+$.

iii) $\bar{F}$ *is a principal* $\bar{B}$*-module.*

PROOF. First we show that for every solid neighborhood V of 0 and every $x \in F_o$ there is $y \in F \cap E_+$ such that $x_+ - y \in V$. Indeed, there is $b \in B \cap [0,e]$ such that $bx \geq 0$ and $(e - b)x_+ \in V$. As $bx \in F$ and

$$x_+ - bx = x_+ - (bx)_+ = x_+ - bx_+ \in V$$

the result follows.

From the above assertion, i) follows at once. To prove ii), take any $x \in (\bar{F})_+$. Let V be a neighborhood of 0 and let W be a solid neighborhood of 0 such that $W + W \subset V$. As $x \in \bar{F}$, there is $y \in F_o$ such that $x - y \in W$. From

$$|x - y_+| = |x_+ - y_+| \leq |x - y|$$

it follows that $x - y_+ \in W$. There is also $z \in F \cap E_+$ such that $y_+ - z \in W$. It follows that $x - z \in V$; as V is arbitrary, x belongs to the closure of $F \cap E_+$.

By i), $\bar{F}$ is a $\bar{B}$-module. To prove it is principal it is enough, according to Theorem 3.2, to show that $\bar{F}$ is B-principal. Let $x_1, x_2 \in \bar{F}$ be such that $x_1 \wedge x_2 = 0$ and let V be a neighborhood of 0. Consider a solid neighborhood W of 0 such that $W + W + W \subset V$. There are $y_1, y_2 \in F_o$ such that $x_i - y_i \in W$, $i = 1, 2$. As

$$|y_1 \wedge y_2| = |y_1 \wedge y_2 - x_1 \wedge x_2| \leq |y_1 - x_1| + |y_2 - x_2|$$

we have that $y_1 \wedge y_2 \in W + W$. There is $b \in B \cap [0,e]$ such that

$$b(y_1 - y_2) \geq 0, \quad (e - b)(y_1 - y_2)_+ \in W.$$

We have

$$y_1 - (y_1 - y_2)_+ = y_1 \wedge y_2 \in W + W.$$

Therefore

$$(e - b)y_1 = (e - b)(y_1 - (y_1 - y_2)_+) + (e - b)(y_1 - y_2)_+ \in W + W + W \subset V.$$

On the other side

$$by_2 = by_1 - b(y_1 - y_2) = by_1 - (b(y_1 - y_2))_+ = by_1 - b(y_1 - y_2)_+ =$$
$$= b(y_1 - (y_1 - y_2)_+) \in W + W \subset V.$$

As V is arbitrary we have obtained that $\bar{F}$ is principal.

4. B-PAIRS

DEFINITION 4.1. By a B-*pair* we shall mean a couple (E,G) formed by a Banach lattice E and a Banach space G ordered by a b-strict cone together with a positive continuous linear map $J : G \to E$.

DEFINITION 4.2. Let (E,G) be a B-pair. A map $U \in C(E)$ is called *G-central* if there is $V \in L(G,G)$ such that $UJ = JV$.

We denote by $C_G(E)$ the set of all G-central maps; $C_G(E)$ is a subalgebra of $C(E)$ containing 1_E.

DEFINITION 4.3. We say that a B-pair (E,G) is *principal* if the $C(E)$-module E is $C_G(E)$-principal.

If (E,G) is a B-pair and F is an order complete Banach lattice we let $LW_r(E,G,F)$ be the subset of all $U \in L_r(E,F)$ such that $|U| J \in LW(E,F)$.

PROPOSITION 4.1. $LW_r(E,G,F)$ *is an order ideal in* $L_r(E,F)$.

PROOF. Let M be the solid hull of $J(B_G)$, B_G being the closed unit ball of G. The assertion will be proved if we show that $LW_r(E,G,F)$ coincides with the set of all $U \in L_r(E,F)$ such that $|U|(M)$ is L-bounded.

Of course if $|U|(M)$ is L-bounded then $U \in LW_r(E,G,F)$. Conversely, let $U \in LW_r(E,G,F)$. As G_+ is b-strict there is $a > 0$ such that for every $x \in B_G$ there is $y \in G$ with $-y \leq x \leq y$ and $\|y\| \leq a$. Let $\varepsilon > 0$. As $|U|(J(aB_G))$ is L-bounded there is $z \in F_+$ such that $\|(||U|(J(x))| - z)_+\| \leq \varepsilon$ for every $x \in aB_G$. Consider $x \in M$; there is $u \in B_G$ such that $|x| \leq |J(u)|$. As $u \in B_G$ there is $v \in aB_G$ for which $-v \leq u \leq v$; it follows that $|J(u)| \leq J(v)$. From

$$||U|(x)| \leq |U|(|x|) \leq |U|(|J(u)|) \leq |U|(J(v))$$

and

$$\|(|U|(J(v)) - z)_+\| \leq \varepsilon$$

we have that $\|(||U|(x)| - z)_+\| \leq \varepsilon$ for every $x \in M$. Hence $|U|(M)$ is L-bounded.

Let E, F be Banach spaces. A bilateral ideal of $L(E,F)$ is a vector subspace I of $L(E,F)$ with the property that $WUV \in I$ whenever $U \in I$, $V \in L(E,E)$ and $W \in L(F,F)$.

We recall that the dual of a Banach lattice has order continuous norm if and only if for every $f \in E'_+$ and every $\varepsilon > 0$ there is $y \in E_+$ such that $f((x - y)_+) \leq \varepsilon$ for every $x \in E$ with $\|x\| \leq 1$. Similarly, E has order continuous norm if for every $x \in E_+$ and every $\varepsilon > 0$ there is $g \in E'_+$ such that $(f - g)_+(x) \leq \varepsilon$ for every $f \in E'$ with $\|f\| \leq 1$ (for the proof of these assertions see [2]).

THEOREM 4.1. *Let* (E,G) *be a principal* B-*pair such that* E' *has order continuous norm. Let* F *be a Banach lattice with order continuous norm. Consider a closed bilateral ideal I in* L(G,F). *Then the set of all* $U \in LW_r(E,G,F)$ *such that* $UJ \in I$ *is a band in* $LW_r(E,G,F)$.

The proof of the theorem will rely on Theorem 2.2 and on the following lemma:

LEMMA 4.1. *Let* E, F *be Banach lattices such that* E' *and* F *have order continuous norms and let* M *be a bounded solid subset of* E. *Consider the solid seminorm* p_M *on* $L_r(E,F)$ *given by*

$$p_M(U) = \sup\{\| \, |U|(x) \| \mid x \in M\}.$$

Let $U \in L_r(E,F)_+$ *be such that* U(M) *is* L-*bounded. Then the solid strong topology is stronger on* [-U,U] *than the topology defined by* p_M.

PROOF. It suffices to prove that for every $\varepsilon > 0$ there is $x \in E_+$ such that $p_M(V) \leq \|V(x)\| + \varepsilon$ whenever $V \in [0,U]$ (because if $V_1, V_2 \in [-U,U]$ then $|V_1 - V_2| \in [0,2U]$ and we may apply the above inequality to $|V_1 - V_2|$).

Without loss of generality we may assume that M is contained in the unit ball of E. Let $\varepsilon > 0$. As U(M) is L-bounded there is $y \in F_+$ such that $\|(U(x) - y)_+\| \leq \varepsilon/3$ for every $x \in M \cap E_+$ with $\|x\| \leq 1$. As F has order continuous norm there is $f_\varepsilon \in F'_+$ such that $(f - f_\varepsilon)_+(y) \leq \varepsilon/3$ for every $f \in F'$ with $\|f\| \leq 1$. As E' has order continuous norm there is $x_\varepsilon \in E_+$ such that $U'(f_\varepsilon)((x - x_\varepsilon)_+) \leq \varepsilon/3$ for every $x \in E$ with $\|x\| \leq 1$. Let $V \in [0,U]$ and let $x \in M \cap E_+$, $f \in F'_+$, $\|f\| \leq 1$. We have

$$f(V(x)) = (f - f_\varepsilon)(V(x)) + f_\varepsilon(V(x - x_\varepsilon)) + f_\varepsilon(V(x_\varepsilon)) \leq$$

$$\leq (f - f_\varepsilon)_+(U(x)) + U'(f_\varepsilon)((x - x_\varepsilon)_+) + f_\varepsilon(V(x_\varepsilon)).$$

But

$$(f - f_\varepsilon)_+(U(x)) = (f - f_\varepsilon)_+(U(x) - y) + (f - f_\varepsilon)_+(y) \leq$$

$$\leq f((U(x) - y)_+) + (f - f_\varepsilon)_+(y) \leq 2\varepsilon/3.$$

Hence

$$f(V(x)) \leq \|V(\|f_\varepsilon\| x_\varepsilon)\| + \varepsilon.$$

As

$$p_M(V) = \sup\{f(V(x)) \mid x \in M \cap E_+, f \in F'_+, \|f\| \leq 1\}$$

the result follows.

PROOF OF THEOREM 4.1. Let B be the set of all $U \in LW_r(E,G,F)$ such that $UJ \in I$. Clearly B is a vector subspace. We prove first that B is an order ideal of $LW_r(E,G,F)$. Let $V,U \in LW_r(E,G,F)$ be such that $|V| \leq |U|$ and $U \in B$. By Theorems 2.2 and 3.1, $L_r(E,F)$ is a principal $\overline{C_G(E)} \bar{\otimes} C(F)$-module. Hence there is a net $(a_\delta) \subset \overline{C_G(E)} \bar{\otimes} C(F)$ such that $a_\delta \in [-1_E \otimes 1_F, 1_E \otimes 1_F]$ and $a_\delta U \to V$ in the solid strong topology; clearly $a_\delta U \in [-|U|, |U|]$.

Let B_G be the closed unit ball of G and let M be the solid hull of $J(B_G)$ in E. By the proof of Proposition 4.1, $|U|(M)$ is L-bounded. By Lemma 4.1 it follows that $p_M(a_\delta U - V) \to 0$. Therefore, for every $\varepsilon > 0$ there is $a \in \overline{C_G(E)} \bar{\otimes} C(F)$ such that $p_M(aU - V) \leq \varepsilon/2$. It is known (see [3]) that $\overline{C_G(E)} \otimes C(F)$ is dense in $\overline{C_G(E)} \bar{\otimes} C(F)$; as $C_G(E)$ is dense in $\overline{C_G(E)}$ it follows that $C_G(E) \otimes C(F)$ is dense in $\overline{C_G(E)} \bar{\otimes} C(F)$. Hence there is $b \in C_G(E) \otimes C(F)$ with $\|b - a\| \leq \varepsilon(2p_M(U) + 1)^{-1}$. Thus

$$p_M(bU - V) \leq p_M((b - a)U) + p_M(aU - V) \leq \|b - a\| \, p_M(U) + p_M(aU - V) \leq \varepsilon \, ;$$

as $B_G \subset M$ the above inequality implies

$$\|(bU)J - VJ\| \leq \varepsilon \, .$$

As I is a bilateral ideal we have $(bU)J \in I$; as I is closed and ε is arbitrary, $VJ \in I$.

Now let (U_δ) be a net in B and $U \in LW_r(E,G,F)$ be such that $0 \leq U_\delta \uparrow U$. As F has order continuous norm, $U_\delta \to U$ in the solid strong topology. By Lemma 4.1, $p_M(U_\delta - U) \to 0$; hence $\|U_\delta J - UJ\| \to 0$. As $U_\delta J \in I$ and I is closed it follows that $UJ \in I$. Therefore B is a band.

COROLLARY 4.1. *Let* (E,G) *and* F *be as in Theorem* 4.1 *and let* U, F : E → F *be such that* $0 \leq U \leq V$. *If* VJ *is compact then* UJ *is also compact.*

If E, F are Banach spaces, a map $U \in L(E,F)$ will be called approximable if it lies in the uniform closure of all finite-rank continuous linear maps. Every approximable map is compact; the set of all approximable maps is a bilateral ideal.

COROLLARY 4.2. *Let* (E,G) *and* F *be as in Theorem* 4.1 *and let* U, F : E → F *be such that* $0 \leq U \leq V$. *If* VJ *is approximable then* UJ *is also approximable.*

We pass now to some examples of B-pairs. In a first place, we consider the case of the Sobolev spaces. Let Ω be a domain in $\mathbf{R}^n$. The Sobolev space $W^{k,p}(\Omega)$ is the space of all p-integrable functions on Ω having p-integrable derivatives (in the sense of distributions) of all orders $\leq k$; the norm is defined by

$$||f|| = \Big(\sum_{0\leq k_1+\dots+k_n\leq k}\int_\Omega |(\partial^{k_1+\dots+k_n}/\partial x_1^{k_1}\dots\partial x_n^{k_n})f(t)|^p dt\Big)^{1/p}.$$

On the Sobolev spaces the following order relation is given: $f \geq 0$ if $f(t) \geq 0$ for almost every $t \in \Omega$. In this way, a structure of ordered Banach space with closed cone is obtained; however, these spaces are far from Banach lattices (for instance, the order intervals are not norm bounded). The most important cases when the positive cone is generating are the following:

a) $k = 1$, in which case $W^{1,p}(\Omega)$ is a vector lattice (though not a Banach lattice).

b) Ω is bounded, has a smooth boundary and $kp > n$. In this case (see [5]), every function in $W^{k,p}(\Omega)$ is bounded; as the constant functions belong to the space, it follows that the positive cone is generating.

Consider a space $W^{k,p}(\Omega)$ with generating cone. Suppose that $1 < p < \infty$. Then the couple $(L^p(\Omega), W^{k,p}(\Omega))$ together with the inclusion map $J : W^{k,p}(\Omega) \to L^p(\Omega)$ form a B-pair satisfying the hypothesis of Theorem 4.1. Indeed, $L^p(\Omega)$ has order continuous dual; to see the pair is principal, note that the multiplication by a C^∞-function with compact support in Ω defines a $W^{k,p}(\Omega)$-central map on $L^p(\Omega)$. Now if $\varepsilon > 0$ is given and $f_1, f_2 \in L^p(\Omega)$ are such that $f_1 \wedge f_2 = 0$, there are compact disjoint subsets K_1, K_2 of Ω such that $\int_{\Omega\setminus K_i} |f_i(t)|^p dt \leq \varepsilon$. We can find a C^∞-function ϕ with compact support in Ω such that ϕ equals 1 on K_1 and 0 on K_2. Then

$$||f_1 - \phi f_1|| \leq \varepsilon, \quad ||\phi f_2|| \leq \varepsilon$$

(the norm being taken in $L^p(\Omega)$). The assertion is proved.

As a second example we shall construct a B-pair (E,G) and a Banach lattice F satisfying the hypothesis of Theorem 4.1 together with a positive linear map $U : E \to F$ such that UJ is compact but not approximable. In this way it will be proved that Corollary 4.2 is not a direct consequence of Corollary 4.1.

We briefly recall the construction of A.Szankowski's reflexive Banach lattice without the approximation property (for details see [9]). Let $\mathcal{B}_n$ be the algebra of subsets of $[0,1]$ generated by the 2^n atoms $[(i-1)/2^n, i/2^n)$, $i = 1,\dots,2^n$. For every n, let ϕ_n be the permutation of $\{1,2,\dots,2^n\}$ defined by $\phi_n(2i) = 2i-1$, $\phi_n(2i-1) = 2i$. The map ϕ_n induces a permutation between the atoms of $\mathcal{B}_n$ and therefore a map (denoted again by ϕ_n) on $\mathcal{B}_n$.

For every $n \geq 2^6$ a partition Δ_n of $[0,1]$ into M_n disjoint $\mathcal{B}_n$-measurable sets of equal measure is constructed. The Szankowski space E is defined to be the space of equivalence classes of measurable functions on [0,1] such that the norm

$$\|f\| = \Big(\sum_{n=2^6}^{\infty} \sum_{B\in\Delta_n} M_n^{\alpha p}\Big(\int_B |f(t)|^r dt\Big)^{p/r}\Big)^{1/p}$$

is finite (α, p and r being certain positive constants).

A subset M in a Banach space E will be called approximable if for every $\varepsilon > 0$ there is $U \in L(E,E)$ such that $\dim U(E) < \infty$ and $\|x - U(x)\| \leq \varepsilon$ for every $x \in M$.

In the Szankowski space E the following nonapproximable compact set is constructed: let E_n be the set of all B_n-measurable functions such that $|f|$ is the characteristic function of $\phi_n(A)$ for some $A \in \Delta_n$. Then $M = \{0\} \cup \bigcup_{n=2^6}^{\infty} \alpha_n E_n$ is a compact nonapproximable set, α_n being some suitable positive numbers. As the sets E_n are finite, M can be disposed in a sequence converging to $0 : M = \{f_0, f_1, \ldots\}$. Let K be the closed convex hull of $\{f_n/\|f_n\|^{\frac{1}{2}} \mid n \geq 0\}$ and let G be the vector subspace generated by K. Define on G the norm which makes K its unit ball. Then G becomes a Banach space and the inclusion map $J : G \to E$ is compact but not approximable (see the proof of Theorem 1.e.4 in [8]). The order relation on E induces an order relation on G. Therefore, if we prove that (E,G) is a principal B-pair we may take F = E and $U = 1_E$ in order to obtain our example.

First of all, let us show that G_+ is b-strict. Observe that K is the set of all elements $\sum_{n=0}^{\infty} a_n g_n$ where $a_n \geq 0$, $\sum_{n=0}^{\infty} a_n \leq 1$ and $g_n = f_n/\|f_n\|^{\frac{1}{2}}$: indeed, the set

$$C = \{(a_n) \mid a_n \geq 0,\ \sum_{n=0}^{\infty} a_n \leq 1\}$$

is a closed subset of the compact space $[0,1]^{\mathbf{N}}$; as $g_n \to 0$, the map $(a_n) \to \sum_{n=0}^{\infty} a_n g_n$ from C into E is continuous, so it takes C into a compact convex subset of E. From the construction of the sets E_n it is obvious that for every n there is n' such that $|g_n| = g_{n'}$. Hence, if

$$f = \sum_{n=0}^{\infty} a_n g_n \in K$$

then

$$\pm f \leq \sum_{n=0}^{\infty} a_n |g_n| \in K$$

which proves our assertion.

Now we show that (E,G) is a principal pair. First we prove that the multiplication by the characteristic function of an atom of $\mathcal{B}_n$ defines a G-central map on E. Indeed, from the construction of the sets E_n it is easy to see that $U(g_n) \in G$ for every $n \geq 0$ and that $U(g_n)$ is a convex combination of the g_j's for sufficiently large n; therefore the restriction of U to G is continuous.

Now let $\mathcal{B} = \bigcup_{n=0}^{\infty} \mathcal{B}_n$. The set of all $\mathcal{B}$-measurable functions (that is, the functions which are $\mathcal{B}_n$-measurable for some $n \geq 0$) is dense in E. By the above argument, the multiplication by such a function defines a G-central map on E. Let $\varepsilon > 0$ be given and let $h_1, h_2 \in E$ be such that $h_1 \wedge h_2 = 0$. We can find $\mathcal{B}$-measurable functions h'_1, h'_2 such that $\| h_i - h'_i \| \leq \varepsilon$, $i = 1, 2$; replacing h'_i by $h'_i - h'_1 \wedge h'_2$ we may assume that $h'_1 \wedge h'_2 = 0$. Now let ϕ be the characteristic function of the support of h'_1. We have

$$\| h_1 - \phi h_1 \| \leq \| (e - \phi)(h_1 - h'_1) \| + \| (e - \phi) h'_1 \| \leq \varepsilon ,$$

$$\| \phi h_2 \| \leq \| \phi (h_2 - h'_2) \| + \| \phi h'_2 \| \leq \varepsilon$$

(e being the function identical one). The proof is complete.

5. THE CASE OF A REFLEXIVE BANACH SPACE WITH CLOSED GENERATING CONE

DEFINITION 5.1. Let E be an ordered vector space. A *latticial extension* of E is a vector lattice $\tilde{E}$ such that E is a vector subspace of $\tilde{E}$ and $E_+ = E \cap \tilde{E}_+$.

Let E be an ordered Banach space and let $\tilde{E}$ be a latticial extension of E. According to § 4, a map $U \in \mathcal{C}(\tilde{E})$ will be called E-*central* if $U(E) \subset E$ and the restriction of U to E is continuous; the subalgebra of all E-central maps will be denoted by $\mathcal{C}_E(\tilde{E})$.

DEFINITION 5.2. We say that an ordered Banach space E has a *principal latticial extension* if there is a latticial extension $\tilde{E}$ of E and a dense vector subspace E_0 of E with the following property: for every $\varepsilon > 0$ and every $x \in E_0$ there is $U \in \mathcal{C}_E(\tilde{E}) \cap [0, 1_{\tilde{E}}]$ and $y \in E$ such that $\| y \| \leq \varepsilon$, $U(x) \geq 0$ and $(1_{\tilde{E}} - U)(x_+) \leq y$.

THEOREM 5.1. *Let* E *be a reflexive Banach space ordered by a closed generating cone and let* F *be an order continuous Banach lattice. Suppose that* E *has a principal latticial extension. Let* I *be a closed bilateral ideal in* L(E,F). *Consider* $U, V : E \to F$ *such that* $0 \leq U \leq V$ *and* $V \in LW(E,F) \cap I$. *Then* $U \in I$.

PROOF. We shall construct a Banach lattice H with order continuous dual such that (H,E) will be a principal B-pair with the following property: for every positive

$U : E \to F$ there is a unique positive $\tilde{U} : H \to E$ such that $U = \tilde{U}J$. Thus an application of Theorem 4.1 will conclude the proof (we remark that E_+, being closed and generating, is also b-strict).

Let $\tilde{E}$ be a principal latticial extension of E and let K be the convex solid hull of the closed unit ball B_E of E in $\tilde{E}$. Denote by G the vector subspace of $\tilde{E}$ spanned by K; G is an order ideal of $\tilde{E}$, hence a vector lattice. As $U(G) \subset G$ for every $U \in C_E(\tilde{E})$, it follows that G is a $C_E(\tilde{E})$-module. Define the solid seminorm p on G by

$$p(x) = \mathbf{inf}\{a \mid a \in \mathbf{R}_+,\ x \in aK\}.$$

By Theorem 3.2 we have that the closure $\bar{E}$ of E with respect to p is a vector sublattice of G, that $(\bar{E})_+$ is equal to the closure of E_+ and $\bar{E}$ is a principal $\overline{C_E(\tilde{E})}$-module. Put $G_0 = \bar{E} \cap p^{-1}(\{0\})$ and let H be the completion of $\bar{E}/G_0$. If $J : E \to H$ is the canonical map, (H,E) is a principal B-pair. Every positive (hence continuous) linear map $U : E \to F$ is continuous for the restriction of p to E; therefore there is a unique continuous linear $\tilde{U} : H \to F$ such that $U = \tilde{U}J$. As H_+ is the closure of $J(E_+)$ it follows that $\tilde{U}$ is positive.

It remains to show that H' has order continuous norm. To this purpose it suffices to prove that for every $\varepsilon > 0$ and every linear positive $f : \bar{E} \to \mathbf{R}$ continuous for p there is $y \in (\bar{E})_+$ such that $f((|x| - y)_+) \leq \varepsilon$ for every $x \in \bar{E}$ with $p(x) \leq 1$.

Let $\bar{E}'$ be the vector lattice of all linear forms on $\bar{E}$ continuous for p and let $(\bar{E}')^\times$ be the vector lattice of all order continuous linear forms on $\bar{E}'$; obviously, $\bar{E}' \subset L_r(\bar{E},\mathbf{R})$. As E_+ is closed, the restriction of every $f \in \bar{E}'$ to E is continuous. Hence the restriction of the map $k : \bar{E} \to (\bar{E}')^\times$ given by $k(x)(f) = f(x)$ to E is continuous for $\sigma(E,E')$ and $\sigma((\bar{E}')^\times,\bar{E}')$. As B_E is $\sigma(E,E')$-compact it follows that $k(B_E)$ is $\sigma((\bar{E}')^\times,\bar{E}')$-compact. Let $\varepsilon > 0$ and let $f \in (\bar{E}')_+$. As $\bar{E}'$ is order complete, 82E and 82G in [4] imply that there is $\phi \in (\bar{E}')^\times_+$ such that $(|k(x)| - \phi)_+(f) \leq \varepsilon$ for every $x \in B_E$. Examining the proof of 81H in [4] we see that ϕ can be taken of the form

$$\sum_{i=1}^{n} |k(y_i)|$$

with $y_i \in E$. As E_+ is generating, for every i there is $z_i \in E_+$ such that $-z_i \leq y_i \leq z_i$. Hence $\phi \leq k(y)$ where $y = \sum_{i=1}^{n} z_i$. By 31C in [4], k is a Riesz morphism. It follows that

$$f((|x| - y)_+) = k((|x| - y)_+)(f) = (|k(x)| - k(y))_+(f) \leq (|k(x)| - \phi)_+(f) \leq \varepsilon$$

for every $x \in B_E$. The set $\{x \mid x \in \bar{E},\ f((|x| - y)_+) \leq \varepsilon\}$ is a closed solid convex set containing B_E; hence it contains every $x \in \bar{E}$ with $p(x) \leq 1$.

COROLLARY 5.1. *Let* E *and* F *be as in Theorem* 5.1 *and let* $U, V : E \to F$ *be such that* $0 \leq U \leq V$. *If* V *is compact then* U *is also compact.*

COROLLARY 5.2. *Let* E *and* F *be as in Theorem* 5.1 *and let* $U, V : E \to F$ *be such that* $0 \leq U \leq V$. *If* V *is approximable then* U *is also approximable.*

We consider now some examples of spaces satisfying the hypothesis of Theorem 5.1. In the first place, take E to be the Sobolev space $W^{k,p}(\Omega)$ such that $1 < p < \infty$, $kp > n$ and Ω is bounded and has a smooth boundary; these conditions ensure that E is reflexive and E_+ is closed and generating. Let $\tilde{E}$ be the vector lattice of all functions on Ω; clearly $\tilde{E}$ is a latticial extension of E. To see it is principal, let $\varepsilon > 0$ and let $f \in E$ be given. Take g to be the function identic equal to c on Ω, c being a suitable positive constant such that $\|g\| \leq \varepsilon$. As f is continuous on $\bar{\Omega}$ ([5], Theorem 5.7.8), the sets $M_1 = \{t \mid t \in \bar{\Omega},\ f(t) \geq c\}$ and $M_2 = \{t \mid t \in \bar{\Omega},\ f(t) \leq 0\}$ are disjoint compact subsets of $\mathbf{R}^n$. Therefore there is a C^∞-function ϕ with compact support on $\mathbf{R}^n$ such that $0 \leq \phi(t) \leq 1$ for $t \in \mathbf{R}^n$, $\phi(t) = 1$ for $t \in M_1$ and $\phi(t) = 0$ for $t \in M_2$. It follows that the map U defined by the multiplication by ϕ has the properties: $U \in C_E(\tilde{E}) \cap [0, 1_{\tilde{E}}]$, $U(f) \geq 0$, $(1_{\tilde{E}} - U)(f_+) \leq g$. These proves our assertion.

As a second example take E to be the space $W_0^{k,p}(0,1)$ with $1 < p < \infty$. It consists of the closure in $W^{k,p}(0,1)$ of the subspace formed by the C^∞-functions with compact support in (0,1). It can be proved that a function f is in $W_0^{k,p}(0,1)$ if and only if f and the derivatives $d^i f / dt^i$ for $1 \leq i \leq k - 1$ are absolutely continuous functions on [0,1] vanishing at 0 and 1 and $d^k f / dt^k$ is in $L^p(0,1)$. The space E is ordered by the closed cone of all functions taking only positive values. For $k \geq 1$, this cone is not normal (though it is latticial for k = 1); for $k \geq 2$, the cone is neither latticial. Nevertheless, we prove that this cone is generating. Indeed, let $f \in E$ be given and let $a > 0$ and $b > 0$ be such that $f(t) < a < b/4^k k!$ for every $t \in [0,1]$. Define the functions $g_0, \ldots, g_k : [0, \frac{1}{4}] \to \mathbf{R}$ by

$$g_k(t) = ((d^k f / dt^k)(t))_+ + b,$$

$$g_{i-1}(t) = \int_0^t g_i(s)ds, \quad 1 \leq i \leq k.$$

By induction we have that $((d^i f / dt^i)(t))_+ \leq g_i(t)$; hence $(f(t))_+ \leq g_0(t)$ for every $t \in [0, \frac{1}{4}]$; on the other side

$$g_0(1/4) \geq b/4^k k! > a.$$

Clearly, $g_0(0) = 0$. In the same way we define the functions $h_0, \ldots, h_k : [\frac{3}{4}, 1] \to \mathbf{R}$ by

$$h_k(t) = ((-1)^k(d^k f / dt^k)(t))_+ + b,$$

$$h_{i-1}(t) = \int_t^1 h_i(s)ds, \quad 1 \leq i \leq k.$$

Clearly $(f(t))_+ \leq h_0(t)$ for $t \in [\frac{3}{4},1]$ and $a < h_0(\frac{3}{4})$. Hence we may find a C^∞-function ϕ on $[0,1]$ such that

$$(d^i\phi / dt^i)(\tfrac{1}{4}) = (d^i g_0 / dt^i)(\tfrac{1}{4}), \quad 0 \leq i \leq k-1,$$

$$(d^i\phi / dt^i)(\tfrac{3}{4}) = (d^i h_0 / dt^i)(\tfrac{3}{4}), \quad 0 \leq i \leq k-1,$$

and $\phi(t) > a$ for $t \in [\frac{1}{4},\frac{3}{4}]$. Then the function f_0 given by

$$f_0(t) = g_0(t), \qquad t \in [0,\tfrac{1}{4}],$$

$$f_0(t) = \phi(t), \qquad t \in [\tfrac{1}{4}, \tfrac{3}{4}],$$

$$f_0(t) = h_0(t), \qquad t \in [\tfrac{3}{4}, 1]$$

is a positive function in E such that $f \leq f_0$.

Now we prove that E has a principal latticial extension; thus, it will provide an example of space satisfying the hypothesis of Theorem 5.1. Indeed, the space $\tilde{E}$ of all functions on $[0,1]$ is a latticial extension of E. As we have already mentioned, the subspace E_0 consisting of C^∞-functions with compact support in $(0,1)$ is dense in E. Let $f \in E_0$ be given. There is $g \in E_0$ such that g is identically 1 on the support of f. Let $\varepsilon > 0$ be given and let $\eta = \varepsilon / \|g\|$. There is $\phi \in E_0$ such that $0 \leq \phi(t) \leq 1$ for $t \in [0,1]$, $\phi(t) = 1$ if $f(t) \geq \eta$ and $\phi(t) = 0$ if $f(t) \leq 0$. The map U given by the multiplication by ϕ is in $C_E(\tilde{E})$. Clearly $0 \leq U \leq 1_{\tilde{E}}$ and $U(f) \geq 0$; on the other side, as $(1 - \phi(t))f(t) \leq \eta g(t)$ for $t \in [0,1]$, it follows that $(1_{\tilde{E}} - U)(f_+) \leq \eta g$. The proof is complete.

REFERENCES

1. **Aliprantis, C.D. ; Burkinshaw, O. :** The components of a positive operator, *Math. Z.* **184**(1983), 245-257.

2. **Dodds, P. ; Fremlin, D.H. :** Compact operators in Banach lattices, *Israel J. Math.* **34**(1979), 287-320.

3. **Fremlin, D.H. :** Tensor products of Archimedean vector lattices, *Amer. J. Math.* **94**(1972), 777-798.

4. **Fremlin, D.H. :** *Topological Riesz spaces and measure theory,* The University Press, Cambridge, 1974.

5. **Fučik, S. ; John, O. ; Kufner, A. :** *Function spaces*, Academia, Prague, 1977.

6. **Kalton, N.J. ; Saab, P. :** Ideal properties of regular operators between Banach lattices, preprint.

7. **Leinfelder, H.:** A remark on a paper of L.D.Pitt, *Bayreuth Math. Schr.* **11**(1982).

8. **Lindenstrauss, J. ; Tzafriri, L. :** *Classical Banach Spaces. I*, Springer, 1977.

9. **Lindenstrauss, J. ; Tzafriri, L. :** *Classical Banach Spaces. II*, Springer, 1979.

10. **de Pagter, B. :** The components of a positive operator, *Indag. Math.* **45**(1983), 219-241.

11. **Peressini, A. :** *Ordered topological vector spaces*, Harper & Row, 1967.

12. **Vuza, D. :** Strongly lattice-ordered modules over functions algebras, INCREST Preprint Series in Math., Bucureşti, **62**(1980).

13. **Vuza, D. :** Extension theorems for strongly lattice-ordered modules and applications to linear operators, INCREST Preprint Series in Math., Bucuresti, **100**(1981).

14. **Vuza, D. :** The perfect M-tensor product of perfect Banach lattices, in *Proceedings of the First Romanian-GDR Seminar*, Lecture Notes in Math. **991**(1983), Springer.

15. **Vuza, D. :** *Modules over commutative ordered rings* (Romanian), Thesis, Bucuresti, 1983.

16. **Vuza, D. :** Principal modules of linear maps and their applications, *Proceedings of the Second International Conference on Operator Algebras, Ideals and their Applications in Theoretical Physics*, Teubner-Texte zur Math. **67**(1984), pp.212-219.

17. **Vuza, D. :** Ideals and bands in principal modules, INCREST Preprint Series in Math. , Bucureşti, **42**(1984); (to appear in *Archiv. der Math.*).

Dan Vuza

Department of Mathematics, INCREST
Bdul Păcii 220, 79622 Bucharest
Romania.

Operator Theory:
Advances and Applications, Vol.17

MORE ABOUT PSEUDODIFFERENTIAL OPERATORS ON BOUNDED DOMAINS

Harold Widom

At the 1983 Operator Theory Conference we presented [2] a formal expansion for pseudodifferential operators on bounded domanins in $\mathbf{R}^n$ and described specific cases in which the expansion represented the quantity in question. This quantity is

$$\text{tr } f(P_\Omega A P_\Omega) \tag{1}$$

where A is a ψ.d.o. of negative order -r on $\mathbf{R}^n$ acting on scalar-valued or vector-valued functions; Ω is a compact set in $\mathbf{R}^n$ with smooth boundary; P_Ω is the projection from $L_2(\mathbf{R}^n)$ to $L_2(\Omega)$; and f is an appropriate but general C^∞ or analytic function. The expansion was obtained by a localization which allowed replacing Ω by $\mathbf{R}^{n-1} \times \mathbf{R}^+$ and then considering $P_\Omega A P_\Omega$ as a ψ.d.o. on $\mathbf{R}^{n-1}$ with symbol valued in the ψ.d.o.'s on $L_2(\mathbf{R}^+)$. A symbolic calculus for operator-valued ψ.d.o.'s then reduced the problem to the evaluation of certain quantities involving (one-dimensional) Wiener-Hopf operators. What resulted finally was an expansion for (1) with terms of two kinds – interior terms, which are integrals over $\Omega \times \mathbf{R}^n$ of expressions involving the symbol of A and its derivatives; and boundary terms which are integrals over the cotangent bundle $T^* \partial\Omega$ of a more complicated nature which involve also the second fundamental form of $\partial\Omega$. The integrands in these integrals are expressed explicitly in terms of the symbol (although more explicitly in the case of scalar symbols than matrix symbols) and so the terms are actually computable. The series is similar to a Taylor series in that it is an expansion "about" an arbitrary principal symbol of A.

Two cases where something was actually proved, in [3], are the following:

Szegö expansion. Here A has amplitude function $\rho(x,y,\varepsilon\xi)$ where ε is a small parameter. There is a formal expansion

$$\sum((-i\varepsilon)^{|\alpha|}/\alpha!)\,\partial_\xi^\alpha \partial_y^\alpha \rho(x,x,\varepsilon\xi)$$

for the symbol of A. Substituting this into the general expansion, using $\rho(x,x,\varepsilon\xi)$ as principal symbol, and making simple variable changes in the integrals which arise leads to a series

$$\sum_{k=0}^{\infty} a_k \varepsilon^{k-n} .$$

This was shown to be a correct asymptotic expansion for (1) as $\varepsilon \to 0$ under natural assumptions on f.

Heat expansion. Here A is a positive (definite) elliptic operator of negative order -r and f is the specific function

$$f(\lambda) = \begin{cases} e^{-t\lambda^{-1}} & \lambda > 0 \\ 0 & \lambda \leq 0 \end{cases}$$

where t is a small positive parameter. It is assumed that A is a classical ψ.d.o. so that its symbol has an asymptotic expansion as $|\xi| \to \infty$

(2)
$$\sigma \sim \sum_{i=0}^{\infty} \sigma_i$$

where σ_i is homogeneous in ξ of degree -r-i. Substituting this into the general expansion, using σ_0 as principal symbol, and making variable changes in the integrals gives a formal series

$$\sum_{k=0}^{\infty} a_k t^{(k-n)/r} .$$

It was shown *in the scalar case*, and *only for* $r < 1$, that the first n + 1 terms of this series are correct in the sense that as $t \to 0+$ we have

(3)
$$\sum e^{-t\lambda_i^{-1}} = \sum_{k=0}^{n} a_k t^{(k-n)/r} + o(1)$$

where the λ_i are the eigenvalues of $P_\Omega A P_\Omega$. Here the coefficient a_0 is given by the interior integral

(4)
$$a_0 = (2\pi)^{-n} r^{-1} \Gamma(n/r) \int_\Omega \int_{S^{n-1}} \sigma_0(x,\omega)^{n/r} d\omega dx$$

where $d\omega$ is the volume element on the unit sphere S^{n-1}. The coefficient a_1 is a sum $b_1 + c_1$ where b_1 is given by the interior integral

(5)
$$b_1 = (2\pi)^{-n} r^{-1} \Gamma((n+r-1)/r) \cdot$$
$$\cdot \int_\Omega \int_{S^{n-1}} \sigma_0(x,\omega)^{(n-r-1)/r} \mathrm{Re}\, \sigma_1(x,\omega) d\omega dx$$

and c_1 is given by a boundary integral involving only σ_o. In the special case where

$$\sigma_o(x,\omega) = s(x)$$

is independent of ω it can be expressed in terms of the hypergeometric function $F={}_2F_1$ as

$$c_1 = ((n-1)\Gamma((n-1)/r))/(2^n\pi^{(n+2)/2}\Gamma((n/2)+1)) \cdot$$

(6)

$$\cdot\int_0^1 [(1-u^{n-r-1})/(1-u^{-r})]/F(\tfrac{1}{2},(n+1)/2,\ (n/2)+1,\ u^2)du \int_{\partial\Omega} s(x)^{(n-1)/r}dx$$

when $n > 1$ and

(6')
$$c_1 = (r/8)|\partial\Omega|$$

when $n = 1$. (It is not until the formula for a_2 that the second fundamental form of $\partial\Omega$ appears.)

It was thought last year that the two restrictions mentioned above were really unnecessary for the truth of (3), that the proof was just not good enough to yield the general result. We still believe that (3) holds for $r < 1$ in the matrix case, but it is definitely false (i.e., not necessarily true) for $r \geq 1$ even in the scalar case. By this we mean, not that there is no result of the form (3), but that our particular a_k may be the wrong ones.

This can be seen most easily by considering the operator A whose kernel is the Green function for the Sturm-Liouville problem

(7)
$$\phi'' + \mu\phi = 0 \qquad \phi(0) + \alpha\phi'(0) = \phi(1) + \beta\phi'(1) = 0\,.$$

If α and β are both nonzero the eigenvalues μ_n $(n = 0,1,\ldots)$ are given by $\mu_n = \pi^2(n + o(1))^2$ and (3) in this case says

$$\sum_{n=0}^{\infty} e^{-t\pi^2(n+o(1))^2} = 1/(2\sqrt{\pi t}) + \tfrac{1}{2} + o(1)$$

and it is correct. But the asymptotics are different if α or β equals 0. In fact for boundary conditions

(8)
$$\phi(0) = \phi(1) = 0$$

the eigenvalues are $\mu_n = \pi^2(n+1)^2$ and

(9)
$$\sum_{n=0}^{\infty} e^{-t\pi^2(n+1)^2} = 1/(2\sqrt{\pi t}) - \tfrac{1}{2} + o(1)\,.$$

In other words (3) is false in this case.

For the problem (7) the kernel (i.e., null space) of the operator $P_\Omega A P_\Omega$ thought of as acting on the space of distributions supported on [0,1] is spanned by $\delta_0 - \alpha\delta'_0$ and $\delta_1 - \beta\delta'_1$ (where δ_x denotes the Dirac distribution centered at x) whereas with boundary conditions (8) the kernel is spanned by δ_0, δ_1. In the latter case therefore, where (3) fails, there is nontrivial kernel in the Sobolev space H_{-s} for any $s > \frac{1}{2}$ while in the former case, where (3) holds, the kernel in H_{-s} is trivial for all $s < 3/2$. This actually accounts for the difference.

Recall that for any s the space $H_s = H_s(\mathbf{R}^n)$ consists of tempered distributions ϕ such that

$$\int (1 + |\xi^2|)^s |\hat{\phi}(\xi)|^2 d\xi < \infty .$$

We shall also (slightly abusing standard notation) denote by $\overset{\circ}{H}_s(\Omega)$ those distributions in H_s which are supported on Ω and by $H_s(\Omega)$ those distributions on the interior of Ω which have an extension to $\mathbf{R}^n$ which belongs to H_s. For any s the operator $P_\Omega A P_\Omega$ represents a bounded operator from $\overset{\circ}{H}_{-s}(\Omega)$ to $H_{r-s}(\Omega)$; here the P_Ω on the right just indicates that the domain of the operator consists of distributions supported on Ω while the P_Ω on the left is the restriction operator from H_{r-s} to $H_{r-s}(\Omega)$. If $|s - (r/2)| < \frac{1}{2}$ then the operator is Fredholm and the kernel of the operator is the same for all choices of s from this interval. (See, for example, [1, § 22].)

Notice that $s = 0$ belongs to this interval precisely when $r < 1$ and our assumption on A (in addition to its acting on scalar-valued functions) was that it was positive definite, so that in particular $P_\Omega A P_\Omega$ had trivial kernel in $L_2(\Omega) = \overset{\circ}{H}_0(\Omega)$. For $r \geq 1$ the triviality of the kernel in the appropriate Sobolev space is not automatic but must be added to the hypotheses.

THEOREM. *For any* $r > 0$ *relation* (3), *with the coefficients* a_k *as determined in* [3], *is correct if* $P_\Omega A P_\Omega$ *has trivial kernel in* $\overset{\circ}{H}_{-s}(\Omega)$ *for some* $s > (r - 1)/2$ *(and hence for all* $s < (r + 1)/2$*).*

The proof of this will appear in a revised version of [3].

Prominent among operators of negative order -r are those with kernel

$$\alpha |x - y|^{r-n} \qquad r - n \neq 0, 2, 4, \ldots$$

where α is a constant with the same sign as $\Gamma((n - r)/2)$. Happily they always do satsisfy the hypothesis of the theorem, as we shall now show. It suffices to show that $P_\Omega A P_\Omega$ has trivial kernel in $\overset{\circ}{H}_{-r/2}(\Omega)$. If ϕ is in this kernel then $(A\phi, \psi) = 0$ for all $\psi \in \overset{\circ}{H}_{-r/2}(\Omega)$.

Now for $r > 0$ the distributional Fourier transform of $|x|^{r-n}$ is a constant with the same sign as $\Gamma((n-r)/2)$ times $|\xi|^{-r}$ interpreted as a distribution. If $r < n$ this distribution applied to a test function $\Phi(\xi)$ is of course just

$$\int |\xi|^{-r}\Phi(\xi)\,d\xi$$

but for $r > n$, if

(10) $$k - 1 < (r - n)/2 < k\,,$$

it is equal instead to

$$\int |\xi|^{-r}\ [\Phi(\xi) - \sum_{|\alpha| < 2k-1} (\xi^{\alpha}/\alpha!)\Phi^{(\alpha)}(0)]\ d\xi$$

(we use the usual multi-index notation) where the integral is interpreted as a principal value at $\xi = 0$. We shall denote the expression in brackets by $R_{2k-1}(\Phi)$, the remainder in the finite Taylor expansion of Φ about $\xi = 0$. It follows from the above discussion that

(11) $$(A\phi\,,\psi) = c\int |\xi|^{-r} R_{2k-1}(\hat{\phi}\bar{\hat{\psi}})d\xi$$

for some positive constant c. (This is proved first if ϕ and ψ are C^{∞} with support contained in the interior of Ω and then for general $\phi, \psi \in \overset{\circ}{H}_{-r/2}(\Omega)$ by the usual density and continuity argument.)

Of course everything is trivial for $r < n$ since then

$$(A\phi\,,\phi) = c\int |\xi|^{-r}|\hat{\phi}|^{2}d\xi$$

and so is positive unless $\phi = 0$. For $r > n$, with k as in (10), we shall show that if

(i) $(A\phi\,,\phi) = 0$,

(ii) $A\phi$ vanishes to order k at $z \in \Omega$,

then ϕ is a linear combination of δ_z (the Dirac distribution centered at z) and its derivatives up to order k-1. (It is easy to see that, conversely, any such linear combination belongs to $\overset{\circ}{H}_{-r/2}(\Omega)$ and satisfies (i) and (ii); in particular (i) alone does not imply $\phi = 0$.) This assertion implies the triviality of the kernel since if ϕ is in the kernel then (i) holds and so does (ii) with two different z's. To prove the assertion consider

$$\{\Phi : \int |\xi|^{-r} R_{2k-1}(\Phi)\,d\xi = 0\}\,.$$

This is clearly a linear space closed under complex conjugation. By (i) (and (11)) the set contains $|\hat{\phi}|^{2}$. By (ii), interpreted as $(A\phi\,, \delta_z^{(\alpha)}) = 0$ for $|\alpha| < k$, the set contais all

$$\hat{\phi}(\xi)\xi^{\alpha}e^{i\xi z} \qquad (|\alpha| < k)$$

and their complex conjugates. The set also contain all ξ^β for $|\beta| < 2k$ since $R_{2k-1}(\xi^\beta) = 0$. Thus the set contains also

$$\Phi_o(\xi) = |\hat{\phi}(\xi)e^{i\xi z} - \sum_{|\alpha|<k} (\xi^\alpha/\alpha!)(\partial_\xi^\alpha \hat{\phi}(\xi)e^{i\xi z})(0)|^2,$$

which is a linear combination of the above. Since Φ_o vanishes to order 2k at $\xi = 0$ we have $R_{2k-1}(\Phi_o) = \Phi_o$ and so

$$\int |\xi|^{-r} \Phi_o(\xi) d\xi = 0$$

whence, since Φ_o is nonnegative, it must be zero. This implies that $\hat{\phi}(\xi)$ is a linear combination of $\xi^\alpha e^{-i\xi z}$ with $|\alpha| < k$ and the assertion is established.

Notice that for $r < n$ all eigenvalues are positive but for $r > n$ there will always be negative eigenvalues; the sum on the left side of (3) must include them also.

Although operators of positive order have not been mentioned the formal expansion of [3] makes sense for them also. To find a "heat expansion" in the traditional sense we take a classical ψ.d.o. A of positive order r, so that its symbol has an expansion (2) where σ_i is homogeneous in ξ of degree $r - i$, assume $\sigma_o > 0$ for $\xi \neq 0$ and set $f(\lambda) = e^{-t\lambda}$. Substituting (2) into the general expansion, using σ_o as principal symbol and making appropriate variable changes, gives a series of precisely the same sort as before,

$$\sum_{k=0}^{\infty} a_k t^{(k-n)/r}, \tag{12}$$

with the a_k given by very similar formulas. Now a_o is given by (4) with the σ_o in the integrand replaced by σ_o^{-1}; and $a_1 = b_1 + c_1$ where b_1 is given by the negative of (5) with the r's in the integrand replaced by -r and c_1 is given by the negative of (6), (6') with each r appearing in an integral of (6) replaced by -r. In particular for the operator $\phi \to \phi''$ on [0,1] the first two terms give

$$1/(2\sqrt{\pi t}) - \tfrac{1}{2} + o(1)$$

which (9) shows is correct as long as we use the boundary conditions (8). This is no accident; the first $n + 1$ terms of (12) are correct as long as Dirichlet boundary conditions are imposed.

THEOREM. *Let $P_\Omega A P_\Omega$, thought of as acting on $\mathring{H}_{r/2}(\Omega)$, have eigenvalues $\lambda_1 \leq \lambda_2 \leq \ldots$. Then with the a_k as described above we have*

$$\sum e^{-t\lambda_i} = \sum_{k=0}^{n} a_k t^{(k-n)/r} + o(1) \qquad (t \to 0+).$$

We must confess that at this writing this "theorem" has not been proved but all indications are that the methods of [3] will go through and it may very well be done by the time this appears in print.

REFERENCES

1. **Eskin, G.I. :** *Boundary value problems for elliptic pseudodifferential equations*, Amer. Math. Soc. Transl. of Math. Monographs, **52** (1981).

2. **Widom, H. :** A spectral asymptotic expansion for pseudodifferential operators, in *Spectral theory for linear operators and related topics*, Birkhäuser Verlag, 1984, pp. 279-289.

3. **Widom, H. :** *Asymptotic expansions for pseudodifferential operators on bounded domains*, Lecture Notes in Math., Springer, to appear.

Harold Widom
Department of Mathematics
University of California
Santa Cruz, California 95064
U.S.A.